Technically-Write!

CANADIAN FIFTH EDITION

Technically-Write!

CANADIAN FIFTH EDITION

Ron Blicq
Lisa Moretto

RGI INTERNATIONAL

PRENTICE HALL CANADA CAREER & TECHNOLOGY
SCARBOROUGH, ONTARIO

Canadian Cataloguing in Publication Data

Blicq, Ron S. (Ron Stanley), 1925–
    Technically-Write!

5th ed.
Includes index.
ISBN 0-13-668468-8

1. Technical writing. I. Moretto, Lisa A. II. Title.
T11.B55 1998   808'.0666   C97-930731-7

 © 1998 Prentice-Hall Canada Inc., Scarborough, Ontario
A Division of Simon & Schuster/A Viacom Company.

Prentice-Hall, Inc., Upper Saddle River, New Jersey
Prentice-Hall International (UK) Limited, London
Prentice-Hall of Australia, Pty. Limited, Sydney
Prentice-Hall Hispanoamericana, S.A., Mexico City
Prentice-Hall of India Private Limited, New Delhi
Prentice-Hall of Japan, Inc., Tokyo
Simon & Schuster Southeast Asia Private Limited, Singapore
Editora Prentice-Hall do Brasil, Ltda., Rio de Janeiro

ISBN 0-13-668468-8

Vice-President, Editorial Director: Laura Pearson
Acquisitions Editor: David Stover
Developmental Editor: Marta Tomins
Copy Editor: Shirley Corriveau
Production Editor: Mary Ann McCutcheon
Editorial Assistant: Ivetka Vasil
Production Coordinator: Julie Preston
Cover Design: David Cheung
Cover Image: Tony Hutchings/Tony Stone Images
Page Layout: Craig Swistun

2 3 4 5   C   02 01 00 99 98

Printed and bound in the United States

Weblinks are included as end-of-chapter references and may occasionally present alternative points of view. Because the World Wide Web is a dynamic medium, Web site addresses may sometimes change over time.

Visit the Prentice Hall Canada Web site! Send us your comments, browse our catalogues, and more. **www.phcanada.com** Or reach us through e-mail at **phabinfo_pubcanada@prenhall.com**

Contents

Chapter 4
Short Informal Reports 85

Chapter 8
Illustrating Technical Documents 269

Chapter 9
Technically-Speak! 297

Chapter 10
Communicating with Prospective Employers 317

Chapter 11
The Technique of Technical Writing 363

About the Authors

Ron Blicq and Lisa Moretto are Senior Consultants with RGI International, a consulting company specializing in oral and written communication. They teach workshops based on the Pyramid Method of Writing presented in this book to audiences all over the world.

Ron is Senior Consultant of RGI's Canadian office. He has extensive experience as a technical writer and editor with the Royal Air Force in Britain and CAE Industries Limited in Canada, and has been teaching technical and business communication since 1967. Ron has authored five books with Prentice Hall and has written and produced six educational video programs, such as *Sharpening Your Business Communication Skills* and *So, You Have to Give a Talk?* He is a Fellow of both the Society for Technical Communication and the Association of Teachers of Technical Writing, and a Life Member of the Institute of Electrical and Electronics Engineers Inc. Ron lives in Winnipeg, Manitoba. In the summer he flies with the Winnipeg Gliding Club; in the winter he has been learning to downhill ski.

Lisa is Senior Consultant of RGI's United States' office. She has experience as an Information Developer for IBM in the U.S. and as a Learning Products Engineer for Hewlett-Packard in the U.K. Lisa holds a BSc in Technical Communication from Clarkson University, New York, and an MSc in User Interface Design from the London Guildhall University. Her specialties include developing on-line interactive information, designing user interfaces and writing product documentation. She is a member of the Society for Technical Communication and the Institute of Electrical and Electronics Engineers Inc. Lisa lives in Myrtle Beach, South Carolina. Although she lives by the ocean, Lisa's heart is in the mountains for she is an ardent skier.

(Photo: Mary Lou Stein)

Preface

This book presents all those aspects of technical communication that you, as a technician, technologist, engineer, scientist, computer specialist, or technical manager, are most likely to encounter in industry. It introduces you to the employees of two technically oriented companies, to the type of work they do, and to typical situations that call for them to communicate with clients, suppliers, and each other.

Since the earlier editions of *Technically-Write!*, changes have occurred in the way that correspondence and reports are written at both companies (H L Winman and Associates in Calgary, Alberta, and Macro Engineering Inc in Toronto, Ontario—see Chapter 2), and these changes are reflected in this fifth edition. Where previously Anna King—the companies' technical editor—edited reports typed from handwritten drafts, now she edits them online, calling them up from files keystroked by the engineers and engineering technicians at their individual computer terminals in both cities. To prepare you for such a writing environment, Chapter 1 suggests techniques you can use to outline and write directly on a computer.

As before, the book contains numerous examples of letters, reports, and proposals, all based on the unique "pyramid" method for structuring a document, a technique that has helped countless technical people overcome "writer's block."

To meet the demands of today's workplace, in Chapter 3 we have introduced new sections on writing effective electronic mail and designing information so that readers' attention is drawn immediately to key issues. In Chapter 10 we have included a section on writing electronic online resumes. Then, to acknowledge the shift to an increasingly global business environment, in Chapter 11 we have inserted a new section on writing for an international audience, an audience for most of whom English is probably a second language. In the same chapter we have also included a section on writing non-gender-specific language, plus the seven guidelines for forming technical abbreviations, etc, that previously were included as a separate chapter with the glossary. The Glossary of Technical Usage is now a stand-alone section at the end of the book.

Along the way, we have very much appreciated the friendly advice and many helpful suggestions from users of the book, both teachers and students. Their ideas have guided us in preparing this fifth edition.

We are also celebrating, for it's 25 years since the first edition of *Technically-Write!* was published, and it's also the first time the book has had two authors.

RB & LM

Supplements

The fifth Canadian edition of *Technically-Write!* is supported by a comprehensive supplements package, which includes:

- Instructor's Manual, with additional exercises, teaching tips, video references, transparency masters, and other useful material, based on the authors' extensive teaching and workshop experience.

- Test Item File (new to this edition), with approximately 30 objective and short answer questions per chapter, designed to reinforce students' understanding of key concepts in the text;

- Computerized Test Item File (new to this edition), available in both Windows and Macintosh versions, with a variety of features allowing instructors to sort questions and add new questions of their own;

- Free Electronic Grammar Workbook, included in the back of the text, which allows students to hone their knowledge of grammar and word usage.

People as "Communicators"

As communicators, most people are lazy and inefficient. We are equipped with a highly sophisticated communication system, yet consistently fail to use it properly. (If our predecessors had been equally lazy and failed to develop their communication "senses" to the level required to ensure survival, many species would be extinct today; indeed, it's even questionable that we would exist in our present form.) This communication system comprises a transmitter and receiver combined into a single package controlled by a computer, the brain. It accepts multiple inputs and transmits in three mediums: action, speech, and writing.

We spend many of our waking hours communicating, half the time as a transmitter, half as a receiver. If, as a receiver, we mentally switch off or permit ourselves to change channels while someone else is transmitting, we contribute to information loss. Similarly, if as a transmitter we permit our narrative to become disorganized, unconvincing, or simply uninteresting, we encourage frequency drift. Our listeners detune their receivers and let their computers (their brains) think about the lunch that's imminent, or wonder if they should rent a video tonight.

As long as a person transmits clearly, efficiently, and persuasively, people receiving the message keep their receivers "locked on" to the transmitting frequency (this applies to all written, visual, and spoken transmissions). Such conditions expedite the transfer of information, or "communication."

In direct contact, in which one person is speaking directly to another, the receiver has the opportunity to ask the transmitter to clarify vaguely presented information. But in more formal speech situations, and in all forms of written and most visual communication, the receiver no longer has this advantage. He or she cannot stop a speaker who mumbles or uses unfamiliar terminology, and ask that parts of a talk be repeated or clarified; neither can the receiver easily ask a writer in another city to explain an incoherent passage of a business letter, or the producer of a video program to describe the point the video is trying to make.

The results of failure to communicate efficiently soon become apparent. If people fail to make themselves clear in day-to-day communication, the consequences are likely to differ from those they anticipated, as Cam Collins has discovered to his chagrin.

Cam is a junior electrical engineer at Macro Engineering Inc, and his specialty is high-voltage power generation. When he first read about a recent extra-high-voltage (EHV) DC power conference, he wanted urgently to attend. In a memorandum to Fred Stokes, the company's chief engineer, Cam described the conference in glowing terms that he hoped would convince Fred to approve his request. This is what he wrote:

Cam's request fails to convince

> Fred
>
> The EHV conference described in the attached brochure is just the thing we have been looking for. Only last week you and I discussed the shortage of good technical information in this area, and now here is a conference featuring papers on many of the topics we are interested in. The cost is only $228 for registration, which includes a visit to the Freeling Rapids Generating Station. Travel and accommodation will be about $850 extra. I'm informing you of this early so you can make a decision in time for me to arrange flight bookings and accommodation.
>
> Cam

Fred Stokes was equally enthusiastic and wrote back:

> Cam
>
> Thanks for informing me of the EHV DC conference. I certainly don't want to miss it. Please make reservations for me as suggested in your memorandum.
>
> Fred

Cam was the victim of his own carelessness: he had failed to communicate clearly that it was *he* who wanted to go to Freeling Rapids!

Elizabeth has a good idea...

Elizabeth Drew, on the other hand, did not realize she had missed a golden opportunity to be first with an innovative computer technique until it was too late to do anything about it. Her story stems from an incident that occurred several years ago, when she was a recently graduated engineer employed by a manufacturer of agricultural machinery. Elizabeth's job was to design modifications to the machinery, and then prepare the change procedure documentation for the production department, service representatives, sales staff, and customers.

"For each modification I had to coordinate three different documents," she explained to us over lunch. "First, there had to be a design change notice to send out to everyone concerned. And then there had to be an 'exploded' isometric drawing showing a clear view of every part, with each part cross-referenced to a parts list. And finally there had to be the parts list itself, with every item labelled fully and accurately."

Elizabeth found that cross-referencing a drawing to its parts list was a tedious, time-consuming task. The isometric drawing of the part was computer generated by the drafting department. The parts list was also keyed into a computer, but by a separate department. However, because the two computer systems were incompatible, cross-referencing had to be done manually.

"And then I hit on a technique for interfacing the two programs," Elizabeth explained. "It was simple, really, and I kept wondering why no one else had thought of it!"

Without telling anyone, she modified one of the computer programs and tested her idea with five different modification kits. "It worked!" she laughed. "And, best of all, I found that cross-referencing could be done in one-tenth of the time."

...it was simple and efficient...

Elizabeth felt her employer should know about her idea: possibly the company could market the program, or even help her copyright it. So the following day she stopped Mr Haddon, the Engineering Manager, as they passed in the hallway, and blurted out her suggestion. This is the conversation that ensued:

Elizabeth	Mr Haddon
Oh! Mr Haddon! You know how long it takes to do the documentation for a new part...?	
	Yes..s..s..?
The problem is in trying to interface between the graphics computer and the parts list...	
	(*Mr Haddon appeared to be listening politely, but internally he was growing impatient.*)
...It has to be done by hand, you see...	
	Doesn't the drafting department do all that?
Oh, yes! They do. I was just trying to help them...to speed up their work a bit.	
	You're working for the chief draftsman now?
Oh, no! It was just an idea I had—to modify one of the computer programs...	

...but Elizabeth doesn't know how to articulate her ideas clearly

I don't remember issuing you a work order...

No. You didn't. I was doing it on my own... (*She meant she was doing it on her own time.*)

You mean the computer people asked you to do it?

Well – uh – no. Not exactly...

But you have been modifying one of the computer programs? Without authority?

(*Reluctantly*) Uh-huh.

I thought I had made it quite clear to all the staff: No projects are to be undertaken without my approval! (*His tone was cold and abrupt.*)

I wanted to try...

That's final! (*And he turned on his heel and continued down the hall.*)

They need to focus their messages

   Elizabeth's simple suggestion had become lost in a web of misunderstanding. By the time she was through explaining what she had been doing, she had given up trying to offer her idea to the company. And so her idea lay dormant for two years, until a major software company came out with a comparable program. Elizabeth knew then that perhaps there had been market potential for her design.

   If Cam Collins and Elizabeth Drew had paused to consider the needs of the people who were to receive their information, they would never have launched precipitously into discourses that omitted essential facts. Cam had only to start his memorandum with a request ("May I have your approval to attend an EHV DC conference next month?"), and Elizabeth with a statement of purpose ("I have designed a computer program that can save us hundreds of dollars annually. May I have a few moments to describe it to you?"), to command the attention of their department heads. Both Mr Stokes and Mr Haddon could then have much more effectively appraised the information.

   Such circumstances occur daily. They are frustrating to those who fail to communicate their ideas, and costly when the consequences are carried into business and industry.

   Bill Carr recently devised and installed a monitor unit for the remote control panel at the microwave relay station where he is the resident engineering technologist. As his modification greatly improved operating methods, Janet Reid, Manager of Technical Services at head office, asked him to submit an installation drawing and an accompanying description. Here is part of his description:

> Some difficulty was experienced in finding a suitable location for the monitor unit. Eventually it was mounted on a locally manufactured bracket attached to the left-handed upright of the control panel, as shown on the attached drawing.

On the strength of Bill's explicit mounting description and detailed list of hardware, Janet instructed technologist Phyllis Walters to convert Bill's description into an installation instruction, purchase materials, assemble 21 modification kits, and ship them to the 21 other relay stations in the microwave link.

Good intentions...

Within a week the 21 resident engineering technologists were reporting to Phyllis that it was impossible to mount the monitor unit as instructed, because of an adjoining control unit. Neither Janet nor Phyllis had remembered that Bill Carr was located at site 22, the last relay station in the microwave link, where there was no need for an additional control unit. Bill had assumed that Janet would be aware that the equipment layout at his station was unique. As he commented afterward: "I was never told *why* I had to describe the modification, or what head office planned to do with my description."

...resulting in confusion!

In business and industry we must communicate clearly and understand fully the implications of failing to do so. A poorly worded order that results in the wrong part being supplied to a job site, a weak report that fails to motivate the reader to take the urgent action needed to avert a costly equipment breakdown, and even an inadequate job application that fails to sell an employer on the right person for a prospective job, all increase the cost of doing business. Such mistakes and misunderstandings are wasteful of the country's labor and resources. Many of them can be prevented by more effective communication—communication that is receiver-oriented rather than transmitter-oriented, and that transmits messages using the most expeditious, economical, and efficient means at our command.

Chapter 1
A Technical Person's Approach to Writing

Engineering technologist Dan Skinner has a report to write on an investigation he completed seven weeks ago. He works for H L Winman and Associates (a consulting engineering firm you will meet in Chapter 2), and he has made several half-hearted attempts to get started. But each time has never seemed to be the right moment: maybe he was interrupted to resolve a circuit problem, or it was too near lunchtime, or a meeting was called, or, when nothing else interfered, he "just wasn't in the mood." And now he is up against the wire. He has neither set pen to paper nor placed his fingers on the keyboard of his personal computer. He hasn't even worked up a writing outline.

Unless Dan is one of those unusual people who can produce only when under pressure, he is in danger of writing an inadequate, hastily prepared report that does not represent his true abilities. He knows he should not have left his report-writing project until the last moment, but he is human like the rest of us and constantly finds himself in situations like this. He does not realize that by leaving a writing task until it is too late to do a good job, and then frantically organizing the work, he is probably inhibiting his writing capabilities even more than necessary.

If Dan were to relax a little, instead of worrying that he has to organize himself and his writing task, he would find the physical process of writing a much more pleasant experience. But first he must change his approach.

Every technical person, from student technician to potential scientist to practising engineer, has the ability to write clearly and logically. But this ability has to be developed. Dan Skinner must first learn some basic planning and writing techniques, and practise using them until he has acquired the skill and confidence that are the trademarks of an effective writer.

Simplifying the Approach

Throughout this book we will be advising you to *tell your readers right away what they most need or want to know*. This means structuring your writing so that the first paragraph (in short documents, the first *sentence*) satisfies their curiosity. Most executives and many technical readers are busy people who have time to read only essential information. By presenting the most important items first, you can help them decide whether they want to read the whole document immediately, put it aside to read later, or pass it along to a specialist in their department.

The writer's pyramid helps you focus your letters and reports.

This reader-oriented style of presentation is known as the "pyramid technique." Imagine that every letter, memorandum, or report you write is shaped like a pyramid, with a small piece of essential information—the key data you want your reader to see right away—at the top, supported on a broad base of details, facts, and evidence. In most letters and short reports the pyramid has only two parts: a brief **Summary** followed by the **Full Development**, as shown in Figure 1-1(a). In long reports an additional part—known as the **Essential Details**—is inserted between the Summary and the Full Development, as in Figure 1-1(b).

Readers normally are not aware when a writer has used the pyramid technique. They simply find the letter or report well organized and easy to read. The key elements are right up front, in the first paragraph, and the full story containing all the technical details is in the remaining paragraphs. For example, in the opening paragraph of his letter report in Figure 1-2, Wes Hillman summarizes what Tina Mactiere most wants to know (whether the training course was a success and what results were achieved). Then in the remainder of the letter he fills in background details, states briefly how the course was run, reports on student participation and reaction, and suggests additional topics that could be covered in future courses.

Every document shown in this textbook has been structured using the pyramid technique. How the pyramid is applied to letters, memorandums, reports, proposals, instructions, descriptions, and even resumes and oral

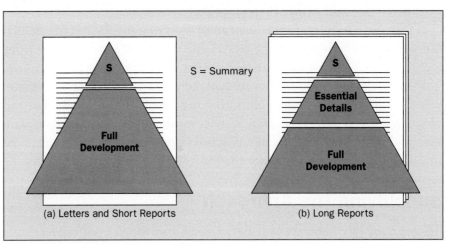

Figure 1-1 The pyramid writing technique.

The Roning Group Inc

Communication Consultants
Box 181 – RPO Corydon
Winnipeg MB R3M 3S7

October 16, 1997

Ms Tina R Mactiere, President
Macro Engineering Inc
600 Deepdale Drive
Toronto ON M5W 4R9

Dear Ms Mactiere

Results of Pilot Report-Writing Course

The report-writing course we conducted for members of your engineering staff was completed successfully by 14 of the 16 participants. The average mark was 63%.

This was a pilot course set up in response to an August 13, 1997, enquiry from Mr F Stokes. At his request, we placed most emphasis on providing your staff with practical experience in writing business letters and technical reports. Attendance was voluntary, the 16 participants having been selected at random from 29 applicants.

Best results were achieved by participants who recognized their writing problems before they started the course, and willingly became actively involved in the practical work. A few said they had expected to attend an "information" type of course, and at first were mildly reluctant to take part in the heavy writing program. Our comments on the work done by individual participants are attached.

Course critiques completed by the participants indicate that the course met their needs from a letter- and report-writing viewpoint, but that they felt more emphasis could have been placed on technical proposals and oral reporting. Perhaps such topics could be covered in a short follow-up course.

We enjoyed developing and teaching this pilot course for your staff, and particularly appreciated their enthusiastic participation.

Sincerely

[signature]

Wesley G Hillman
Course Leader

enc

A full block letter, spare on punctuation (also see page 67)

Figure 1-2 A letter report written using the pyramid technique.

presentations is described in Chapters 3 through 7, and 9 and 10. For the moment, just remember that using the pyramid is the simplest, fastest, most effective way to plan and write any document, regardless of its length. If Dan Skinner had known about the pyramid technique, his difficulty in getting started would have been considerably eased.

Planning the Writing Task

The word "planning" seems to imply that report writers must start by thoroughly organizing both themselves and their material. We disagree. Organizing too diligently, or too early in the writing process, for many people inhibits rather than accelerates writing. The key is to organize your information in a spontaneous, creative manner, allowing your mind to freewheel through the initial planning stages until the topics have been collected, scrutinized, sorted, grouped, and written into a logical outline that will appeal to the reader.

"Disorganize" the writing task!

We recommend that at first Dan Skinner neither make an outline nor take any action that resembles organization. Instead, he should work through seven simple planning stages that are less structured and therefore less confining. These stages are shown in Figure 1-3 and described in detail below.

1. Gather Information

Dan's first step should be to assemble all the documents, results of tests, photographs, samples, computer data, specifications, and other supporting material that he will need to write his report, or that he will insert into it. He must gather everything he will need now, because later he will not want to interrupt his writing to look for additional facts and figures.

Pay primary attention to the ultimate reader

2. Define the Reader

Next, Dan must clearly identify his audience. This is probably the most important part of his planning, for if he does not, he may write an

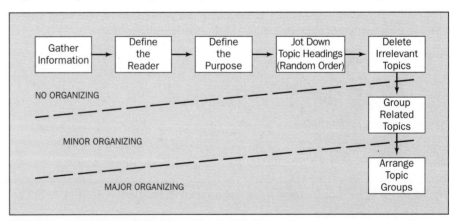

Figure 1-3 The seven planning stages. In practice, these stages can overlap.

unfocused report that misses its mark. He must conjure up an image of the person or people who will read his report by asking himself six questions:

1. *Who, specifically, is my reader?* It may be someone he knows, in which case his task is simplified, or it may be someone he is not acquainted with (such as a customer in an out-of-town firm), for whom he must imagine a persona.
2. *Is he or she a technical person?* Dan needs to know whether he can use or must avoid technical terms.
3. *How much does the reader know about the subject I will be describing?* This will give Dan a starting point, since there is no need to cover information a reader already knows.
4. *What does the reader want to know or expect to be told?* Dan must be able to anticipate whether the reader will be receptive or hostile to the information he has to present.
5. *Will more than one person read my report?* If so, Dan must repeat questions 2 through 4 for the additional readers.
6. *Who is my primary reader?* The primary reader is the person who will make a decision or take specific action after reading Dan's report. Often it is the person to whom the report is directed, but on occasion may be one of the secondary readers. For example, a report may be addressed to a department manager, but the person who will use it or do something about it will be an engineer on the manager's staff.

Dan's inability to identify his reader was one of the reasons he had difficulty getting started on his report-writing task.

3. Define the Purpose
Now that he has identified his reader, Dan has to ask himself one, and possibly two, more questions:

7. *Why am I writing to this person (or these people)?* He needs to decide whether his objective is to pass along information (to *inform* the reader about something), or to convince the reader to act or react (to **persuade** the reader to do something; for example: to reply, make a decision, or approve a request). If his purpose is to inform, Dan must now identify the key information he wants his reader to know. If his purpose is to persuade, then he needs to ask one more question.

   Decide: Why am I creating this message?

8. *What action do I want the reader to take?* At this point Dan is really asking himself: What do I want my memo, letter, or report to achieve? With this answer in mind, he will be able to fashion a *focused* writing plan.

4. Jot Down Topic Headings

Now that Dan has a reader in mind he can start making notes, either with a pen and paper or at a computer keyboard. At this third stage he must "loosen up" enough to generate ideas spontaneously. He needs to let his mind brainstorm, so that it comes up with ideas and pieces of information quickly and easily. He must not stop to question the relevance of this information (that will come later); his role for the moment is purely to collect it.

Normally, at the outlining stage, a technical person will key in or write down a set of familiar or arbitrary headings, such as "Introduction," "Initial Tests," and "Material Resources," and arrange them in logical order. (These seem to be standard headings in many reports.) But we want our report writer to be different. We want Dan Skinner to free his mind of the elementary organized headings, and even to refuse to divide his subject mentally into blocks of information. Then we want him to key into his computer the series of main topics he plans to discuss, writing only brief headings rather than full sentences. He must do this in random order, making no attempt to force the topics into groups (although it is quite possible that grouping will occur naturally, since many interdependent topics are likely to occur to him in logical order). The topics he knows best will spring readily to mind, then there will be a gradual slowdown as he encounters less familiar topics.

When he finishes his initial list, he should scroll up the screen and examine each topic to see if it suggests less obvious topics. As additional topics come to mind he must key them in, still in random order, until he finds he is straining to find new ideas.

Dan must not try to decide whether each topic is relevant during this spontaneous brainstorming session. If he does, he will immediately inhibit his creativity because he will become too logical and organized. He must list all topics, regardless of their importance and eventual position in the final report.

At the end of this session Dan's list should look like Figure 1-4. He can now take a break, knowing that the first details of his report are on disk. What he may not yet realize is that he has almost painlessly produced his first outline.

5. Delete Irrelevant Topics

The fourth stage calls for Dan to print a hard copy to work on, and then to examine his list of headings with a critical eye, dividing them into those that bear directly on the subject and those that introduce topics of only marginal interest. His knowledge of the reader—identified in stage 2—will help him decide whether each topic is really necessary, so he can delete irrelevant topics as has been done in Figure 1-5.

Loosen up: delve deeply into brainstorming

Start grouping your topics into compartments

```
        Building OK – needs strengthening
        Elevators – too slow, too small
        Talk with YoYo – elev mfr (10% discount)
        Waiting time too long – 70 sec.
        Shaft too small
        How enlarge shaft?
                Remove stairs?
        Talk with fire inspector
        Correspondence – other elev mfrs
        Talk with Merrywell – Budget $800 000
        Sent out questionnaire
        Tenants' preferences –
                Express elev        No stop – 2nd flr
                Executive elev      Faster service
                Prestige elev       No stop – ground flr
                Freight elev
        Freight elev – takes up too much space
        Shaft only 10.85 × 2.5 m (when modified)
        Big freight elev – omit basement
        Tenants "OK" small freight elev
                (YoYo "C" – 2.45 m)
        YoYo – has office in Montrose
        Basement level has loading dock
        Service reputation – YoYo?
                – Others?
```

Figure 1-4 Initial list of topic headings, typed in random order.

Let the initial outline develop naturally, loosely

6. Group Related Topics

The headings that remain should be grouped into "topic areas" that will be discussed together. Dan can do this simply by coding related topics with the same symbol or letter. In Figure 1-5, letter (A) identifies one group of related topics, letter (B) another group, and so on.

Now start pulling the pieces together

7. Arrange the Topic Groups

It's at this stage that we encourage Dan to take his first major organizational step, which is to arrange the groups of information in the most suitable order and at the same time sort out the order of the headings within each group. He must consider:

Let the final outline evolve from the subject matter...

- which order of presentation will be most interesting,

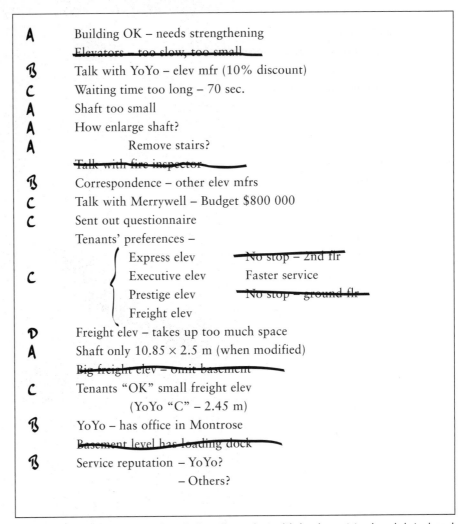

Figure 1-5 The same list of topic headings, but with irrelevant topics deleted and remaining topics coded into subject groups (A–structural implications; B–elevator manufacturers; C–tenants' preferences; D–freight elevator).

- which will be most logical, and
- which will be simplest to understand.

The result will become his final writing plan or report outline. Figure 1-6 shows Dan's final writing plan. Depending on how he prefers to work, Dan can use a computer-printed version of his outline or work online, directly from a word-processed file.

A final comment about outlining: If you have already developed an outlining method that works well for you, or you are using outlining software successfully, then we suggest you continue as you have been doing. The outlining method suggested here is for people who are seeking a simpler, more creative way to develop outlines than the one they are currently using.

```
Building condition:
        OK – needs strengthening (shaft area)
        Existing elev shaft too small
        Remove adjoining staircase
        Shaft size now 10.85 × 2.5 metres

Tenants' needs:
        Sent out questionnaire
        Identified 5 major requests
        Requests we must meet:
                Cut waiting time: 32 sec (max)
                Handle freight up to 2.3 m long
        Requests we should try to meet:
                Express elev to top 4 floors
                Deluxe models (for prestige)
                Private elev (for executives)


Budget: must be within $800 000


Elevator manufacturers:
        Researched 3
        Only YoYo Co. offers discount
        Only YoYo Co. has Montrose office
```

Figure 1-6 Topic headings arranged into a writing outline.

Writing the First Draft

As we sit at our desks, the heading "Writing the First Draft" at the top of Ron's computer screen and "Focus the Letter" at the top of Lisa's (Lisa is working on Chapter 3), we find we are experiencing the same problem that every writer encounters from time to time: an inability to find the right words—*any* words—that can be strung together to make coherent sentences and paragraphs. The ideas are there, circling around inside our skulls, and the outlines are there, so we cannot excuse ourselves by saying we have not prepared adequately. What, then, is wrong?

> Write where you won't be disturbed: no telephone; no pager; no cellphone

The answer is simple. Ten minutes ago the telephone rang and Jack, a neighbor, announced he would shortly bring over a "Block Neighborhood Watch" plan for Ron to sign. Ron paused to switch on the coffee, for we know that Jack will expect a cup while we converse, and now we can hear the percolator grumbling away in the distance. We cannot concentrate when we know our continuity of thought is so soon to be broken.

Continuity is the key to getting one's writing done. In our case, this means writing at fairly long sittings during which we *know* we will not be disturbed. We must be out of reach of the telephone, visiting friends, and even family, so we can write continuously. Only when we have reached a logical break in the writing, or have temporarily exhausted an easy flow of words, can we afford to stop and enjoy that cup of cappuccino!

It is no easier to find a quiet place to write in the business world. The average technical person who tries to write a report in a large office cannot simply ignore the surroundings. A conversation taking place a few desks away will interfere with one's creative thought processes. And even a co-worker collecting money for the pool on that night's NHL game between the Edmonton Oilers and the Toronto Maple Leafs will interrupt writing continuity.

The problem of finding a quiet place to write can be hard to solve, particularly now that most people key their reports into a computer and so cannot move away from their desks (unless they are fortunate enough to own a portable computer). In business we recommend a short walk around the premises to find a hidden corner, or an office which is temporarily unoccupied. Then perhaps, if you are lucky, you can just "disappear" for a while. For technical students, who frequently have to work on a tiny writing space in a crowded classroom, or in a roomful of computer terminals, conditions are seldom ideal. Outlining in the classroom, followed by writing in the seclusion of a library cubicle, is a possible alternative.

Some people prefer to use a pen and paper for writing reports and then type only the final copy into the computer. Others prefer to key directly into the computer. If you do, before you start you have to consider aspects that normally do not concern you when you write with a pen; items that a typist normally looks after. For example, you have to define the page layout, making the following decisions:

- What font you will use, and whether it should be serif or sans serif. A serif type has tops and tails on its ascenders and descenders (this book is set in Sabon, which is a serif type). A sans serif type is much plainer (**Helvetica** is a typical sans serif type).
- Whether you will print the report in 10 or 12 point type (i.e. with 10 or 12 characters to the linear inch). Generally, 11 or 12 point is better for serif fonts, and 10 or 11 point is better for sans serif fonts.
- The number of lines you want on a page, and the width of your planned typing lines.
- The width of the margins you want on either side of the text and at the top and bottom of the page.

Write whatever way works best for you

- Whether you want the right margin to be justified (straight) or ragged. Research shows that paragraphs set with a ragged right margin are easier to read than paragraphs set with a justified right margin.
- Where you want the page numbers to be positioned (top or bottom of the page, and either centred or to one side of the page); on most systems page numbers are printed automatically, but you can select where they are to appear.
- The line spacing you want (single or double), and how many blank lines you want between paragraphs (normally one or two).
- Whether the first line of each paragraph is to be indented or set "flush" with the left margin; and, if indented, how long the indentation is to be.
- For long words at the end of a line, whether you or the computer will decide where the word is to be hyphenated (you can also select no hyphenation).
- The levels of headings you will use, and how you will use different font sizes and boldface type to differentiate between them. (See page 368 and Figure 11-2 of Chapter 11 for guidelines.)

You have to set up page parameters only once; the first time

Most popular word-processing programs provide default settings for these options, but you should be aware of them and how to customize the page layout for your particular needs. (A "default" is a condition, such as the position of page numbers, that is set automatically.) You should consult the documentation that comes with your word-processing software for specific instructions on how to change an option. Every program is different.

When Dan Skinner has found a peaceful location out of reach of the telephone, his friends, and even his boss, he can make a fresh start. But this is exactly where a second difficulty may occur. Equipped with an outline on his left and a pad of paper or a keyboard in front of him, he may find that he does not know where to begin. Or he may tackle the task enthusiastically, determined to write a really effective introduction, only to find that everything he writes sounds trite, unrealistic, or downright silly. Discouraged after many false starts, he deletes his most recent attempt from the screen.

We frequently advise technical people who encounter this "no start" block that the best place for them to start writing is at paragraph two, or even somewhere in the middle. For example, if Dan finds that a particular part of his project interests him more than other parts, he should write about that part first. His interest and familiarity with the subject will help him write those few first words, and keep him going once he has started. The most important thing is to *start* writing, to put any words down, even if they are not exactly the right words, and to let them lead naturally into the next group of ideas.

Tips for combatting "writer's block"

Write without stopping
to revise; that comes
later

This is where continuity becomes essential. When you have found a quiet place to work, and have allowed sufficient time to write a sizable chunk of text, don't interrupt the writing process to correct a minor point of construction. If you are to maintain continuity, don't strive to write perfect grammar, find exactly the right word, fiddle with page layout, or construct sentences and paragraphs of just the right length. That can be done later, during revision. The important thing is to keep building on that rough draft (no matter how "rough" it is), so that when you stop for a break you know you have written something you can work up into a presentable document.

If, as he writes, Dan cannot find exactly the word he wants, he should jot down a similar word and key in a question mark enclosed in brackets immediately after it, as a reminder to change the word when the first draft is finished. Similarly, if he is not sure how to spell a certain word, he should resist the temptation to turn to a dictionary, for that will disrupt the natural flow of his writing. Again, he should draw attention to the word as a reminder that he must consult his dictionary later. See Figure 1-7.

We cannot stress too strongly that writers should not correct their work as they write. Writing and revising are two entirely separate functions, and they call for different approaches; they cannot be done simultaneously without the one lessening the effectiveness of the other. Writing calls for creativity and total immersion in the subject so that the words tumble out in a constant flow. Revision calls for lucidity and logic, which force a writer to reason and query the suitability of the words he or she has written. The first requires exclusion of every thought but the subject; the second demands an objectivity that constantly challenges the material from the reader's point of view. Writers who try to correct their work as they write soon become frustrated, for creativity and objectivity are constantly fighting for control of their fingers.

Most technical people find that, once they have written those first few difficult sentences, their inhibitions begin to fall away and they write more naturally. Their style begins to change from the dull, stereotyped writing we expect from the average technical person, to a relaxed narrative that sounds as though they are telling a co-worker about the subject. As a writer becomes interested in the topic, speed builds up and he or she has difficulty keying the words in fast enough. They may not be exactly those of the final report, but when the writer starts checking the first draft he or she will be surprised at the effectiveness of many of the sentences and paragraphs written earlier.

The length of each writing session will vary, depending on the writer's experience and the complexity of the topic. If a document is reasonably short, it should be written all at one sitting. If it is long, it should be divid-

Tenants' Needs

To find out what the building's tenants most needed in elevator service, we asked each company to fill out a questionaire (sp?). From their answers we were able to identify 5 factors needing consideration:

1. A major problem seems to be the length of time a person must wait for an elevator. Every tenant said we must cut out lengthy waits. A survey was carried out to find out how long people had to wait (during rush hours). This averaged out at 70 sec, more than twice the 32 sec established by Johnson (Ref?), before people get impacient (sp?). From this we calculated we would need 3 or 4 passenger elevators.

2. At first it seemed we would be forced to include a full-size freight elevator in our plan. Two companies (which?) both carry large but light displays up to their floors, but both later agreed they could hinge them, and if they did this they would need only 2.3 metres width (maximum). They also said they did not need a freight elevator all the

Use question marks as a search tool

Figure 1-7 Part of the author's first draft. Note that the author has not stopped to hunt up minor details. Several revisions were made between this first draft and the final product (see pages 214 to 216).

ed into several medium-length sessions that suit the writer's particular staying power.

In Dan's case, at the end of each session he should glance back over his work, note the words he has circled or questioned, and make a few necessary changes (Figure 1-7 is a page from a typical first draft). He must not yet attempt to rewrite paragraphs and sentences for better emphasis. Such major changes must be left until later, when enough time has elapsed for him to read his work objectively. Only then can he review his work as a complete document and see the relationship among its parts. Only then can he be completely critical.

Taking a Break

When we have written the final paragraph of a long report, we have to resist the temptation to start revising it immediately. There are sections we know are weak, or passages we are not happy about, and the desire to correct them is strong. But we know from experience it's too soon. For the moment we pass the draft through a spell-checker, make a safety copy onto a 3.5-inch disk, and print out the pages (we create a double-spaced draft, so we will have room to write in revisions when we are editing). Then we staple them together and set them aside while we tackle a completely unrelated task.

Reading without a suitable waiting period encourages writers to look at their work through rose-tinted spectacles. Sentences that normally they would recognize as weak or too wordy appear to contain words of wisdom. Gross inaccuracies that under normal circumstances they would pounce upon go unnoticed. Paragraphs that may not be understood by a reader new to the subject, to their writers seem abundantly clear. Their familiarity with their work blinds them to its weaknesses. The remedy is to wait.

Reading with a Plan

Read all the way
through without a pen in
your hand

Dan Skinner's first reading should take him straight through the draft without stopping to make corrections, so that he can gain an overall impression of the report. Subsequent readings should be slower and more critical, with changes written in as he goes along. As he reads he should check for clarity, correct tone and style, and technical and grammatical accuracy.

Checking for Clarity

Checking for clarity means searching for passages that are vague or ambiguous. If the following paragraph remained uncorrected, it would confuse and annoy a reader:

Muddled Paragraph   When the owners were contacted on April 15, the assistant manager, Mr Pierson, informed the engineer that they were thinking of advertising Lot 36 for sale. He has however reiterated his inability to make a definite decision by requesting his company to confirm their intentions with regard to buying the land within two months, when his boss, Mr Davidson, general manager of the company, will have come back from a business tour in Europe. This will be June 8.

The only facts you can be sure about after reading this paragraph are that the owners of the land were contacted on April 15 and the general manager will be returning on June 8. The important information about the possible sale of Lot 36 is confusing. Probably the writer was trying to say something like this:

<table>
<tr><td>Revised
Paragraph</td><td>The engineer spoke to the owners on April 15 to enquire if Lot 36 was for sale. He was informed by Mr Pierson, the assistant manager, that the company was thinking of selling the lot, but that no decision would be made until after June 8, when the general manager returns from a business tour in Europe. Mr Pierson suggested that the engineer submit a formal request to purchase the land by that date.</td><td>Clear!</td></tr>
</table>

The more complex the topic, the more important it is to write clear paragraphs. Although the paragraph below is quite technical, it would be generally understood even by nontechnical readers:

<table>
<tr><td>Clear
Paragraph</td><td>A sound survey confirmed that the high noise level was caused mainly by the radar equipment blower motors, with a lesser contribution from the air-conditioning equipment. Tests showed that with the radar equipment shut down the ambient noise level at the microphone positions dropped by 10 dB, whereas with the air-conditioning equipment shut down the noise level dropped by 2.5 dB. General clatter and impact noise caused by the movement of furniture and personnel also contributed to the noisy working conditions, but could not be measured other than as sudden sporadic peaks of 2 to 5 dB.</td><td>Technical, but still clear</td></tr>
</table>

Here paragraph unity has added much to readability. The writer has made sure that

- the topic is clearly stated in the first sentence (the topic sentence),
- the topic is developed adequately by the remaining sentences, and
- no sentence contains information that does not substantiate the topic.

If any paragraph meets these basic requirements, its writer can feel reasonably sure the message has been conveyed clearly.

Writers who know their subject thoroughly may find it difficult to identify paragraphs that contain ambiguities. A passage that is abundantly clear to them may be meaningless or offer alternative interpretations to a reader unfamiliar with the subject. For example:

<table>
<tr><td>Our examination indicates that the receiver requires both repair and recalibration, whereas the transmitter needs recalibration only, and the modulator requires the same.</td><td>Muddled writing...</td></tr>
</table>

This sentence plants a question in the reader's mind: Does the modulator require both repair and recalibration, or only recalibration? The technician who wrote it knows, because he has been working on the equipment, but the reader will never know unless he or she writes, phones, or emails the technician. The technician could have clarified the message easily, simply by rearranging the information:

Clear writing...

> Our examination indicates that the receiver requires both repair and recalibration, whereas the transmitter and modulator need only recalibration.

Sometimes ambiguities are so well buried they are surprisingly difficult to identify, as in this excerpt from a chief draftsperson's report to a department head:

> The drafting section will need three Nabuchi Model 700 CAD computers. The current price is $3175 and the supplier has indicated his quotation is "firm" for three months. We should therefore budget accordingly.

The department head took the message at face value and inserted $3175 for CAD computers into the budget. But two months later the company received an invoice for $9525. Unable by then to return two of the three computers, the department head had to overshoot his budget by $6350. This financial mismanagement was caused by the chief draftsperson, who had omitted to insert the word "each" immediately after "$3175."

Many ambiguities can be sorted out by simple deduction, although a reader should not have to interpret a writer's intentions. It's the writer's job to make reading a document as easy and free of stress as possible, by eliminating confusing statements and alternative meanings.

Occasionally an ambiguity can provide a humorous note, as in these extracts from an equipment failure report and a field trip report:

> A A circuit board was inserted into slot 2014 by a technician with bent pins.
>
> B High grass and brush around the propane tanks impeded the technicians' progress and should be cleared before they grow too dense.

Unintentionally funny writing

So that readers will not mistakenly think the technician in sentence **A** has bent legs (pins), the reference to bent pins must be linked more closely with the circuit board:

> A A technician inserted a circuit board with bent pins into slot 2014.

Similarly, so that readers will not think the technicians in sentence **B** are growing too dense, the two thoughts in the sentence should be separated:

B   High grass and brush around the propane tanks impeded the technicians' progress. This undergrowth should be cleared before it grows too dense.

Checking for Correct Tone and Style

How do you know when your writing has the right tone? One of the most difficult aspects of technical writing is establishing a tone that is correct for the reader, suitable for the subject, and comfortable for you, the writer. If you know your subject well and have thoroughly researched your reader, you will most likely write confidently and, often, will automatically establish the correct tone. But if you try to set a tone that does not feel natural, or if you are a little uncertain about the subject and the reader, your reader will sense unsureness in your writing. And no matter how skilfully you edit your work, the hesitancy will show up in the final sentences and paragraphs.

Finding the Best Writing Level

To check that he has set the correct tone, Dan Skinner must assess whether his writing is suitable for both the subject matter and the reader. If he is writing on a specific aspect of a very technical topic and knows that his reader is an engineer with a thorough grounding in the subject, he can use all the technical terms and abbreviations that his reader will recognize. Conversely, if he is writing on the same topic for a nontechnical reader who has little or no knowledge of the subject, Dan may have to write a simplified narrative rather than state specific details, explain technical terms that he would normally expect to be understood, and generally write more informatively.

Keep coming back to your readers: plant yourself in their shoes

Writing on a technical subject for readers who do not have the same technical knowledge as yourself is not easy. You have to be much more objective when checking your work and try to think in the same way as the nontechnical reader. You can use only those technical words you know will be recognized, yet you must avoid oversimplified language that may irritate the reader.

For example, when engineer Rita Corrigan wrote the following in a modification report, she knew her readers would be electronics technicians at radar-equipped airfields.

> We modified the MTI by installing a K-59 double-decade circuit. This brightened moving targets by 12% and reduced ground clutter by 23%.

Although this statement would be readily understood by the readers for whom it was intended, to any reader not familiar with radar terminology it would mean very little. So when Rita reported on the same subject to the airport manager, she wrote this:

We modified the radar set's Moving Target Indicator by installing a special circuit known as the K-59. This increased the brightness of responses from aircraft and decreased returns from fixed objects on the ground.

Adjust the level of writing to suit the reader

For this reader Rita included more descriptive details and eliminated specific technical details that might not be meaningful to the airport manager. In their place she has made a general statement that aircraft responses were "increased" and ground returns "decreased." She also knew that the airport manager would be familiar with terms such as *Moving Target Indicator*, *responses*, and *returns*.

Now suppose that Rita also had to write to the local Chamber of Commerce to describe improvements in the airport's air traffic control system. This time her readers would be entirely nontechnical, so she would have to avoid using *any* technical terms:

We have modified the airfield radar system to improve its performance, which has helped us to differentiate more clearly between low-flying aircraft and high objects on the ground.

Keeping to the Subject

Having established that he is writing at the correct level, Dan Skinner must now check that he has kept to the subject. He must take each paragraph and ask: Is this truly relevant? Is it direct? And is it to the point?

If Dan prepared his outline using the method described earlier, and followed it closely as he wrote his report, he can be reasonably sure that most of his writing is relevant. To check that his subject development follows his planned theme, he should identify the topic sentences of some paragraphs and check them against the headings in his outline. If they follow the outline, he has kept to the main theme; if they tend to diverge from the outline or if he has difficulty in identifying them, he should read the paragraphs carefully to see whether they need to be rewritten, or possibly even eliminated. (For more information about topic sentences, see Chapter 11.)

Technical writing should always be as direct and specific as possible. Technical writers should convey just enough information for their readers to understand the subject thoroughly. Technical writing, unlike literary writing, has no room for details that are not essential to the main theme. This is readily apparent in the following descriptions of the same equipment.

Literary Description The new cabinet has a rough-textured dove grey finish that reflects the sun's rays in varying hues. Contrary to most instruments of this type, its controls are grouped artistically in one corner, where the deep black of the knobs provides an interesting contrast with the soft grey and white background. A cover plate, hardly noticeable to the layperson's inexperienced eye, conceals a cluster of unsightly adjustment screws that would otherwise mar the overall appearance of the cabinet and would nullify the esthetic appeal of its surprisingly effective design.

Technical Description
The grey cabinet is functional, with the operator's controls grouped at the top right-hand corner where they can be grasped easily with one hand. Subsidiary controls and adjustment screws used by the maintenance crews are grouped at the bottom left-hand corner, where they are hidden by a hinged cover plate.

Technical writing is functional *writing*

Comparison of these examples shows how a technical description concentrates on details that are important to the reader (it tells *where* the controls are and why they have been so placed); and it maintains an efficient, businesslike tone.

Using Simple Words

A writer who uses unnecessary superlatives sets an unnaturally pompous tone. The engineer who writes that a design "contains ultrasophisticated circuitry" seems to be justifying the importance and complexity of his or her work rather than saying that the design has a very complex circuit. The supervisor who recommends that technician Johannes Schmitt be "given an increase in remuneration" may be understood by the company controller but will only be considered pompous by Johannes. If the supervisor had written that Johannes should be "given a raise," he would have been understood by both. Unnecessary use of big words, when smaller, more generally recognized, and equally effective synonyms are available, clouds technical writing and destroys the smooth flow that such writing demands.

Don't use a 90-cent word when an equally suitable 25-cent word exists

Removing "Fat"

During the reading stage Dan should be critical of sentences and paragraphs that seem to contain too many words. He should check that he has not inserted words of low information content, that is, phrases and expressions that add little or no information. Their removal, or replacement by simpler, more descriptive words, can tighten up a sentence and add to its clarity. Low-information-content words and phrases are often hard to identify because the sentences in which they appear seem to be satisfactory. Consider this sentence:

For your information, we have tested your spectrum analyser and are of the opinion that it needs calibration.

The expressions "for your information" and "are of the opinion that" are words of low information content. The first can be deleted, and the second replaced by "consider," so that the sentence now reads:

> We have tested your spectrum analyser and consider it needs calibration.

The same applies to this sentence:

> If you require further information, please feel free to telephone Mr Thompson at 489-9039.

Weed out unnecessary expressions

The phrase "if you require," although not wrong, could be replaced by the single word "for"; but "please feel free to" is archaic and should be eliminated. The result:

> For further information please telephone Mr Thompson at 489-9039.

See Tables 11-2 and 11-3 in Chapter 11, which contain lists of low-information-content words and wordy expressions.

Inadvertent repetition of information can also contribute to excessive length. For example, Dan may write:

> We tested the modem to check its compatibility with the server. After completing the modem tests we transmitted messages at low, medium, and high baud rates. The results of the transmission tests showed...

If he deletes the repeated words in sentence 2 ("After completing the modem tests") and sentence 3 ("...of the transmission tests..."), he can achieve a much tighter paragraph.

> We tested the modem to check its compatibility with the server, and then transmitted messages at low, medium, and high baud rates. The results showed...

Checking for Accuracy

Maintain top-level quality control

Nothing annoys readers more than to discover that they have been presented with inaccurate information (particularly if they have been using the information for some time before they discover the error). Readers of Dan's report will assume that he knows his facts and has checked that they have been correctly transcribed into the report. Discovering even a single error in his report can undermine Dan's credibility in readers' minds.

There is no way to prevent some errors from occurring when quantities and details are copied from one document to another. Therefore, Dan must carefully check that facts, figures, equations, quantities, and extracts from other documents are all copied correctly. He must do this regardless of whether he keyed in the report at a computer terminal or a typist typed it from his handwritten draft. Even though he may feel the typist is responsible for proofreading the report, the ultimate responsibility for checking its accuracy is his.

Checking for accuracy also means ensuring that grammar, punctuation, and spelling have not been overlooked. Dan must check spelling with care, because his familiarity with the subject may blind him to obvious errors. (How many of us have inadvertently written "their" when we intended to write "there"? And "too" when we meant "two"?)

Dan has to recognize that spell-check programs are not 100% reliable. He may use a word—particularly a technical word—that is not in the spell-checker's memory, or he may key in a word inaccurately and inadvertently form another word that the spell-checker recognizes. For example, he may have planned to key in "departure" but accidentally keyed in "department," which the spell-checker would not recognize as an error. (Neither would it flag "their" and "too" as errors.) A spell-check program does not comprehend the context of the words; it simply examines each word and compares it to its master list. If it finds a "match," it takes no action; if it does not find a match, it highlights the word and sometimes also emits an audible warning.

Be wary when using spell-check programs

Revising Your Own Words

We do not recommend that Dan read and revise online, i.e. directly on his computer screen. He needs to see his words in the same form that his ultimate readers will see them: as printed words on a page. (True, if Dan sends his report through the Internet it may be read on screen, but even then most readers will make a hard copy when they download a long report.) Our experience, and also of many other report writers, is that when you read text online you miss some of the typographical errors that you would much more readily identify on a printed page. Consequently, we suggest that Dan marks up a hard copy of his report and then, as a separate step, transfers the changes to the online document.

Proofread on hard copy

Now, and each time he makes revisions online, Dan must perform two more "housekeeping" checks:

1. Proofread the hard copy printouts of the revised sections very carefully (he need not proofread sections he has not changed).
2. Again pass the report through the spell-check program.

Each time he reads his report, Dan should continually ask himself five questions:

1. **Can my readers understand me?**
   Will the person I am writing for be able to read my report all the way through without becoming lost?
   What about other readers who might also see my report: Will they understand it?
2. **Is the focus right?**
   Is my report reader-oriented?
   Are the important points clearly visible?
   Have I summarized the main points in an opening statement that the reader will see right away?
3. **Is my information correct?**
   Is it accurate?
   Is it complete?
   Is all of it necessary?
4. **Is my language good?**
   Is it clear, definite, and unambiguous?
   Are there any grammatical, punctuation, or spelling errors?
   Does every paragraph have a topic sentence (preferably at the start of the paragraph)?
   Have I used any big "overblown" words where simpler words would do a better job?
   Are there any low-information-content words and phrases?
5. **Have I kept my report as short as possible while still meeting my readers' needs and covering the topic adequately?**

Make yourself a checklist, then use it!

By now Dan's draft should be in good shape. Any further reading and revising will be final polishing, and the amount will depend on the importance of the report. If his report is for limited or in-company distribution, a standard-quality job normally will suffice. But if it is to be distributed outside the company, or submitted to an important client, then Dan will spend as much time as necessary to ensure that it conveys a good image of both himself and H L Winman and Associates (his employer).

Reviewing the Final Draft

The final step occurs when Dan feels that his report is ready to be issued. Before instructing the computer to print a letter-quality copy, he should ask himself four more questions:

1. Does the report achieve the purpose I established earlier?
2. What reaction will it incur from the intended reader(s)?
3. Is this the reaction I want?
4. Would I want to read what I have just written?

Now he must be extremely self-critical, ready to doubt his ability to be truly objective. If his answers are at all hesitant, he should ask an independent reviewer to read his report. Ideally, this person will be technically and mentally equivalent to the eventual reader, but not too familiar with the project. The reviewer must be able to criticize constructively, reading the report completely rather than scanning it, and making notes describing possible weaknesses and ambiguities. The reviewer should take time to discuss the report with Dan, and explain to him why certain sentences or paragraphs need to be reworked.

Ask for a second opinion

Dan Skinner will now be able to issue his report with confidence, knowing that he has fashioned a good product. The approach described here will not have made report writing a simple task for him, but it will have assisted him through the difficult conceptual stages, and helped him to read and revise more efficiently. When Dan has to write his next report, he will be less likely to put it off until it is so late that he has to do a rush job.

ASSIGNMENTS

Exercise 1.1
Describe why the "pyramid" method of writing will help you become a better presenter of information.

Exercise 1.2
Which do you feel is the better way for you to develop an outline for a report: the organized method or the "random" method? Explain why.

Be comfortable with your writing method

Exercise 1.3
    (a)   What are the seven stages advocated for planning a report?
    (b)   Which is the most important stage? Explain why.
    (c)   Must the stages be followed exactly in the sequence listed?

Exercise 1.4
If a report is likely to be read by several people, how would you identify which is your primary reader?

Exercise 1.5
If you had a long project to write next month, describe where you would write it and why you would write it there.

Exercise 1.6
Is it better to write a report without stopping to "clean up" the construction along the way, or to write a page at a time and to edit it before going on to the next page? Explain why.

Exercise 1.7

What two factors will help you write more confidently, and probably help you set the right tone?

Exercise 1.8

Describe what is meant by low-information-content words and expressions, and explain why they should be avoided in your writing.

Exercise 1.9

From the list of five main questions that you, as a writer, should ask yourself during the revision stage (see the boldface questions on page 22), which do you think is most important? Explain why.

Exercise 1.10

What kind of person is best equipped to do a final review of a report you have written?

WEBLINKS

Technical/Professional Communication: Summary
www.rpi.edu/~decemj/teach/techcomm/tpsum.html

In this essay, Professor John December of the Rensselaer Polytechnic Institute summarizes his ideas about the nature of technical/professional communication, focusing on viewing technical communication as a process of shaping information.

Technical Communication Quarterly
rhet.agri.umn.edu:80../~tcq/

Technical Communication Quarterly is the journal of the Association for Teachers of Technical Writing. Abstracts and an index to the contents of the journal are available at the association's home page.

Something about Technical Writing
logic.csc.cuhk.edu.hk/~s946014/tech.html

This article gives a brief description of technical writing, lists its salient characteristics, then discusses why students must have some knowledge of it.

Revision in Business Writing
owl.english.purdue.edu/Files/90.html

This handout from Purdue University's Online Writing Lab covers the importance of revising your writing. The principles discussed here are relevant to technical communications: language, conciseness, and clarity; tone, attitude, and audience; organization; and grammar, punctuation, and spelling.

CoreCOMM's Worst-Technical-Writing-Sample-of-the-Month Contest
www.corecomm.com/worst.html
CoreCOMM started this contest to allow technical writers to display some of the writing that makes their jobs secure. The site includes the contest rules and publishes the (disguised) winning entries.

Chapter 2
Meet "H L Winman and Associates"

Many of the projects you read or write about in this book are based upon events affecting two companies: H L Winman and Associates, a Calgary, Alberta, firm of consulting engineers, and its Ontario affiliate, Macro Engineering Inc. To help you identify yourself in this working environment, this chapter contains short histories and partial organization charts of the companies, plus brief biographical sketches of some of the men and women who have brought them to their current position and who most influence how their engineers, scientists, technologists, and technicians write. The chapter also describes what your role might be if you were employed by either company.

Corporate Structure

H L Winman and Associates was founded by Harvey Winman some 30 years ago. Starting with a handful of engineers working in a small office in one of the older sections of Calgary, Harvey gradually built up his company to where it now has 920 employees: 390 at head office in Calgary, 350 in Toronto, 40 in a United States' branch in Cleveland, Ohio, and 140 in smaller offices throughout Canada. Some are branches of H L Winman and Associates (there may be one in your area), while others are local engineering firms such as Macro Engineering Inc that Harvey has purchased. Figure 2-1 illustrates the company's two main locations and larger branches.

Head Office: H L Winman and Associates

The head office of H L Winman and Associates is located in a modern building at 475 Lethbridge Trail, a major thoroughfare in Calgary. A partial organization chart in Figure 2-2 shows Martin Dawes as president, with nine department managers who are responsible for the day-to-day operation of the company reporting directly to him.

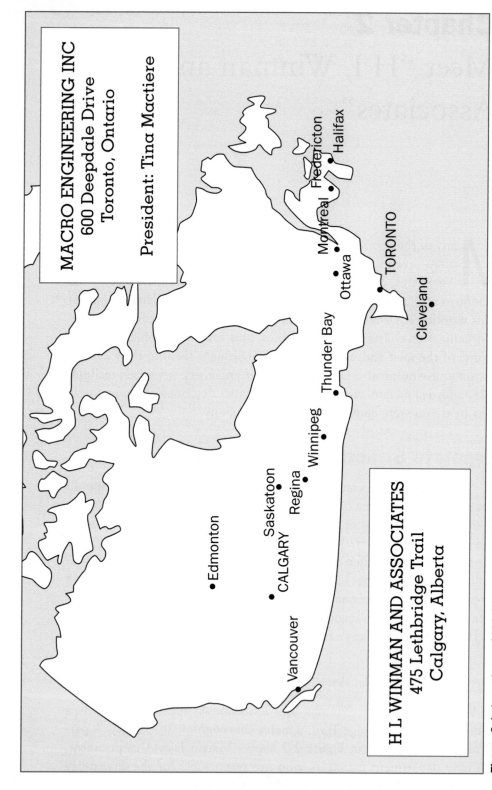

Figure 2-1 Locations of H L Winman and Associates' offices.

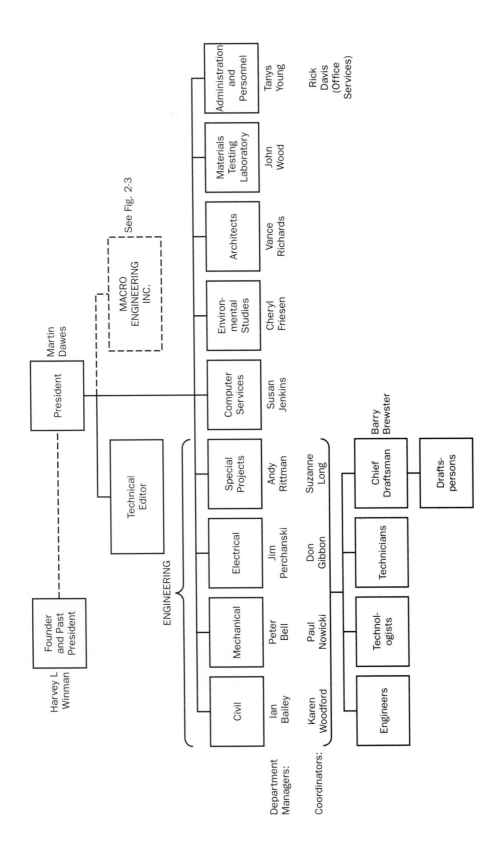

Figure 2-2 Partial organization chart—H L Winman and Associates.

Harvey Winman's
company started
small...

...but with good man-
agement grew into
a multiple-city
organization

Harvey retired early to
create upward mobility
within the firm

Over the years H L Winman and Associates has developed an excellent reputation among its clients for the management of large construction projects in remote areas of Canada and in developing countries. It has managed the construction of whole townsites, airfields, and dams for large hydroelectric power generating stations; engineered, designed, and supervised construction of major traffic intersections, shopping complexes, bridges, and industrial parks; and designed and supervised numerous large buildings such as hotels, schools, arenas, and manufacturing plants. More recently, the company has entered the systems engineering field. Some of its current studies involve problems in air pollution, protection of the environment, and the effects of oil exploration and pipeline construction on arctic tundra. To handle so many diverse tasks, Harvey has built a highly adaptable, flexible staff with representation from many scientific and technical disciplines.

Although he is past retirement age, Harvey Winman maintains an active interest in his company's business. Short, with a round face topped by a bald head with a fringe of hair, he is often seen strolling from department to department. He stops frequently to talk to his staff (most of whom he knows by their first name) and to enquire both about the projects they are working on and their home lives. Well liked and thoroughly respected by all, from senior management down to the newest clerk, he is a manager of the old school, a type seldom seen in today's brusque business atmosphere.

The company's chief executive is Martin Dawes, to whom Harvey has handed full operating control. Martin is a civil engineer who developed a fine reputation for designing and constructing hydroelectric dams in remote, mountainous country. An ardent fisherman, he has often combined field projects with trips into the surrounding countryside, from which he has returned laden with salmon or trout. Before becoming president, he was vice president for two years and, before that, head of the civil engineering department for 15 years.

Both Harvey Winman and Martin Dawes write well, and they expect their staff also to be good writers. They recognize that the written word is the means that most often conveys

Martin Dawes. (Photo: Bob Wick)

an image of their company to customers. So, to ensure that all letters, reports, and company publications are of top quality, they have introduced a technical editor into their management team.

The editor is Anna King, who graduated from McGill University with a Bachelor of Science in electrical engineering. She worked for the Southern Ontario Power and Light Company for nine years, first in the relay department and later in development engineering, where her capabilities both as engineer and report writer became evident.

Computer technology has significantly changed a technical editor's role

Eventually Anna moved west and became H L Winman and Associates' technical editor. "Yours will be an exacting job," Martin Dawes said on her first day. "You will be setting writing standards for the whole company. I don't want my engineers to think you are here simply to do their writing for them. Your role is to encourage them to become better writers."

Ontario Affiliate: Macro Engineering Inc

To increase his company's presence in Eastern Canada, Harvey Winman purchased two companies in the Toronto area and amalgamated them to form Macro Engineering Inc (MEI). One was Robertson Engineering Company (Wayne G Robertson, president), which had a well-established reputation for designing and custom-manufacturing high-technology electronic, nucleonic, and electro-mechanical instruments. The other was Mactiere Microelectronics (Tina Mactiere, president), which specialized in designing computer interface systems.

When he amalgamated the two companies, Harvey combined the first three letters of Mactiere Microelectronics with the first two letters of Robertson Engineering to form Macro Engineering Inc. Tina and Wayne agreed to head the new company jointly, with Tina as president and chief executive officer (CEO) and Wayne as general manager and marketing manager, as shown in the company's partial organization chart in Figure 2-3.

Macro Engineering Inc has approximately 350 employees, of whom 225 are engineers, technologists, and computer science specialists. The company occupies the first two floors of

Tina R Mactiere, president of Macro Engineering Inc. (Photo: Bob Wick)

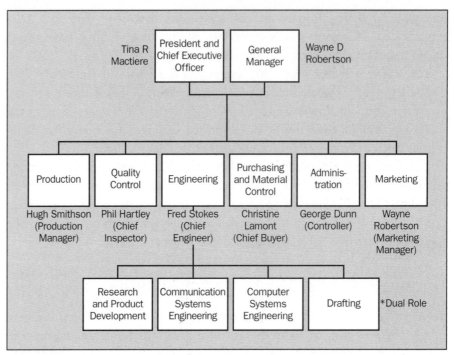

Figure 2-3 Partial organization chart showing engineering department of Macro Engineering Inc.

the Wilshire Building, an office and light-manufacturing complex in Toronto, owned by the Wilshire Insurance Company.

Acquiring Macro Engineering Inc brought a high-tech component into Harvey Winman's company

Like Harvey Winman, both Tina and Wayne recognize the importance of communication in their business, and particularly stress that the operating instructions and training manuals that accompany their products be abundantly clear. To this end, they employ a small technical writing group that can write manuals not only in English, but also in French, German, and Spanish. Occasionally they have Anna King fly in from Calgary to prepare particularly important technical proposals.

Many of the letters and reports in this book are based on work done by these two companies. Similarly, many of the end-of-chapter assignments assume you are employed by one of the companies and that the letter or report you are to write occurs naturally as part of your work.

Your Role as a Technical Employee

As you read the remaining chapters in this book, you will now be able to identify your "employer" and visualize yourself as an "employee." But you still need to know what your role will be as an engineer, technologist, or computer specialist working for either H L Winman and Associates or Macro Engineering Inc.

Working for a Consulting Firm

Consultants provide specialist help to other companies or government organizations. Most clothing manufacturers, for example, are unlikely to have engineering help on staff, other than perhaps an electrical and a mechanical technician to deal with day-to-day maintenance. A clothing manufacturer who wants to build an extension to the plant, introduce robotics into the manufacturing operation, or correct an in-plant air pollution problem, does not have the expertise to do the work. Conceivably, the manufacturer could go directly to a builder and ask for a design and construction estimate, or to a manufacturer of robotics equipment and ask for a proposal, or to an air-conditioning company and request improved air circulation, but in each case the clothing manufacturer would be buying limited, rather biased, expertise. Often, too, a nontechnical company does not want to develop construction specifications, request and screen proposals, or coordinate and supervise contractors' work ("That can be one big headache," most agree). They would rather leave all these details to a consultant.

What will your role be, after you graduate?

By calling in an engineering consultant the manufacturer is in effect saying, "We have a problem and we want you to use all the resources at your command to investigate it, recommend how it can be corrected, and (sometimes) manage the corrective action for us." Because consultants are not "tied in" with particular builders, robotics manufacturers, air-conditioning installers, and so on, they have an unbiased perspective from which to view their clients' needs. Some consultants specialize in a particular field, such as civil engineering or environmental control, while others offer a fairly wide range of consulting services, as does H L Winman and Associates. And if a general consultant does not specialize in a particular area in which expertise is needed, the consultant will collaborate with another consultant to provide the necessary services for a client.

The steps in a client-engineering consultant relationship are generally as described below and illustrated in Figure 2-4. There are two phases: the initial contract between the client and consultant, followed by a study or investigation; and the coordination of a construction, renovation, or service contract that may evolve from the investigation.

Phase 1:
1. The client approaches the consultant, describes what is required or the problem that needs to be corrected and, if the project is likely to be extensive, asks for a proposal and cost estimate. (If a project is small or straightforward, the client may dispense with the proposal and instruct the consultant to go ahead with the project.)
2. The consultant examines what is involved and then prepares a proposal outlining

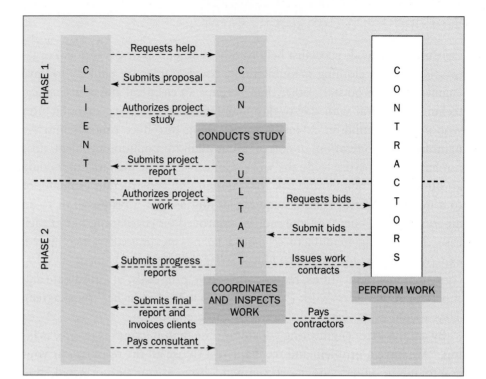

Figure 2-4 A simplified view of communication between (1) the client and consultant, and (2) the consultant and contractors.

If you work for a firm of engineering consultants...

...you could have one of several roles

- the consultant's understanding of the problem or project, and the work involved,
- what human and material resources will be used,
- any reports that will be prepared,
- the consultant's previous experience with similar projects (for the client to assess the consultant's capabilities), and
- the costs for (a) carrying out the investigation and (b) carrying out the work.

3. If the consultant's proposal is accepted, the client writes a letter or issues a purchase order authorizing the consultant to go ahead with the project.
4. The consultant conducts the investigation and then prepares a detailed report outlining what was found out and what corrective action needs to be taken. (If the project is lengthy, periodically the consultant will send a progress report to the client.)

Some client-consultant arrangements stop here. If, however, the client wants the consultant to manage the ensuing construction, installation, renovation work, or service, the relationship continues.

Phase 2:

5. The client authorizes the consultant to manage the project through to completion.
6. The consultant then

  - prepares a work plan and job specification,
  - calls for bids from contractors,
  - issues contracts to the most suitable contractors,
  - coordinates and monitors the contractors' work,
  - deals with problems,
  - submits regular progress reports to the client,
  - inspects completed work, asks for correction of unsatisfactory work, pays the contractor, and
  - prepares a job completion report and invoice for the whole project, and submits both to the client.

As an engineer, technologist, technician, or computer specialist, you may be involved in any of these stages, and you will be expected to write reports at the appropriate times both for your own management and for your company's clients. Good communication is an essential component in any client-consultant relationship, because clients who are "kept in the picture" subconsciously feel that the consultant is effectively looking after their interests.

Working for an Engineering or Technical Firm

If you are employed by a technical firm, you may still provide services to clients and customers. You may be a technical service representative who installs, repairs, modifies, and maintains electronic, computer, or mechanical equipment leased or sold by your company. As such you will be expected to be an expert not only on the technical aspects but also in interpersonal relations. A service representative is both a technical specialist and a salesperson.

If you work for a technical company...

As a technical service representative you will write reports describing both routine maintenance and special conditions. After each service visit you will have to describe what you observed about the equipment, what repairs or adjustments you made, and what components you installed or replaced. Generally, much of this information will be written onto a preprinted form that has appropriate spaces for filling in details. But when special visits are made or problems occur, you will more likely have to write a short investigation report in which you describe a problem and how you resolved it, or an inspection report in which you describe the condition of equipment or a system. (For examples of such reports, see Chapter 4.)

...your role may deal directly with equipment or a system

Alternatively, you may be employed by a contractor to direct, coordinate, and supervise the work of installation teams, and to provide engineering assistance when problems occur. Much of the time you will work at the job site, and will keep management at head office informed of progress by submitting reports describing the teams' progress and problems you encountered. (In this capacity, you may be on the payroll of one of the contractors providing services to the consultants described earlier.)

Or you may even be employed by a design company or a manufacturer, for whom you will provide engineering or technical services that may include

- designing new products or processes, new computer software, or modifications to existing systems,
- investigating manufacturing problems or assembly line "glitches,"
- proposing improved manufacturing methods,
- installing and testing prototype equipment and components,
- instructing operators and users of new or modified equipment, and
- inspecting work performed by assemblers and junior technicians.

In such a capacity your reporting will have to be comprehensive, for you will be expected to describe what you have seen, present management with an analysis or justification, and sometimes recommend corrective action. You will have to use all the report-writing skills suggested in Chapters 4 through 6 for written reports, and Chapter 9 for spoken reports.

WEBLINKS

Technical Communication Overview
www.vsl.ist.ucf.edu/~deej/itoverview.html

A detailed description of the role of the technical communicator is available at this site.

Society for Technical Communication
stc.org/

With more than 20 000 members worldwide, STC is the largest professional organization serving the technical communication profession. The society's diverse membership includes writers, editors, illustrators, printers, publishers, educators, students, engineers, and scientists employed in a variety of technological fields.

Chapter 3
Technical Correspondence

When you write a personal letter to a friend or relative, you probably do not worry whether your letter is too long or contains too much information. You assume your reader will be pleased to hear from you, so you launch into a general discourse, inserting comments and items of general news without concerning yourself very much about organization.

But when you write a business letter you have to be much more disciplined. Now your readers are busy people who want only the details that specifically concern them. Information they do not need irks them. For these people your letters must be focused, well planned, brief, and clear.

Focusing the Letter

Letter writing is like photography: you have to centre the image in the viewfinder and then focus the image sharply before you snap the picture. But first you have to decide exactly what you want to depict; otherwise viewers will wonder why they are looking at the picture.

Identifying the Main Message

If you write your letters pyramid-style, you will automatically focus the reader's attention on your main message. Before you place your fingers on the keyboard, fix clearly in your mind *why you are writing* and *what you most want your reader to know*. Then focus sharply on this information by placing it right up front, where it will be seen immediately.

If you begin a letter with background information rather than the main point, your reader will wonder why you are writing until he or she has read well into the letter. Don McKelvey's letter to Jim Connaught is a typical example of an unfocused letter.

Readers want to know *right away* what you most need to tell them

Dear Mr Connaught

I refer to our purchase order No. 21438 dated April 26, 1997, for a Vancourt microcopier model 3000, which was installed on May 14. During tests following its installation your technician discovered that some components had been damaged in transit. He ordered replacements and in a letter dated May 20 informed me that they would be shipped to us on May 27 and that he would return here to install them shortly thereafter.

It is now June 10, and I have neither received the parts nor heard from your technician. I would like to know when the replacement parts will be installed and when we can expect to use the microcopier.

Sincerely

Don McKelvey

Readers don't want to plough through para-graphs of background information before they encounter your main message

Jim had to read more than 70 words before he discovered what Don wanted him to do. If Don had written pyramid-style, starting with a main message, Jim would have known immediately why he was reading the letter:

Dear Mr Connaught

We are still unable to use the Vancourt 3000 microcopier we purchased from you on April 26, 1997. Please inform me when I can expect it to be in service.

And placing the main message up front would have helped Don write a shorter explanation that would have been simpler to follow:

The microcopier was ordered on P.O. 21438 and installed on May 14. During tests, your technician discovered that some components had been damaged in transit. He ordered replacements, then in a letter dated May 20 informed me that they would be shipped to us on May 27, and that he would return here to install them. To date, I have neither received the parts nor heard from your technician.

Sincerely

Don McKelvey

Unfortunately, knowing you should open every letter with a main message is not enough. You also need to know how to find exactly the right words to put at the top of the pyramid. And that is where many technical people have trouble.

Getting Started

To overcome this block, try using a technique recommended by Anna King, H L Winman and Associates' technical editor. She suggests that when you start a letter, first write these six words:

I want to tell you that...

And then finish the sentence with what you *most* want to tell your reader. For example:

Dear Ms Reynaud

*I want to tell you that...*the environmental data you submitted to us on October 8 will have to be substantiated if it is to be included with the Labrador study.

Then, when your sentence is complete, *delete* the first six words (the *I want to tell you that...* expression). What you have left will become a focused opening statement:

Dear Ms Reynaud

The environmental data you submitted to us on October 8 will have to be substantiated if it is to be included with the Labrador study.

Often you can use an opening statement formed in this way just as it stands when you remove the six "hidden" words. At other times, however, you may feel the opening statement seems a bit abrupt. If so, you can soften it by inserting a few additional words. For example, in the letter to Ms Reynaud, you might want to add the expression "I regret that...":

Dear Ms Reynaud

I regret that the environmental data you submitted to us on October 8 will have to be substantiated if it is to be included with the Labrador study.

Figure 3-1 depicts this convenient way to start a letter and concurrently create a main message. It also shows that in business letters the main message is more often referred to as the **Summary Statement**.

Here are three more examples of properly formed Summary Statements:

Dear Colonel Watson

We will complete the XRS modification on June 14, eight days earlier than scheduled.

This proven technique will never fail you!

Writing newspaper-style will help you get started

The writer's pyramid
helps draw attention to
the most important
information

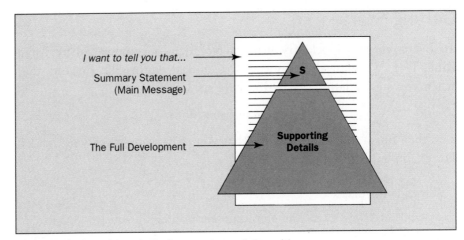

Figure 3-1 Creating a letter's summary statement.

Dear Ms Mohammed

Your excellent paper "Export Engineering" arrived just in time to be included in the program for the Pacific Rim Conference.

Dear Mr Voorman

Seven defective castings were found in shipment No. 308.

(You can check that *I want to tell you that...* was used to form these three opening sentences by mentally inserting the six hidden words at the start of each sentence.)

Avoiding False Starts

If you do not use the six hidden words to start a letter, you may inadvertently open with an awkwardly constructed sentence that seems to be going nowhere. For example:

Dear Mr Corvenne

In answer to your enquiry of December 7 concerning erroneous readouts you are experiencing with your Mark 17 Analyser, and our subsequent telephone conversation of December 18, during which we tried to pinpoint the fault, we have conducted an examination into your problem.

Anna King refers to a long, rambling opening like this as "spinning your wheels," because such a sentence does not come to grips with the topic early enough. She has prepared a list of expressions (see Figure 3-2)

WHEN YOU WRITE A LETTER...

Never start with a word that ends in "ing":
> *Referring...*
>
> *Replying...*

Never start with a phrase that ends with the preposition "to":
> *With reference to...*
>
> *In answer to...*
>
> *Pursuant to...*
>
> *Due to...*

Never start with a redundant expression:
> *I am writing...*
>
> *For your information...*
>
> *This is to inform you...*
>
> *The purpose of this letter is...*
>
> *We have received your letter...*
>
> *Enclosed please find...*
>
> *Attached herewith...*

IN OTHER WORDS...

DON'T SPIN YOUR WHEELS!

Figure 3-2 Anna King's suggestion to H L Winman engineers.

Try inserting *I want to tell you that...* in front of these openings: it doesn't work!

that can easily cause you to write complicated, unfocused openings. In their place she recommends starting with the *I want to tell you that...* expression, which will help you focus your reader's attention on the main message. If the letter referring to the Mark 17 Analyser had started this way, it would have been much more direct:

Dear Mr Corvenne

(*I want to tell you that...*) The problem with your Mark 17 Analyser seems to be in the extrapolator circuit. Following your enquiry of December 7 and your subsequent description of erroneous read-outs, we examined... (etc).

A "getting down to business" start

Planning the Letter

When you have identified and written the main message, your next step is to select, sort, and arrange the remaining information you want to convey to your reader. These details should amplify the bare message you have already presented and provide evidence of its validity. For example, when Paul Shumeier wrote the following Summary Statement he realized he would be presenting his reader with costly news:

> Dear Mr Larsen
>
> Tests of the environmental monitoring station at Wickens Peak show that 60% of the instruments need to be repaired and recalibrated at a cost of $7265.

He also realized that Mr Larsen would expect the remainder of the letter to tell him why the repairs were necessary, exactly what needed to be done, and how Paul had derived the total cost. To provide this information, Paul first had to identify what questions would be foremost in Mr Larsen's mind after he had read the Summary Statement. This meant asking himself six questions, all based on *Who?*, *Where?*, *When?*, *Why?*, *What?* and *How?*:

Who (was involved)?
Where (did this happen)?
When (did this happen)?
Why (are the repairs necessary)?
What (repairs are needed)?
How (were the costs calculated)?

To insert the answers to these questions into his letter, Paul now had to open up the lower part of the writer's pyramid. This becomes the **Full Development** (or *supporting details*) shown in Figure 3-1.

Opening Up the Pyramid

To help Paul—and you—organize a letter's Full Development, the lower part of the pyramid is stretched vertically and divided into three compartments known as the **Background**, **Facts**, and **Outcome** (see Figure 3-3).

The **Background** covers *what* has happened previously, *who* was involved, *where* and *when* the event occurred or the facts were gathered and, sometimes, for whom the work was done. Paul wrote:

> Our electronics technicians examined the Wickens Peak Monitoring station on May 16 and 17, in response to your May 10 request to Patrick Friesen.

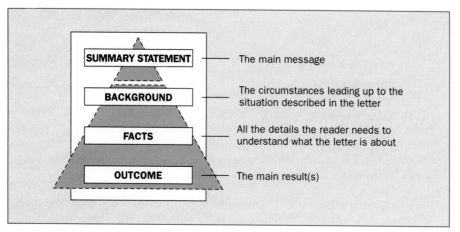

Figure 3-3 Basic writing plan for a business letter or interoffice memorandum.

The **Facts** amplify the main message. They provide specific details the reader needs to fully understand the situation or to be convinced of the need to take further action. Here, Paul wrote:

> Most of the damage was caused by a tree northwest of the site that fell onto the station during a storm on April 23 and damaged parts of the roof and north and west walls. Instruments along these walls were impact-damaged and then soaked by rain. Other instruments in the station were affected by moisture.
>
> Major repairs and recalibration are required for the 16 instruments listed in attachment 1, which describes the damage and estimated repair cost for each instrument. This work will be done at our Shepperton repair depot for a total cost of $4485. Minor repairs, which can be performed on site, are necessary for the 27 instruments listed in attachment 2. These on-site repairs will cost $2780.

Use attachments to simplify a letter

These two paragraphs clearly answer the *Why?*, *What?*, and *How?* questions. Note that, rather than clutter the middle of his letter with a long list, Paul placed the details in two attachments and summarized only the main points in the body of his letter. (The attachments are not shown here.)

The **Outcome** describes the result or any effect the facts have had or will have. If the letter is purely informative and the reader is not expected to take any action, the Outcome simply sums up the main result. Paul would have written:

> I have obtained Ms Korton's approval to perform the repairs and a crew was sent in on May 23. They should complete their work by May 31.
>
> Sincerely

MACRO
ENGINEERING INC.

600 Deepdale Drive, Toronto ON M5W 4R9

FROM: Kevin Toshak   DATE: October 22, 1997
TO: Tina Mactiere   SUBJECT: Monitor Installation at WRC

I have installed a TL-680 monitor unit in room 215 at the Wollaston Research Centre, as instructed in your memorandum of October 15.

The unit was installed without major difficulties, although I had to modify the equipment rack to accept it as illustrated in the attached sketch. Post-installation tests showed that the unit was accepting signals from both the control centre and the remote site.

Kevin T

Figure 3-4 An informative letter in memorandum format.

But if the reader is expected to take some action, or approve somebody else taking action (often, the letter writer), then the Outcome becomes a *request for action*. Because Paul wanted an answer, he wrote:

> If these repair costs are acceptable, please telephone, fax, or email your approval so I can send in our repair crew.
>
> Sincerely

These three parts can help you arrange the Full Development of any letter or memorandum into a logical, coherent sentence. Before starting, however, you have to decide whether you are writing an informative or a persuasive letter.

Write an *Action Statement* if you want your reader to act or react

Writing to Inform

Letters and memorandums that purely inform, with no response or action required from the reader, normally can be organized around the basic Summary Statement-Background-Facts-Outcome writing plan shown in Figure 3-3. Kevin Toshak's memo to Tina Mactiere, in Figure 3-4, falls into this category.

Another example is a confirmation letter, in which the writer confirms previously made arrangements. In the following Macro Engineering Inc memorandum, general manager Wayne Robertson ensures that he and chief buyer Christine Lamont both understand the arrangements that will evolve from a decision made at a company meeting:

	Christine
Statement	I am confirming that you will represent both Macro
Summary	Engineering Inc and H L Winman and Associates at the Materials Handling conference in Montreal on May 15 and 16, 1998, as
Background	agreed at the Planning Meeting on March 23. At the conference you will
Facts	• take part in a panel discussion on packaging electronic equipment from 10:00 to 11:15 a.m. on May 15, and
	• host a wine-and-cheese reception for delegates from 5:00 to 7:00 p.m. on May 16.
	Janet Kominsky is making your travel and hotel reservations, and the catering arrangements for the reception. Anna King will provide brochures from Calgary, and my secretary will make up packages for you to distribute.
Outcome	I'll brief you on other details before you leave.
	Wayne

An informative letter tells *the reader what has been done or what has to be done...*

...It doesn't expect the reader to respond

Although the basic writing plan for letters has four compartments (see Figure 3-3), you do not have to write exactly four paragraphs. As both Wayne's and Kevin's memos show, you may combine two compartments into a single paragraph, or let one compartment be represented by several paragraphs.

Writing to Persuade

In a persuasive letter you expect your reader either to respond to your letter or to take some form of action. Consequently, the writing plan's Outcome compartment is renamed **Action**, as shown in Figure 3-5, to remind you to end a persuasive letter with an action statement. A request and a complaint are typical examples of persuasive letters, and so is the informal proposal described in Chapter 5.

A persuasive letter sells *the reader to take some form of action*

Making a Request

Many technical people claim that placing the message at the start of a letter is not a problem until they have either to ask for something or to give the reader bad news. Then they tend to lead gently up to the request or the unhappy information.

Many writers hesitate to open with a request

Bill Kostash is no exception. He is service manager for Mechanical Maintenance Systems Ltd, and he has to write to customers to ask if they will accept a change in the preventive maintenance contracts his company has with them. He starts by writing to Ms Bea Nguyen, the contracts administrator for Multiple Industries in Cambridge, Ontario:

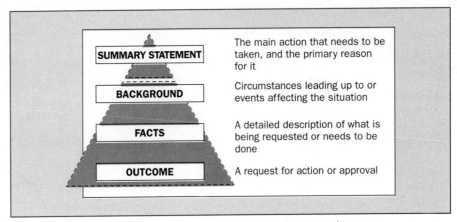

Figure 3-5 Writing plan for a persuasive letter or memorandum.

June 18, 1997

Dear Ms Nguyen

I am writing with reference to our contract with you for the preventive maintenance services we provide on your RotoMat extruders and shapers. Under the terms of the current contract (No. RE208) dated January 2, 1997, we are required to perform monthly inspection and maintenance "...on the 15th day of each month or, if the 15th falls on a weekend or holiday, on the first working day thereafter."

(Bill has got off to a bad start. Instead of opening with a Summary Statement he has inserted all the background details first, with the result that Bea Nguyen does not yet know why he has written to her. He has also opened with one of the expressions Anna King lists as an awkward start in Figure 3-2. Let's see how he continues.)

An unfocused, meandering request letter

Our problem is that almost all of our clients ask that we perform their maintenance service between the 5th and 25th of each month, to avoid their end-of-month peak accounting periods. This in turn created difficulties for us, in that our service technicians experience a peak workload for 20 days and then have virtually no work for 10 days.

(Bea Nguyen still does not know why he is writing.)

Consequently, to even out our workload, I am requesting your approval to shift our inspection date from the 15th to the 29th of each month. If you agree to my request, I will send our technician in to service your machine on June 29—a second time this month—rather than create a six-week period between the June and July inspections. Could you let me know by June 25 if this change of date is acceptable?

Sincerely

William J Kostash

(*Now* Bea knows why Bill has written to her—but she had to read a long way to find out. And probably she had to reread his letter to fully understand the details.)

If Bill had used the writing plan in Figure 3-5 to shape his letter, his request would have been much more effective. The revised letter is shown in Figure 3-6, in which

1 is his **Summary Statement** (he states his request and what the effect will be),

2 contains the **Background** (the contract details),

Writing with a plan creates a coherent request

3 contains the **Facts** (it describes the problem), and

4 is the **Action** statement, in which he mentions *two* actions: what he wants Bea to do (to call him) and what he will do (schedule a second visit).

Registering a Complaint

The approach is the same if you have to write a letter of complaint or ask for an adjustment. You can use the writing plan for a persuasive letter shown in Figure 3-5, inserting the following information into each compartment:

- **Summary Statement.** State the problem and say what you want the reader to do about it:

Complaint letters, particularly, need to be structured pyramid-style

> Dear Mr Bruyere
>
> The Nabuchi 700 portable computer you recently sold me had a defective lithium-ion battery that had to be replaced while I was outside Canada. Consequently I am requesting reimbursement of the repair expenses I incurred.

(Often, it is better to generalize what action is needed in the Summary, and then to state exactly what has to be done in the Action compartment.)

- **Background.** Identify the circumstances leading up to the event, quoting specific details:

> I bought the computer and a Nabuchi 701PC international power converter from your Willows Mall store on September 4, 1997, on Sales Invoice No. 14206A.

Set the scence,...

(If there are few Background details, you may combine them with either the Summary Statement or the Facts rather than place them in a very short independent paragraph.)

MECHANICAL MAINTENANCE SYSTEMS LTD
2120 Cordoba Avenue
London, Ontario N6B 2G6

June 18, 1997

Ms Bea Nguyen
Contracts Administrator
Multiple Industries Limited—Manufacturing Division
18 Commodore Bay
Cambridge ON N1R 5S2

A focused, definite, direct request

Dear Ms Nguyen

1 I am requesting your approval to change the date of our monthly preventive maintenance visits to service your RotoMat extruders and shapers to the 29th of each month. This will help spread my technicians' workload more evenly and so provide you with better service.

2 Our contract with you is No. RE208 dated January 2, 1997, and it requires that we perform a monthly inspection and maintenance on the 15th day of each month. Unfortunately, almost all of our clients ask that we perform their **3** maintenance service between the 5th and the 25th. This creates a problem for us in that our service technicians experience a peak workload for 20 days and then have very little work for 10 days.

4 Could you let me know by June 25 if you can accept the change? Then I will send a technician to your plant on June 29 for a second visit this month, rather than create a six-week space between the June and July inspections.

Sincerely

William J Kostash

William J Kostash
Service Manager

Figure 3-6 A request letter written pyramid-style.

- **Facts.** Describe exactly what happened, in chronological order, so that the reader will understand the reason for your complaint or request for adjustment:

  > The computer worked satisfactorily for the first six weeks, but during that time I had no occasion to use it solely on battery power.

  > On October 25 I left for Europe, first giving the batteries an 18-hour charge as recommended in the operating instructions. While using the computer in flight, after only 35 minutes the low-battery lamp lit up and the screen warned of imminent failure. I recharged the batteries the following day, in Rheims, France, but achieved less than 25 minutes of operating time before the batteries again became fully discharged.

  > As the Nabuchi line is neither sold nor serviced in France, I had to buy and install a replacement lithium-ion battery (a Mercurio Z7S), which has since worked admirably. I enclose the defective battery, plus a copy of the sales receipt for the replacement battery I purchased from Lestrange Limitée, Rheims.

  ...offer the details,....

- **Action.** Identify specifically what action you want the reader to take, or that you will take:

  ...and end with a firm Action Statement

  > Please send me a cheque for $244.30, which at the current rate of exchange is the Canadian equivalent of the 1190 francs shown on the sales receipt.

  > Sincerely

  > *Suzanne Dumont, P.Eng*

Creating a Confident Image

Readers react positively to letters and memorandums in which the writer conveys an image of a confident person who knows the subject well and has a firm idea of what he or she plans to do, or expects the reader to do. Such an image is conveyed by both the quality of the writing and the physical appearance of the piece of correspondence.

Be Brief

For technical business correspondence, brevity means writing short letters, short paragraphs, short sentences, and short words.

Short Letters

The key word here is "short"

A business reader will tend to react readily to a short letter, viewing its writer as an efficient provider of information. In contrast, the same reader may view a long letter as "heavy going" (even before reading it) and tend to put it aside to deal with it later. A short letter introduces its topic quickly, discusses it in sufficient depth, and then closes with a concluding statement, its length dictated solely by the amount of information that has to be conveyed.

We know of a company in which the managing director has ruled that no letter or memorandum may exceed one page. This is an effective way to encourage staff to be brief, and it works well for many people. But for letter writers who have more to say than they can squeeze onto a single page, the limitation can prove inhibiting. For them, we suggest borrowing a technique from report writing. Instead of placing all their information in the letter, they should change the letter into a semiformal report and then summarize the highlights—particularly the purpose and the outcome—into a one-page letter placed at the front of the report (so that the report becomes *an attachment* to the letter, as depicted in Figure 3-7).

If you use this device, in the body of the letter both refer to the attachment *and* insert a main conclusion drawn from it, as has been done here:

A short cover letter is like an executive summary (see page 186)

During the second week we measured sound levels at various locations in the production area of the plant, at night, during the day, and on weekends. These readings (see attachment) show that a maximum of 55 dB was recorded on weekdays, and 49 dB on weekends. In both cases these peaks were recorded between 5 and 6 p.m.

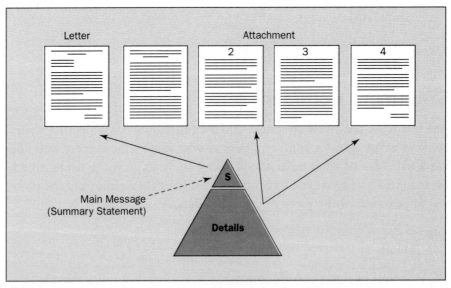

Figure 3-7 A short letter with attachments is an adaptation of the pyramid method of writing. An example can be seen in Figure 5-4 (page 135).

Short Paragraphs

Novelists can afford to write long paragraphs because they assume they will have their readers' attention, and their readers have the time and patience to make their way through leisurely descriptions. But in business and industry readers are working against the clock and so want bite-size paragraphs of easy-to-digest information.

Think of a paragraph as a miniature pyramid

Let the first sentence of each paragraph introduce just one idea, then make sure that subsequent sentences in that paragraph develop the idea adequately and do not introduce any other ideas. In technical business writing the first sentence of each paragraph should be a "topic sentence," so that the reader immediately knows your main idea. Consequently, a busy reader who skims your document by reading just the first sentence of every paragraph will still gain a good understanding of the main points. Let the first sentence of each paragraph summarize the paragraph's contents, and the remaining sentences support the first sentence by providing additional information:

We have tested your 15 Vancourt 801 CD-ROM drives and find that 11 require repair and recalibration. Only minor repairs will be necessary for 6 of these drives, which will be returned to you next week. Of the 5 remaining drives, 3 require major repairs which will take approximately 20 days, and 2 are so badly damaged that repairs will cost $180 each. Since this is more than the maximum repair cost imposed by you, no work will be done on these drives.

If an idea you are developing breeds an overly long paragraph, try dividing the information into a short introductory paragraph and a series of subparagraphs, as has been done here:

My inspection of the monitoring station at Freedom Lake Narrows revealed three areas requiring attention, two immediately and one within three months:

1. The water stage manometer is recording erratic readouts of water levels. A replacement monitor needs to be flown in immediately so that the existing unit can be returned to a repair depot for service.

2. The tubing to the bubble orifice is worn in several places and must be replaced (30.5 metres of 8 mm tubing will be required). This work should be done concurrently with the monitor replacement.

3. The shack's asphalt roof is wearing and will need resurfacing before freeze-up.

Paragraphs that are longer than 8 or 9 printed lines are too long

We are not suggesting that your letters should contain a series of small, evenly sized paragraphs, which would appear dull and stereotyped.

Paragraphs should vary from quite short to medium-long to give the reader variety. How you can adjust paragraph and sentence length to suit both reader and topic, and also to place emphasis correctly, is covered in Chapter 11.

Short Sentences

Convoluted sentences create the impression their writer is confused

If you write short, uncomplicated sentences, you will be helping your readers quickly grasp and readily understand each thought you present. Sometimes expert literary writers can successfully build sentences that develop more than one thought, but such sentences are confusing and out of place in the business world. Compare these two examples of the same information:

Complicated There has been intermittent trouble with the vacuum pumps, although the flow valves and meters seem to be recording normal output, and the 18 cm pipe to the storage tanks has twice become clogged, causing backup in the system.

Clear There has been intermittent trouble with the vacuum pumps, and twice the 18 cm pipe to the storage tank has become clogged and caused backup in the system. The flow valves and meters, however, seem to be recording normal output.

The first example is confusing because it jumps back and forth between trouble and satisfactory operation. The second example is clear because it uses two sentences to express the two different thoughts.

Short Words

Short words are especially important for readers whose first language is not English

Some engineers and engineering technologists feel that the technical environment in which they work, and the complex topics they have to write about, demand that they use long, complex words in their correspondence. Similarly, some people feel that long words build credibility and respect for their position; the opposite is true. They write "an error of considerable magnitude was perpetrated," rather than simply "we made a large error." In so doing, they make a reader's job unnecessarily difficult.

Because the engineering and scientific worlds encompass many long and complex terms that have to be used in their original form, to make your correspondence more readable surround such technical terms with simple words. Be aware, too, that in today's global society, your readers may read and write English as a second language. Long words that are not in the average person's vocabulary may cause confusion and misunderstanding. Chapter 11 has more information about writing for an international audience.

Be Clear

A clear letter conveys information simply and effectively, so that the reader readily understands its message. To write clearly demands ingenuity and attention to detail. As a writer you must consider not only how you will write your letters, but also how you will present them.

Create a Good Visual Impression

Experienced writers know that a nicely laid out letter impresses the reader. Subconsciously it seems to be saying "my neat appearance demonstrates that I contain quality information that is logically and clearly organized."

The appearance of a letter tells much about the writer and the company he or she represents. If a letter is sloppily arranged or contains strikeovers, visible erasures, or spelling errors, then its readers imagine a careless individual working in a disorganized office. But if readers are presented with a neat letter, placed in the middle of the page and printed by a quality printer, then they imagine a well-organized individual working for a forward-thinking company noted for the quality of its service. Most people prefer to deal with the latter company and will read its correspondence first.

Clarity depends on appearance as well as clarity of expression

Develop the Subject Carefully

The key to effective subject development is to present the material logically, progressing gradually from a clear, understood point to one that is more complex. This means developing and consolidating each idea for the reader's full understanding before attempting to present the next idea. The sections on paragraph unity and coherence in Chapter 11 (see pages 371 and 372) provide examples of coherent paragraphs. Using the pyramid technique also will help you structure your information in a logical sequence.

Be Definite

People who think better with their fingers on a keyboard or a pen in their hand sometimes make decisions as they write, producing indecisive letters that are irritating to read. These writers seem to examine and discard points without really grappling with the problem. By the time they have finished a letter, they have decided what they want to say, but it has been at the reader's expense.

Know clearly what you want to say before you start writing

Decision making does not come easily to many people. Those of us who hesitate before making a decision, who evaluate its implications from all possible angles and weigh its pros and cons, may allow our indecisiveness to creep into our writing. We hedge a little, explain too much, or try to say how or why we reached a decision before we tell our reader what the decision is. This is particularly true when we have to tell readers something unfavorable or contrary to their expectations.

The key, as before, is to use the pyramid:

1. Decide exactly what you want to say (i.e. develop your main message), and then
2. Place the main message right up front (use *I want to tell you that...* to get started).

If you also write primarily in the active voice, you will sound even more decisive. Active verbs are strong, whereas passive verbs are weak. For example:

Write directly from person to person, and name the "doer"

These passive expressions	*Should be replaced with*
it was our considered opinion	we considered
it is recommended that	I recommend
an investigation was made	we investigated
the outage was caused by a defective transformer	a defective transmitter caused the outage

For hints on how to use the active voice, see "Emphasis" in Chapter 11 (pages 380 to 382).

Use Information Design to Draw Attention

You can help your readers understand and access information by the appearance of your documents. We call this *Information Design*. The suggestions we provide here will help the information jump off the page and make it easier to read and understand. This doesn't mean, however, that the content and structure of your document aren't important. Information Design will just enhance the readability.

Computerized word processing makes it much easier to incorporate Information Design techniques into all types of documents: letters, memos, reports, faxes, notices.

Insert Headings as Signposts

In longer letters, and particularly those discussing several aspects of a situation, help your reader by inserting headings. Each heading must be informative, summarizing clearly what is covered in the paragraphs that follow. (Figure 11-2 on page 369 offers advice on creating and inserting headings.) If, for instance, we had replaced the heading preceding this paragraph with the single word *Headings*, it would not have summarized adequately what this paragraph describes.

Pick Only One Font

A font such as Century OldStyle or Helvetica is a set of printing type consisting of the same features. Some fonts are called serif (they have a slight finishing stroke - T) and some are called sans serif (they don't have a finishing stroke - T). The font you choose will project an image of you, your company and your document. Statistics show that a serif font like Century OldStyle is easier to read because the serifs lead the eye from letter to letter, and so are more suitable for longer documents. (This text is printed in a serif type called Sabon.) Sans serif fonts, like Helvetica or Franklin Gothic are clean, clear, and portray a neat and modern image, yet are not as easy to read and so are suitable for shorter documents.

Daniel Thomashewski chose the sans serif font Helvetica to type the agenda for the Electronic Facsimile Research Committee meeting in Figure 9-3 on page 308). Morley Wozniak, however, used Times New Roman for his report evaluating proposed landfill sites in Figure 5-4 on page 135).

Once you decide on a font, stay with it. Don't switch to a different font to show emphasis. Instead, use **bold,** *italic* or a larger character size to emphasize particular sections of text. Notice, however, that Morley's cover letter and report are printed on H L Winman and Associates company letterhead which uses a sans serif font. Letterhead and logos are excluded from the "maintain one font" guideline.

Make sure that the character size you choose is appropriate for your document and audience. In a one-page letter or memo we sometimes use 10 pt (point size) to help keep the document to only one page. (The smaller the font, the less space it takes.) In a longer document we use 12 pt because the type will be slightly larger and the reader's eye won't tire so easily. If you know your audience is older, use a larger font—at least 12 pt—to help accommodate their eyesight.

Justify Only on the Left

Word processors make it easy to justify both the left and right margins, which permits you to create lines of exactly the same length. We recommend you justify text only at the left margin and leave the text at the right margin "ragged," otherwise the computer will generate spaces between words and characters to force the right margin to be straight. Unless you are using very sophisticated word-processing software, such as has been used for this book, the uneven spaces may prove stressful for the readers' eyes since they constantly have to adjust to the unevenness. It may be only a subtle difference, yet it's something you as an author can control and so make the reader's task more pleasant.

Morley Wozniak has used a ragged right margin for his report on landfill sites in Figure 5-4 (see page 135), whereas Karen Woodhouse has used

The availability of numerous type styles is not an invitation to mix and match

Select character size to suit font style and readers' visual acuity

Let careful use of white space focus readers' attention

a justified right margin for her formal report on radiant heating in Figure 6-8 (see pages 197 to 207).

Use Wider Margins to Draw Attention

Many technical people hesitate to change the standard settings that come with word-processing packages. Yet once they see the value of being unique, they are easily convinced to try a new way. For example, Anna King, the Technical Editor at H L Winman and Associates, encourages the company's engineers to use a wider left margin in their longer documents and to place the headings all the way to the left. She explains that this helps draw readers' attention to the headings and so helps them retrieve information faster.

Figure 3-8 shows a page from a letter proposal presented with normal paragraphs and headings (a), and a page with a wider left margin and left-justified headings (b). The latter may use more space, but it makes the information much more accessible.

Adding white space or blank areas in your text is another valuable Information Design technique. You can also use diagrams and figures to break up long passages of text and to complement the message. A simple flow chart or table makes a nice diversion for the reader and makes the information more visual and concrete. For example, Bob Walton's memorandum-form incident report in Figure 4-3 on page 88 describes an accident in which he was involved. He could have described the positions of the vehicles prior to the accident, but instead he sketched a diagram and attached it to his memo.

Use Subparagraphs to Present Ideas

Anna King encourages H L Winman engineers to use bulleted lists to break up text and make it visually accessible. Morley Wozniak effectively

Shaping your information can encourage readers to keep reading

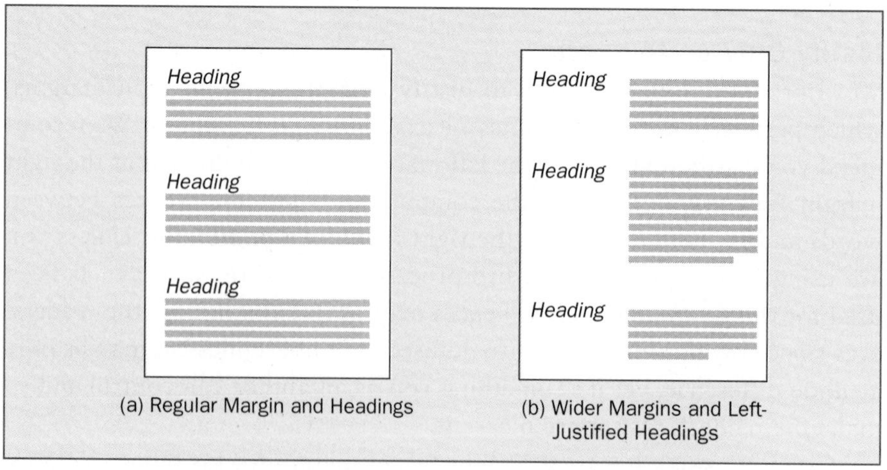

(a) Regular Margin and Headings

(b) Wider Margins and Left-Justified Headings

Figure 3-8 Using standard margins and a wide left margin for text.

used white space in his report evaluating proposed landfill sites (see Figure 5-4 on page 135, and particularly the "chunks" of information on its fourth page). Here's an excerpt of a report Morley wrote before he consulted with Anna:

I have analysed our present capabilities and estimate that we can increase our commercial business from $20 000 to $30 000 per month. But to meet this objective we will have to shift the emphasis from purely local customers to clients in major centres. To increase business from local customers alone will require extensive sales effort for only a small increase in revenue, whereas a similar sales effort in a major centre will attract a 30% to 40% increase in revenue. We will also have to increase our staff and manufacturing facilities. The cost of additional personnel and new equipment will in turn have to be offset by an even larger increase in business. Properly administered, such a program should result in an ever-increasing workload. And, third, we will have to create a separate department for handling commercial business. If we remove the department from the existing production organization it will carry a lower overhead, which will result in products that are more competitively priced.

L-o-n-g paragraphs
build reader resistance

Anna made a simple suggestion, "If you break up the second long sentence by inserting a colon after *we will have to*, and then make a numbered list of the actions you need to take, the information will be much easier to read and understand. Watch what happens visually."

I have analysed our present capabilities and estimate that we can increase our commercial business from $20 000 to $30 000 per month. But to meet this objective we will have to:

Bite-size paragraphs
build reader acceptance

1. Shift the emphasis from purely local customers to clients in major centres. To increase business from local customers alone will require extensive sales effort for only a small increase in revenue, whereas a similar sales effort in a major centre will attract a 30% to 40% increase in revenue.

2. Increase our staff and manufacturing facilities. The cost of additional personnel and new equipment will in turn have to be offset by an even larger increase in business. Properly administered, such a program should result in an ever-increasing workload.

3. Create a separate department for handling commercial business. If we remove the department from the existing production organization it will carry a lower overhead, which will result in products that are more competitively priced.

Table 3-1 Writing plan for a request.

A table can create distinct bite-size compartments of information

Compartment	What goes in it
Summary	A brief description of your request and a request for approval
Background or Reason	The circumstances leading up to the request
Request Details	A detailed explanation of what your request entails, what will be gained if the request is approved, any problems the request may cause, and what the cost will be.
Action	A statement that identifies clearly what you want the reader to do after reading your request.

Use Tables to Capture Information

Displaying text in a table is an alternative way to design information for maximum impact. Many technicians reserve tables for numerical data, but we suggest you also try using tables for presenting text. A table can compartmentalize information into easy-to-find chunks, as Anna King demonstrates in Table 3-1. She uses the table to describe the writing compartments for a request letter. There are other examples in Table 8-3 on page 275 and Table 11-1 on page 373.

Use Columns to Increase Readability

Although columns are rarely suitable for a letter itself, they certainly can be used for attachments to a letter. The value of using columns is described in Figure 3-9.

Use Good Language

It hardly seems necessary to tell you to use good language, but in this case we mean language that you know your readers will understand. Use only those technical terms and abbreviations they will recognize immediately. If you are in doubt, define the term or abbreviation, or replace it with a simpler expression. Chapter 11 provides special guidelines for abbreviating technical and nontechnical terms. In comprehensive letters and proposals, you may find it helpful to attach a Glossary of Technical Terms that defines new or unusual terminology.

Two columns help make
information seem more
accessible

Setting Text in Two Columns Increases Readability

Using a newspaper-style, two-column printed text area helps guide the reader. The line length is forced to be shorter so a reader's eye can follow the line easily and therefore is less stressed. Imagine how difficult it would be to read a newspaper that was printed in one long six-inch wide column!

We don't recommend this format for letters or short reports, but it can have a good impact for longer reports and particularly technical proposals.

This column format is especially useful when you insert graphics into a document. Text and graphics can be clearly integrated by wrapping text around the image. The two are then visually linked.

Graphics used in two-column format also help to balance the page. Often a graphic is inserted and seems to be isolated because there is too much white space around it. With two-columns, each column offsets the other.

Another benefit of the two-column format is that it forces the author to write shorter paragraphs. Otherwise the column would be just one big block of text without any breaks.

Figure 3-9 Setting text in two columns.

Close on a Strong Note

You may feel you should always end a letter with a polite closing remark, such as: *I look forward to hearing from you at your earliest convenience,* or *Thanking you in advance for your kind cooperation.* In contemporary business correspondence—and particularly in technical correspondence—such closing statements are not only outmoded but also weaken your impact on the reader. Today, you should close with a strong, definite statement.

In effect, the Outcome part of the letter provides a natural, positive close, as illustrated by the final sentences in the letters to Mr Larsen (page 41) and Ms Nguyen (Figure 3-6). You should resist the temptation to add a polite but uninformative and ineffective closing remark. Simply sign off with "Regards" or "Sincerely."

Let your *Outcome* or
Action Statement be
your last word

The factors we have described so far are mostly manipulative details that can be learned. Armed with this knowledge and the basic letter formats illustrated later in this chapter, you will have some ground rules to follow in the practical aspects of business letter writing. But you still have to acquire the more difficult technique of letting your personality appear in your letters without letting its presence become too obvious.

Adopting a Pleasant Tone

Sincerity and tone are intangibles that defy close analysis. There is no quick and easy method that will make your letters sound sincere, nor is there a checklist that will tell you when you have imparted the right tone. Both qualities are extensions of your own personality that cannot be taught. They can only be shaped and sharpened through knowledge of yourself and which of your attributes you most need to develop.

To achieve the right tone, your correspondence should be simple and dignified, but still friendly. Approach your readers on a person-to-person basis, following the five suggestions below.

Know Your Reader

Reminder: Know who you are writing to!

If you have not identified your reader properly, you may have difficulty setting the correct tone. You need to know you reader's level of technical knowledge and whether he or she is familiar with the topic you are describing. Without this focus you may unwittingly seem condescending to a knowledgeable reader because you explain too much and use overly simple words, when the reader clearly expects to read technical terms. Conversely, you may just as easily seem overbearing to, or even overwhelm, a reader who has only limited technical knowledge if you confront him or her with heavy technical details.

Ideally you should select just the right terminology to hold the reader's interest and perhaps offer a mild challenge. By letting readers feel they are grasping some of the complexities of a subject (often by using analogies within their range of knowledge), you can present technical information to nontechnical readers without confusing or upsetting them.

Sometimes you will know the person you are writing to, and then you will probably find it much easier to adopt a pleasant tone. Be careful, though, not to make your correspondence *too* chatty or informal. In business and technical writing you should consistently sound professional. You can never tell when your letter may be passed on to someone else!

Be Sincere

Care about both your topic and your reader

At one time it was considered good manners not to permit one's personality to creep into business correspondence. Today, business letters are much less formal and, as a result, much more effective.

Sincerity is the gift of making your readers feel that you are personally interested in them and their problems. You convey this by the words you use and the way you use them. A reader would be unlikely to believe you if you came straight out and said, "I am genuinely interested in your project." The secret is to be so involved in the subject, so interested by it, that you automatically convey the ring of enthusiasm that would appear in your voice if you were talking about it.

Be Human

Too many letters lack humanity. They are written from one company to another, without any indication that there is a human being at the sending end and another at the receiving end. The letters might just as well be sent from computer to computer.

Do not be afraid to use the personal pronouns, "I," "you," "he," "she," "we," and "they." Let your reader believe you are personally involved by using "I" or "we," and that you know he or she is there by using "you." Contrary to what many of us were told in school, letters may be started in the first person. If you know the reader personally, or you have corresponded with each other before, or if your topic is informal, let a personal flavor appear in your letters by using "I" and the reader's first name:

Dear Ben

I read your report with interest and agree with all but one of your conclusions.

Let your presence be apparent

If you do not know your reader personally and are writing formally as a representative of your company, then use the first person plural and the person's surname:

Dear Mr Wicks

We read your report with interest and agree with all but one of your conclusions.

Avoid Words That Antagonize

If you use words that imply the reader is wrong, has not tried to understand, or has failed to make himself or herself understood, you will immediately place the reader on the defensive. For example, when field technician Des Tanski omitted to send motel receipts with his expense account, Andy Rittman (his supervisor) had to write to Des and ask for them. Andy wrote:

You have failed to include motel receipts with your expense account.

This subtly antagonized Des, because the words *you have failed* seemed to imply that he is something of a failure! Andy should have written:

Please send motel receipts to support your expense account.

Other expressions that may annoy readers or put them on the defensive are listed below:

You have *neglected*...
You *ought to known*...
You seem to have *overlooked*...
You have *not understood*...
⎫ *Words that make a reader feel guilty*

You *must* return the instrument...
Your *demand* for warranty service...
We *insist* that you...
I am sure *you will agree*...
⎫ *Words that provoke a reader*

We *have to assume*...
I *must request*...
I *simply do not understand* your...
You must understand our position...
Undoubtedly you will...
⎫ *Words that "talk down" to a reader*

When a reader has to be corrected, the words you use should clear the air rather than electrify it. Tell readers gently if they are wrong, and demonstrate why; reiterate your point of view in clear terms, to clarify any possibility of misunderstanding; or ask for further explanation of an ambiguous statement, refraining from pointing out that his or her writing is vague.

When your goal is to achieve some sort of action or response from the reader, using words that may antagonize will hinder communication. You can still be clear and direct without using these words, and you will find you are more likely to get the result you expect.

Know When to Stop

When a letter is short, you may feel it looks too bare and be tempted to add an extra sentence or two to give it greater weight. If you do, you may inadvertently weaken the point you are trying to make. This is particularly true of short letters in which you have to apologize, to criticize, to say "thank you," or to pay a compliment (i.e. to "pat the reader on the back").

In all of these cases the key is to be brief: Know clearly what you want to say, say it, and then close the letter *without repeating what you have already said*. The following writer clearly did not know when to stop:

Dear Mr Farjeon

I want to say how very much we appreciated the kind help you provided in overcoming a transducer problem we experienced last
month. We have always received excellent service from your organization in the past, so it was only natural that we should turn to you
again in the hour of our need. The assistance you provided in helping us to identify an improved transducer for phasing in our standby
generator was overwhelming, and we would like to extend our
heartfelt thanks to all concerned for their help.

Sincerely

Paul Marchant

Simple words are much
more meaningful than
flowery expressions

Paul's letter would have been much more believable if he had simply said
"thank you":

Dear Mr Farjeon

Thank you for your prompt assistance last month in identifying an
improved transducer for phasing in our standby generator. Your help
was very much appreciated.

Regards

Paul Marchant

If a writer says too much when saying thank you or apologizing, the
reader begins to doubt the writer's sincerity. You cannot set a realistic tone
if you overstate a sentiment, if you gush, or if you overwhelm your reader with the intensity of your feelings.

Writing Electronic Mail

Electronic mail (email) has become one of the fastest and most widely used
means of communication. The criterion for writing email remains the same
as for all other forms of written communication: keep the message brief but
clear. Never allow an overzealous desire for brevity to cloud your message,
because it can cost more to question an obscure communication than it
would have cost to write a slightly longer but clearer message in the first
place. You want to avoid having your reader make a telephone call or email
you to find out exactly what you were trying to say.

When Mike Toller in Vancouver, BC, opened up a shipment of parts
from Carlson Distributors, he found the order was incomplete and contained some items he had not ordered. He made a note of the deficiencies,
sat at his keyboard, and typed in this message:

The fast new way to
communicate...

> To: Carlson Distributors, Montreal, Quebec:
>
> Your inv 216875 Oct 19, our P.O. W1634. Short-shipped 10 toolsets MKV, 4 801 sockets plus 2 doz mod 280A lathe bits unordered. Advise.
>
> M Toller
> Crown Manufacturing, Vancouver

...needs just as much care and attention

In Montreal, Chantal Goulet puzzled over the message on her computer screen, and then typed this brief reply:

> To: M Toller, Crown Manufacturing, Vancouver
> From: C Goulet, Carlson Distributors, Montreal
>
> The message you sent regarding our invoice 216875 and your P.O. W1634 was difficult to understand. Please explain your concerns.
>
> Chantal Goulet

Mike was surprised: he thought his message was crystal clear. So he again sat at his keyboard and wrote:

What Mike Toller should have written the first time

> To: Chantal Goulet, Carlson Distributors, Montreal
> From: Mike Toller, Crown Manufacturing, Vancouver
>
> My message was quite clear: You short-shipped us 10 toolsets type MKV and 4 No. 801 sockets. You also shipped 2 dozen model 280A lathe bits we did not order. Please ship the missing items and advise how you want the bits returned.
>
> Mike Toller

Chantal replied in three words:

> Mike:
>
> Thanks. Will investigate.
>
> Chantal

If Mike had been more explicit when he wrote his original message, both he and Chantal would have saved time.

Email Netiquette

The word "Netiquette" means "the etiquette of writing electronic mail on the net (the Internet)." There are no established guidelines for Netiquette, but we can give you some suggestions that will help you be an efficient email communicator.

First, email does not give you a licence to

- write snippets of disconnected information,
- write incorrectly constructed sentences,
- forget about using proper punctuation,
- ignore misspelled words, or
- be abrupt or impolite.

Keep the message short but keep it clear

Neither, however, is it a forum for telling long stories and anecdotes. You can use regular mail—"snail mail"—for that.

Write "Pyramid Style"

You can use the pyramid method for writing email messages, just as you do for ordinary mail:

1. Start with what you most want your reader to know and, if appropriate, what action you want the reader to take.

2. Follow with any background information the reader may need to understand the reason for your message, and provide details about any point that may need further explanation.

Check that each message contains *only* the information your reader will need to respond or to act—and no more. That is, take care to separate the essential *need to know* information from the less important *nice to know* details.

Proofread with Care

Proofread email *very* carefully: the informality of the medium and the speed with which you can create and answer messages can invite carelessness. It may sound contradictory to suggest that you print your email messages and edit them on hard copy before you send them, but we recommend you do so if a message is long or its contents are particularly important. This is especially true if you are replying to a message immediately after you have read it.

Reread what you have typed, even for a one-sentence reply

Be Prudent

If you are annoyed or irritated by a message you receive, *wait* before replying. Let your irritation cool down. Email is ideal for transmitting facts; it's the wrong medium for sending emotionally charged messages.

Remember that email is not a good medium for conveying confidential information, and particularly is not a medium for making uncomplimentary remarks about other people. Email messages can too easily be forwarded or copied to other readers, and then you have no control over who else may see what you have written. Be just as professional as you are when writing regular letters and memorandums.

Similarly, be just as sensitive when deciding to copy a message to another person. Be sure in your mind that the original sender would want his or her message distributed to a wider audience.

Email Guidelines

Here are some suggestions that will help you write more effective email messages:

- Remember that busy readers who receive many messages want them to be concise yet complete. Feed their needs.

- If you are writing to multiple readers, consider sending *two* messages rather than a single all-embracing message. Write

  1. a short summary, which you send to readers who are interested only in the main event and the result, and

  2. a detailed message, which you send to readers who need all the details.

Limit how many readers are to receive your message

Avoid Overloading the System

Be selective when replying to a multiple-reader message. It may be tempting to simply click the "Reply" button rather than take the time to address your reply to only those readers who need it, but if you do your reply will go to everyone. And if other people reply in the same way, the system—and everyone else's In Basket—will quickly become overloaded.

When accessing email, download it immediately into your In Basket so that you remain online only briefly. Then read and answer your mail offline (i.e. when you are not connected to the service). But avoid letting messages accumulate for too long in your In Basket. If you want to keep a message or may need to refer to it later, store it in an electronic "project folder" in the "filing cabinet" (or an electronic receptacle of a similar name, depending on the service you are using).

Avoid routinely printing copies of messages you want to keep: creating extra paper defeats the aim of email!

Help Identify the Originator

When replying to a message, particularly if your reply will go to multiple addressees, quote a line or two from the original message to help put your

Make sure the originator's name is evident

reply in context. Identify the excerpt by placing a ">" sign before each line, like this:

Dan Reitsma wrote on May 12:

> The Society's constitution was last updated in
> 1984 and needs amending.

I agree, but first we need to check how much editing was done by Karen Ellsberg before she retired in 1997.

Write your name at the foot of every message you create, even though your name appears in the "To-From" list at the top. If a recipient decides to forward the message to other people, frequently only the text will be forwarded and then recipients will not be able to identify the originator.

Avoid Complex Formatting

Use only simple formatting. Write short paragraphs with line lengths of no more than 60 characters, and separate each paragraph with a blank line. Avoid creating columns and indenting subparagraphs, because what you see on screen most likely will not be what your readers see. For example, your screen may look like this:

Facilities are located as follows:

Facility	Location	Distance
Master Control	Calgary, AB	28.6 km south of transmitter
Remote Site 1	Regina, SK	Downtown
Remote Site 2	Thunder Bay, ON	2.5 km north of university

Most email systems do not transmit tables and charts well

But your readers may see something like this:

```
Facilities are located as follows:

Facility Location Distance

Master Control Calgary, AB 28.6 km south of transmitter

Remote Site 1 Regina, SK Downtown

Remote Site 2 Thunder Bay, ON 2.5 km north of university
```

If you need to format columns, consider creating the message as a word-processor file and sending it as an attachment to an email message. (This can be done well with some programs, less effectively with others.)

Indicate Emphasis with Care

Because with most email software you cannot insert boldface or italic type, if you want to emphasize a word you may insert an asterisk on both sides of the word or expression:

This service is available *only* to first-time software buyers.

Don't shout!

Particularly avoid bellowing! Use upper and lower case letters, just as this sentence has been written (not like the one below).

PARAGRAPHS COMPOSED OF ALL CAPITAL LETTERS ARE HARD TO READ. YOU CANNOT EASILY IDENTIFY WHICH ARE THE KEY WORDS.

Finally, avoid inserting "cute" graphics or humorous remarks into your email. They make you appear unprofessional.

Using a Businesslike Format

There are many opinions of what comprises the "correct" format for business correspondence. Most popular word-processing packages include templates for writing business letters, memos, faxes, and proposals. Some are good and easy to use; others are less practicable. The examples illustrated here are those most frequently used by contemporary technical organizations.

Letter Styles

There are two letter formats: the full block and the modified block (see Figures 3-10 and 3-11). Full block is more widely used and is the format Anna King has adopted for H L Winman and Associates' correspondence (there are examples in the letter report in Figure 5-4 of Chapter 5, and the cover letter preceding the first report of Chapter 6). Anna also is aware that letter styles are continually changing. Some companies now omit the salutation and complimentary close (e.g. "Sincerely"), write dates European style (e.g. day-month-year: "27 January 1998"), omit all punctuation from names and addresses (e.g. "Ms Jayne K Tooke"), and use interoffice memorandums and email for informal correspondence. If these trends become more firmly established, Anna will adapt H L Winman and Associates' letter format to reflect the changes.

Most business letters in North America are written full-block style

The modified block style is more conservative and is used primarily by individuals for their personal correspondence and by some small businesses.

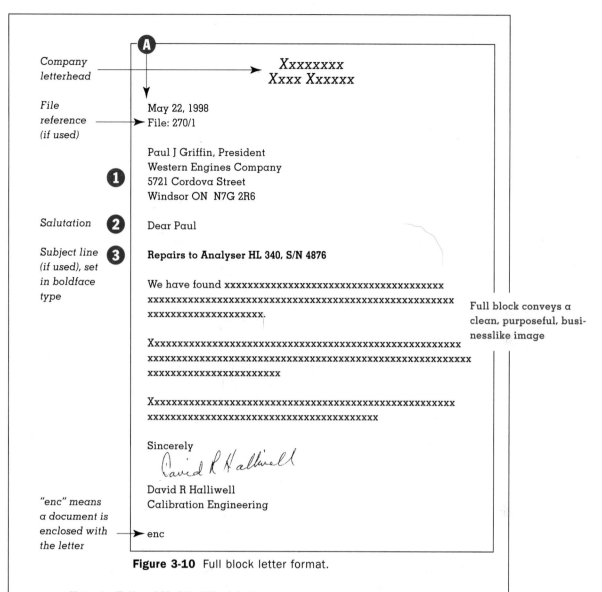

Figure 3-10 Full block letter format.

Notes for Full and Modified Block Letters

1 All but essential punctuation is eliminated from the address, salutation, and signature block (more punctuation is used in the US than in Canada). There should be one blank space between the city and the province or state, and then two spaces between the province and the postal code (or the state and the zip code). The province or state is always printed as two capital letters (e.g. "ON" for Ontario).

2 Today's trend toward informality encourages writers to use first names in the salutation: "Dear Jack" instead of "Dear Mr Sleigh."

3 Subject lines should be *informative* (not just "Production Plan" or "Spectrum Analyser"); they may be preceded by *Subject:*, *Ref:*, or *Re:*. They should be set in boldface type and *not* underlined.

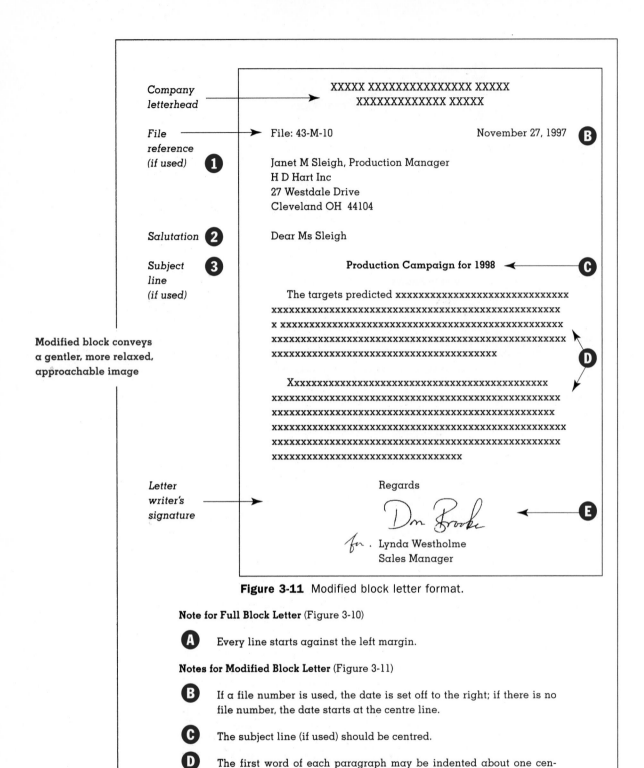

Company letterhead

File reference (if used) **1**

B

XXXXX XXXXXXXXXXXXXX XXXXX
XXXXXXXXXXXXX XXXXX

File: 43-M-10   November 27, 1997

Janet M Sleigh, Production Manager
H D Hart Inc
27 Westdale Drive
Cleveland OH 44104

Salutation **2**

Dear Ms Sleigh

Subject line (if used) **3**

Production Campaign for 1998   ◄——— **C**

The targets predicted xxxxxxxxxxxxxxxxxxxxxxxxxxxxxxx
xxxxxxxxxxxxxxxxxxxxxxxxxxxxxxxxxxxxxxxxxxxxxxxxxxxxxx
x xxxxxxxxxxxxxxxxxxxxxxxxxxxxxxxxxxxxxxxxxxxxxxxxxxxxx
xxxxxxxxxxxxxxxxxxxxxxxxxxxxxxxxxxxxxxxxxxxxxxxxxxxxxxx
xxxxxxxxxxxxxxxxxxxxxxxxxxxxxxxxxxxxxxxx

Xxxxxxxxxxxxxxxxxxxxxxxxxxxxxxxxxxxxxxxxxxxxxx
xxxxxxxxxxxxxxxxxxxxxxxxxxxxxxxxxxxxxxxxxxxxxxxxxxxxxxx
xxxxxxxxxxxxxxxxxxxxxxxxxxxxxxxxxxxxxxxxxxxxxxxxxxxxxx
xxxxxxxxxxxxxxxxxxxxxxxxxxxxxxxxxxxxxxxxxxxxxxxxxxxxxxx
xxxxxxxxxxxxxxxxxxxxxxxxxxxxxxxxxxxxxxxxxxxxxxxxxxxxx
xxxxxxxxxxxxxxxxxxxxxxxxxxxxxxxxx

D

Modified block conveys a gentler, more relaxed, approachable image

Letter writer's signature

Regards

Don Brooke

for . Lynda Westholme
Sales Manager

E

Figure 3-11 Modified block letter format.

Note for Full Block Letter (Figure 3-10)

A   Every line starts against the left margin.

Notes for Modified Block Letter (Figure 3-11)

B   If a file number is used, the date is set off to the right; if there is no file number, the date starts at the centre line.

C   The subject line (if used) should be centred.

D   The first word of each paragraph may be indented about one centimetre, or started flush with the left margin.

E   The signature block should start at the centreline. Here the letter writer has signed "for" his manager.

H L WINMAN AND ASSOCIATES

INTER-OFFICE MEMORANDUM

1 To: A Rittman From: R Davis
Date: December 3, 1997 Subject: Early Mailing of Pay Cheques for Field Personnel 4

2 I need to know who you will have on field assignments during the week before Christmas so that I can mail their pay cheques early. Please provide 3 me with a list of names, plus their anticipated mailing addresses, by December 8.

Cheques will be mailed on Tuesday, December 14. I suggest you inform your field staff of the proposed early mailing.

5

Figure 3-12 Interoffice memorandum.

Notes for Memorandum (Figure 3-12)

1. The informality of an interoffice memorandum means titles of individuals (such as Office Manager and Senior Project Engineer) may be omitted.
2. No salutation or identification is necessary. The writer can jump straight into the subject.
3. Paragraphs and sentences are developed properly. The informality of the memorandum is *not* an invitation to omit words so that sentences seem like extracts from telegrams.
4. The subject line should offer the reader some information; a subject entry such as "Pay Cheques" would be insufficient.
5. The writer's initials are sufficient to finish the memorandum (although some organizations repeat the name in type beneath the initials). Some people prefer to write their initials beside their name on the "From" line, instead of signing at the foot of the memo.

Interoffice Memorandum

The simplest of reporting mediums, the memo is slowly being replaced by email

The memorandum is a flexible document normally written on a prepared form similar to that shown in Figure 3-12. Formats vary according to the preference of individual companies, although the basic information at the head of the form is generally similar. Examples of memorandums appear throughout Chapters 3, 4, and 5.

Fax Cover Sheet

A fax cover sheet may carry a message in addition to being a transmittal document

Any document sent by facsimile machine is normally preceded by a single-page fax cover sheet that identifies both addressee and sender, and their respective fax and telephone numbers (see Figure 3-13). The cover sheet usually has a space for the sender to write a short explanatory note. A sender who has only a short message to send may write the message directly onto the fax cover sheet and then transmit just the single page.

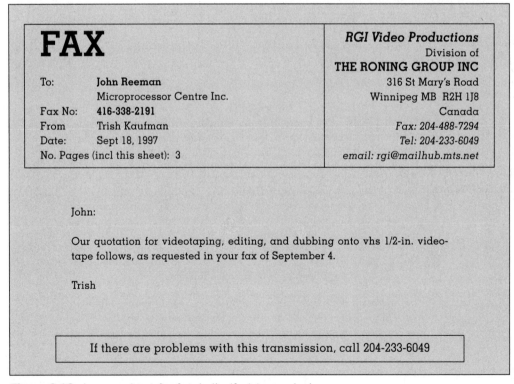

Figure 3-13 A cover sheet for facsimile (fax) transmissions.

Most of the letter and memo writing projects below include all the details you need to do the assignment. You are encouraged, however, to introduce additional factors if you feel they will increase the depth or scope of your letter.

Project 3.1: Unexpected Software Expenses

You are an engineering assistant at the local branch of H L Winman and Associates. One week ago, branch manager Vern Rogers asked you to research a software program called *Amaze 2.3*, and to have it in-house within a week. (With *Amaze*, the company will be able to create graphics for computer-generated slides to be projected through an LCD panel.)

You identified Cottonwood Computers Ltd at 333 Main Street as the local supplier, where Alicia told you that *Amaze* costs $395 and delivery takes about three weeks (the program has to be special-ordered from the manufacturer in the US).

"No. That's too long," you replied. "We need it within three days." So you agreed to have the software couriered to you overnight, at a cost of $40, and specified that it be supplied on 3.5-inch disks.

Amaze arrived three days ago. However, when you tried to install it the first disk (of seven) would not run. Your screen contained the message: "The disk in Drive A contains corrupted data—cannot access." So you called customer service at Carlton Software Associates (CSA) in Chicago—the manufacturer—where you were put on hold for 13 minutes until service representative Kevin came on the line, to whom you explained the problem. Kevin said: "We'll send you another disk. It'll take about 10 days."

A shipment of damaged disks creates a problem...

You asked Kevin to courier the disk to you: your need had now become urgent. He agreed, and said there would be a service charge of $30 in US funds. You explained that there should be no charge because you paid for the original program to be couriered to you, but Kevin said that could be done only if the original order had been placed directly with CSA. You would have to work through the local supplier. Exasperated, you agreed to pay the charge.

Today an envelope arrives, not with the replacement 3.5-inch disk you had requested, but with a CD-ROM disk.

...a problem that keeps on growing

As a CD-ROM drive has not been installed at your office, you again phone CSA (at your expense, because they don't list either an 800 or an 888 number). This time the customer service representative is Andrea, and you wait 17 minutes for her to come on line. You explain the problem and that you had specifically ordered 3.5-inch disks.

"We don't supply our programs on 3.5-inch disks any more," Andrea replies. "Only on CD-ROM. Can you borrow a CD-ROM drive? Or perhaps connect to our Internet site and download the files you need?"

You arrange to download the files, but you are considerably annoyed by the inconvenience (Kevin should have explained this) and the costs you have incurred:

- Two couriered shipments, one at $40 Canadian and one at $30 US.
- Two long-distance telephone calls at $13.47 and $16.20 Canadian.

You decide to write to the head of customer service at CSA and request a refund of some of your expenses. Their address: 2120 Ferguson Loop, Chicago, Illinois, 60546.

Project 3.2: Revising a Letter

At 4:15 p.m. Norm Behouly comes to you with a problem. "I'm going on vacation tomorrow," he announces, "and I'll be away for three weeks. The trouble is, I've keyed two letters into the computer, and now the system has gone down and I can't get them out!"

Norm asks you to print them when you come in tomorrow morning, and mail them for him. He gives you two file names: SURVEY.TXT and FENCE.TXT. "You'll have to sign them for me," he adds, "and I would appreciate it if you would take the time to read them first, just in case there is a typographical error I have missed."

Now it is 9:15 a.m. of the following morning and the computer system is again operational. You key Norm's two letters onto your screen and immediately see that they need much more than just a cursory check for typographical errors.

Revise or correct each letter. Insert a full address for each recipient, including the name of your city and a hypothetical postal code.

Part 1: File SURVEY.TXT

Dear Mr Antony

Simplify a letter that has too many words

In response to your letter of June 7, 1997, and our meeting at your residence at 960 Bidwell Street on June 14, when you showed me the plan of your Lot (Lot 271-06) and the position of the fence bordering the Lot to the south, at 964 Bidwell Street, which is Lot 271-07. You claimed there is a discrepancy between the city site plan and the physical position of the fence, and asked me to do a survey of your Lot so as to establish the correct position.

Your Lot was surveyed by me and an assistant on June 21 and while there I hammered in two markers to delineate the southeast and southwest corners. (No markers were placed on the north side because the position of that fence is not in question.) Your neighbors to the south—Mr and Mrs Beamish—will not be happy when they find out that the fence between Lots 06 and 07 encroaches on your property. You will note from the positions of the markers that the east end of the fence is 0.37 metres inside your territory, but is angled toward the south so that at the west end, where it stops at the garage, it is correctly positioned.

It is assumed that you recognize that the south fence is yours, and the fence to the north is the responsibility of your neighbor to the north. Consequently you have the right to move the fence if you wish or to leave the fence where it now stands until repairs are necessary and then rebuild it in its correct position. As obviously you are aware, the fence is in good condition.

As per your request, I am writing to your neighbors today to inform them of the discrepancy and attaching our invoice.

Yours sincerely

Part 2: File FENCE.TXT

Dear Mr and Ms Beamish

As I am sure you must have been aware, a survey of Lot 271-06 was done recently, on June 14, to determine the exact borders of the Lot at 960 Bidwell Street, to your north. While the survey was being done, markers were positioned at the southeast and southwest corners of the Lot, to establish the exact dividing line between your Lot and that of Mr and Ms Antony at 960 Bidwell Street. No markers were placed at the northeast and northwest corners of the Antony's Lot.

Unfortunately the fence is incorrectly positioned between your Lot (No. 271-07) and Lot 271-06. At the southeast corner of Lot 271-06 the fence is 0.37 metres too far to the north and so encroaches onto your neighbors' Lot. (Actually, the fence slants toward your property as it progresses westward and at the garage end is properly positioned.)

Make this letter more direct and easily understood

I can only assume that you are unaware of this discrepancy, so at Mr Antony's request I am writing to you so that you will know of the circumstances should Mr Antony choose to reposition his fence. I am equally sure that you and the Antonys can come to an amiable agreement.

Please feel free to contact me at your convenience if you need more information concerning this matter.

I remain, yours truly,

Part 3: File GARAGE.DFT

Norm calls you from the airport: "I forgot to tell you," he says. "There's a third file—GARAGE.DFT. It's some notes about the garages on the Antonys' and the Beamishes' Lots, and I think the owners should know about them. Could you write to each of them, for me? It shouldn't wait until I return."

From the notes in file GARAGE.DFT you gather that:

Create a letter from notes

1. The two garages are parallel to each other and the space between the adjacent walls is only 0.53 metres.
2. There is a pile of lumber stacked between the garages to a height of 1.3 metres.
3. City by-law 216, subparagraph 2(c) stipulates that garages must be a minimum of 0.60 metres apart.
4. City by-law 216, subparagraph 2(h) requires that passageways between garages must be accessible, for fire safety reasons.

You feel the homeowners could ignore the separation discrepancy for the moment, but should do something about the stacked lumber (the city inspectors may never notice the too-narrow distance between the garages, but they almost certainly will eventually notice that access between the garages is blocked and this may lead them to measure the separation distance).

Write a letter to Mr and Ms Antony informing them of the problem. Tell them you are sending an identical letter to the Beamishes next door.

Project 3.3: Correcting a Billing Error

Today you receive a credit card statement from WorldCard, covering last month's purchases. There are eight debit entries, three personal and five for expenses incurred during a business trip you made to Wapiti Paper Mill between the 8th and the 12th. (You are an engineering technologist employed by the local branch of H L Winman and Associates, and you went to the mill to investigate and rectify a problem in the process control system.) The five business expenses are:

Nearly everyone has had to correct a billing error

Item	Date	Vendor/Location	Control No.	$
3	08	River Motel, Burntwood Lake	0134652	73.90
4	09	Burntwood Auto Service	0147162	305.60
5	10	Wapiti Autos	0203916	38.66
6	12	Wapiti Inn	0205771	256.50
7	12	Burntwood Auto Service	0211606	31.58

Item 4 puzzles you. You know you purchased gasoline three times and stayed one night on the road in a motel and three nights at another motel near the mill. But you could not have bought $305 of gasoline (your car's tank would not hold that much!).

Fortunately, you always keep a travel log and in it you recorded these entries:

9th	–	51.02 L	@	59.9 c/L
10th	–	63.17 L	@	61.2 c/L
12th	–	52.63 L	@	59.9 c/L

You do not have the credit card vouchers because you attached them to the expense account you handed in to branch manager Vern Rogers on the 19th, and he has sent them on to head office in Calgary. But from your records you can work out what the error is and can guess that it occurred during data entry at WorldCard's data centre in Toronto.

Write to the manager of customer accounts at the credit card company, inform him or her of the error, and ask for an adjustment. WorldCard's address is: Suite 2160 – 24 Henderson Avenue, Toronto, Ontario, M6J 2B5.

Project 3.4: Letter of Thanks

Last night you attended a talk delivered by Ms Tina Mactiere to the local chapter of the Inter-Provincial Engineering Association (IPEA). Today you have to write a letter of thanks to Ms Mactiere, expressing your and the IPEA chapter's appreciation. (You are the chapter's technical program coordinator, and you arranged for Ms Mactiere to give the talk.) Some details you may need are:

1. You are employed by Hogan Consultants Ltd at 212 Broad Avenue of your city, where your company president, Gavin Hogan, encourages his technical staff to participate in IPEA activities.
2. Tina Mactiere is president and chief executive officer of Macro Engineering Inc (see Chapter 2).
3. Her talk was given in the Prairie Room of the Chelmsford Hotel. The event was the Annual General Meeting (AGM) of the local IPEA Chapter. The program included a formal dinner at 6:30 p.m., Tina Mactiere's address at 8:15 p.m., and the AGM at 9:15 p.m. The affair concluded at 10:15 p.m.

Say thank you: elegantly but briefly

4. Tina's talk was titled "Look After the P's and Q's." Her main thrust was that technical people are so concerned with keeping abreast of new technology that they omit other essential aspects of their professional development. She cited, for example, the need for scientists, engineers, and technologists to attend courses or seminars in supervisory management, interpersonal relations, and oral and written communication—topics she referred to as "people skills."
5. Tina proved to be a dynamic speaker. She used slides and a humorous three-minute videotape that neatly underscored the points she was making.

6. There were numerous questions from the audience after her talk, and a strong round of applause.
7. Many people came up to you after the AGM and congratulated you on your choice of speaker and the appropriateness of her topic.
8. Seventy-six IPEA members attended the dinner and meeting.

Project 3.5: Email — Proposal Readiness

You are an engineer with H L Winman and Associates in their Toronto office and have been asked to answer a Request for Proposal (RFP) from the City of Lakeville. The city has decided to build an additional recycling collection location. The site you have selected is at the northwest corner of Magyar Street and Wellington Avenue (Wellington Avenue is a major road between the Lakeville city centre and a prime residential district known as Somerville Estates).

An email to introduce a "product"

You've prepared the proposal complete with architectural diagrams, a cost analysis and a time schedule. Before dropping it into the mail to the Lakeville City Council, you decide to email the contact person to introduce the proposal and highlight the major points. From the RFP you have the following information:

Send proposals and questions to Ellen Johnson

> Lakeville City Council
> 143 Cedar Street
> Lakeville ON M2J 1K7
> (905) 626-1427 (tel)
> (905) 626-1425 (fax)
> email: E.Johnson@LCC.com

Write the email message.

Project 3.6: Portable Computer Problems

After several years working for Macro Engineering Inc (MEI) in Toronto, you decide to "go independent." You resign from MEI and set up your own business, registering your company's name as Pro-Active Consultants Limited. You set up an office in your home, and use your home telephone also as your business telephone. You buy a fax machine which can sense when a fax is coming in on the telephone line.

On July 1 of this year you buy a new computer to replace a computer destroyed by a power surge (the circumstances are described in Project 4.1 on page 109). The new machine is a Nabuchi model 300CDT, and it cost $3360. It has a sound card, an internal CD-ROM drive, a 120 MHz Pentium processor, 32 Mbyte of RAM, and a 2 Gbyte hard drive. You install Windows NT plus a host of software programs you expect to use as a consultant.

However, a problem occurs within a week: every now and then the mouse button freezes and the cursor cannot be moved. The only remedy is to reboot the computer by depressing the CTRL-ALT-DEL keys simultaneously, but that means you lose whatever you have been working on during the last 15 minutes. After several such episodes, you wait for the next mouse-freeze and then carry the computer to Westside Computer Centre, where you bought it. Yet when you demonstrate the problem to technician Michel Olenick, the fault has rectified itself. You leave the computer with Michel for 24 hours but, no matter how much he tries, the fault does not recur. "Intermittent problems can be far harder to diagnose than a full failure," Michel comments when you pick up the computer.

A new computer with too many problems

Ten days later another fault occurs: the whole keyboard suddenly freezes up and there is *nothing* you can do to reboot it. You can't even switch the computer off! The only remedy is to disconnect the power cord and battery. When you reconnect them, the fault is resolved. You visit Westside Computer Centre again, and again the fault refuses to show up.

This week you are experiencing still another problem: the internal power source transformer is immediately beneath the internal PCMCIA modem card. It tends to overheat the modem and cause the modem to malfunction: when it is hot, it will not dial the server's phone number!

"Oh, that's a recurring problem," Michel says. "You just have to remove the modem, allow it to cool down to near-ambient temperature, then reinsert it. Then it will work for you."

Michel is right, of course, but you are still annoyed by this most recent symptom. It's inconvenient and you feel you really shouldn't have to fiddle around like that, and particularly as the other problems are still occurring intermittently. So you decide to write to the computer manufacturer. In your letter you describe the problems and request they supply you with a fully functional, new computer. Nabuchi Electronics' head office in Canada is at 2830 rue St Augustine, Montreal, Quebec, H3B 2S2. Address your letter to the customer service manager.

Project 3.7: Mis-ticketed for Flights

As an independent consultant you find yourself frequently travelling to different client sites. Most of your work is done remotely from your home office, but there are times when an important meeting or presentation requires that you see people in person. You realize the value of developing a relationship with your clients. Even with all of today's technology you find the best way is still face-to-face.

Since you just recently established your own company, Pro-Active Consultants Limited, you don't have the resources or luxury of having a secretary to make your travel arrangements so you have to do it yourself.

The first step—booking the flights—was easy...

When you called Canada Jet Express Airlines (you called them directly because you thought you might get a better price than if you used a travel agent) you spoke with a friendly representative named Joyce. You explained to her that you want to fly to Little Rock, Arkansas, on Sunday, June 7, because you have a business meeting at the new site June 8 to 10, and then on Thursday, June 11, you want to fly from Little Rock to Mobile, Alabama, to visit a friend, returning to your city on Sunday, June 14.

"Wow," Joyce said. "Have I got a deal for you. I can get you to where you want to go for a total of $790.00. That's a great price considering it's not a straight, round-trip ticket but what we in the airline industry call an open-jaw ticket."

You said you need to confirm your plans with the site manager and talk to your friend in Mobile, to make sure they are going to be available before you give her your credit card number and pay for the flights.

"No problem," Joyce said. "I can hold these flights for 24 hours. Just call back before midnight tomorrow."

After a series of answering machine messages back and forth, you finally got in touch with the site manager and your friend: the dates and times you discussed with Joyce at CJE Airlines were fine. When you called the airline to provide your credit card details and secure the flights, you were connected with a different representative named Jonathan.

"I'm sorry, but I can't find your reservations," he replied. "Are you sure you phoned back within 24 hours?"

"Yes, I'm sure," you said. " This is all I've spent my time on in the past 24 hours!" At this point you were getting a little annoyed. Every phone call seemed to eat away 30 to 40 minutes of your time.

"Oh, wait a minute. There it is. It appears your reservations have been cancelled," Jonathan said. "I don't know why but they have gone."

Luckily you wrote down the exact dates and flights that Joyce quoted. Here's what she found:

Sunday, June 7	LV your city	1:00 p.m.	Flt. 832
	AR Chicago, IL	3:00 p.m.	
	LV Chicago, IL	4:26 p.m.	Flt. 808
	AR Little Rock, AK	6:18 p.m.	
Thursday, June 11	LV Little Rock, AK	6:30 p.m.	Flt. 2430
	AR Mobile, AL	8:30 p.m.	
Sunday, June 14	LV Mobile, AL	3:41 p.m.	Flt. 81
	AR Chicago, IL	5:13 p.m.	
	LV Chicago, IL	7:02 p.m.	Flt. 2160
	AR your city	9:17 p.m.	

Jonathan was patient and, although he couldn't get you the great deal Joyce did, he was able to get you on the exact same flights for only $38.00 more.

...then frustration set in, one telephone call at a time

"Fine," you said, "I'll give you my credit card details to guarantee these flights. I'll put them on my company VISA card number 4321 1238 7898 5000, expiration date 9/99.

"OK," said Jonathan. "They'll be in the mail to you today."

Today, five days later, the tickets arrive and when you open the envelope you are shocked. "Unbelievable!" you shout out loud. "CJEA hasn't included the June 11 leg from Little Rock to Mobile!"

So, you make *another* phone call to the airline (another hour of your time) and speak with a representative called Ashley, who isn't as friendly as the first two representatives. She explains that your only option is to purchase a one-way ticket from Little Rock to Mobile for $134.50.

"But that's $172.50 more than my original quote!"

"Well", says Ashley with a tone of sarcasm, "You could always take a bus from Little Rock to Mobile, couldn't you?"

With little choice you agree to purchase the additional ticket but you are not very pleased or impressed. So you decide to write to the airline and express your dissatisfaction, and ask Ashley for a name and address to write to. Here's the information she gives you:

Donavan Johnson
Director of Consumer Affairs
Canada Jet Express Airlines
6001 Airport Highway
Winnipeg MB R3C 2A6

Now it's time to write for an adjustment

Write the letter. Ask for compensation for the trouble you have experienced and the expenses you have incurred.

Project 3.8: Request to Attend TCI

Assume that you are employed by Macro Engineering Inc of Toronto, that today is the second Monday of the current month, and that for the past four weeks you have been on a field assignment to Gillam, Manitoba, where you are installing a computer control system at Manitoba Hydro's power-generating station. You are assisted by technicians Denise Stockman and Graham Holtz, and your project is four days ahead of schedule. The task is to be completed by the 10th of next month.

Today you are shown a brochure describing a Technical Communication Institute (TCI) to be held at the Fort Garry Hotel in Winnipeg from the 26th to the 29th of this month. The brochure provides the following details:

- TCI is held annually, and this is its fifth year.
- It presents seven courses over four days, some of two days duration, some one day.
- Courses this year are:
      Planning and Designing Multimedia
      Designing Online Documentation
      Usability Testing
      Project Management
      Developing a Web Site
      Preparing World-Ready Information
      Human Factors for Technical Communicators

A course worth
attending

- The cost is $785 plus GST ($54.95).
- Hotel cost is $79 single or double occupancy (plus GST)
- Classes are held at the University of Manitoba and Red River Community College downtown extension divisions.
- Attendance is limited to 55 people, with no more than 15 in computer-based courses and no more than 20 in other courses.
- Classes are from 8:30 a.m. to 5:00 p.m. daily.
- Instruction is principally "hands on."
- There's a reception and banquet on the third night (28th)
- Applications are to be sent to TCI at Box 181 RPO Corydon, Winnipeg, R3M 3S7.

Because your work contains both technical and management aspects, and requires that you do a lot of report writing, you consider that TCI would be ideal for you. Terri Molineux, a technical writer at Gillam (she gave you the brochure), attended last year's TCI: "The quality of instruction, and the dedication of the faculty are excellent," she says. "It was well worth the price. I learned a lot."

From the course descriptions in the brochure, you identify the two-day *Usability Testing* and *Project Management* courses as the most useful for you (although *Designing Online Documentation* also appeals to you, and you decide to list it as an alternative choice, just in case one of the other courses is full).

Ask for approval to go

Write a memo to your department head, Fred Stokes, in which you

- describe the courses (convince him they are good),
- ask if you can attend,
- ask for the company to pay the tuition fee,
- ask to be spared from the Manitoba Hydro installation project for five days, one of which is for travel time (be convincing),
- explain that the hotel cost will be offset by the Manitoba Hydro charge for accommodation at their field house ($85 per night, including all meals),

- ask for a quick reply, because time is short. (Because attendance is limited, you're afraid the courses may be filled before you apply!

Calm Air flies between Gillam and Winnipeg, once a day. The airfare is $627 incl taxes.

Send your request by the fastest possible means.

Project 3.9: A Faulty CD Player

Ten days ago you were in the city of Montrose, 1400 km from your home, at the end of a driving vacation. In Montrose you visited Sheila Wilson and Gary Schultz, and they showed you a five-disc in-car CD player Sheila had bought for Gary's birthday. It fitted neatly into the glove compartment and connected to their car radio.

On your third day in Montrose you visited Madison Music Centre, the store where Sheila purchased the CD player, and explained to store owner George Madison that you wanted to buy an identical unit.

A music source you like...

He shook his head: "That's the Regent 501: it's last year's model. I don't have any left." Instead, he showed you a Regent 601. "This is their new model," he said. "It came in just yesterday. You can put it in the glove compartment just like the 501. The main difference is that it plays six discs and they claim it has better interference-free circuitry."

You enquired about the price. He said $374.50 plus taxes, to which Sheila said: "That's $50 more than I paid!"

George Madison shrugged, and you turned away, ready to leave the store. Then suddenly he said: "I'll give you a deal: I'll let you have a voucher for free installation, over at Marine Autos. Installation normally costs $35. You can't beat that!"

You would have preferred to have it installed in your home city, but the free offer appealed to you and so you bought the CD player on Madison Music Centre's sales invoice No. 2324. The following day you had the CD player installed on your way out of Montrose (it took 35 minutes), but had to pay both federal and provincial tax on the $35 "free" installation charge. Then you drove home: a 15-hour journey.

The following day you took a stack of CDs out to the car, loaded them into the Regent 601, and enjoyed the music. Great! Until the player reached the end of the first disc. Then it stopped. You fiddled with the controls and got it to start playing again, but it was the same CD. No matter how hard you tried, you could not get it to play any of the other discs—only the top disc—even though you tried shuffling them and reinserting them.

In exasperation, you telephoned around to find who had the service contract for the Regent line of home electronics: no one! So you took the

...proves to have a problem

601 into Kelvin Electronics at 241 Marchand Avenue. When you picked it up today, technician Merv Halverson handed you an electronic microchip mounted on a tiny circuit board. "There's your problem," he said. "I've replaced it and the set works fine." And he presented you with the defective microchip and a repair bill (Invoice 1731) for $87.50.

"Shouldn't that be under warranty?" you asked. "The player's only one week old."

"Not with us, I'm afraid," Merv answered. "I guess you can get your money back from the manufacturer, or the place you bought it." He tells you that the Regent line is made in China, but he couldn't find a mailing address.

You decide to write to Madison Music Centre in Montrose (street address: 611 Garrick Street). Tell George Madison what has happened and ask for a refund of $.... (you decide how much).

Project 3.10: Acknowledging a College Award

Expressing personal thanks is not always easy

Assume that you are in the second year of the course you are enrolled in, and that three weeks ago the head of the department came to you and announced that you have been selected to be this year's winner of the Inter-Provincial Engineering Association (IPEA) scholarship for "proficiency in technical studies." Yesterday you attended an awards luncheon attended by other scholarship winners and representatives of the firms donating the scholarships. You sat next to Calvin Wycks, vice-president of the local chapter of IPEA, who presented the award to you.

Today, you write to IPEA to thank the Association for the award. Use these details:

- Address your letter to the president, Marjorie McIvor.
- IPEA's address is 710 Durham Drive of your city.
- The award is a cheque for $500 and a wall plaque inscribed with your name.

How to Write Business Letters That Get Results
www.smartbiz.com/sbs/arts/bly48.htm

Robert W. Bly, president of the Center for Technical Communication, provides valuable advice about writing correspondence. "Failure to get to the point, technical jargon, pompous language, misreading the reader—these are the poor stylistic habits that cause others to ignore the letters we send." He recommends that the writer "determine the action you want your letter to generate and tell the reader about it."

Strategies for Writing Persuasive Letters
www.wuacc.edu:80/services/zzcwwctr/persuasive-ltrs.wm.txt

This step-by-step guide covers the purpose of the persuasive letter, prewriting questions for the writer, writing strategies, and revision tips. Examples are included.

Persuasive Communications: Using You-Attitude and Reader Benefit
www.wuacc.edu:80/services/zzcwwctr/you-attitude.txt

Receivers of communications usually are more concerned about themselves than about the writer or the company that person represents. This article describes how to use the "you-attitude" and show reader benefit in your persuasive communication.

Audience Analysis
www.vsl.ist.ucf.edu/~deef/itaudience.html

This site elaborates on the golden rule of technical communication—to help readers process and understand the information they need to know.

Memos
www.rpi.edu/dept/llc/writecenter/web/text/memo.html

Rensselaer Polytechnic Institute's Writing Center offers suggestions about writing effective memos. Some references are included.

A Beginner's Guide to Effective Email
www.webfoot.com/advice/email.top.html?Yahoo

This useful guide includes an introducton to email and a discussion about why it differs from ordinary correspondence. Other sections describe email context, page layout, intonation, jargon, and acronyms. Included are links to other sites about email.

Chapter 4
Short Informal Reports

W hen you hear that someone has just finished writing a technical report, you may imagine a nicely bound formal document, tastefully typed and printed. In some cases you would be correct, but most of the time you would be wrong. Far fewer formal reports are issued than informal reports, which reach their readers as letters and memorandums. This chapter describes the short informal reports you are likely to write as a technologist, engineer, or engineering technician.

Appearance

The memorandum report is the least formal technical report, and it can be transmitted by fax, mail, or email. Normally an interoffice or interdepartmental communication, its length and tone can vary considerably. It can be short and direct, it can develop its topic in great detail to present a convincing case, or it can lie anywhere in between.

The letter report, although still basically informal, can vary in formality according to its purpose, the type of reader, and the subject being discussed. Some letter reports may be as informal as a memorandum report, particularly if they are conveying information between organizations whose members know each other well or have corresponded frequently. Others may be more formal, presented as business letters conveying technical information from one company to another.

Although there are many types of informal reports, all are based on the writing plan outlined in Figure 4-1. Each report contains

1. a brief statement describing what the reader most needs to know,
2. a short introduction to the problem,
3. a discussion of the data, situation, or problem, and what has been done or could be done about it, and
4. a conclusion that sums up the results and possibly recommends what should be done next.

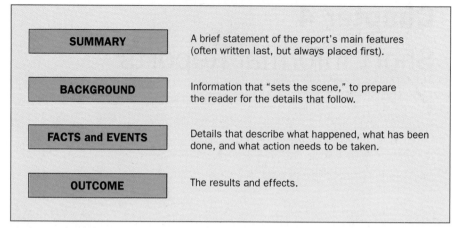

SUMMARY	A brief statement of the report's main features (often written last, but always placed first).
BACKGROUND	Information that "sets the scene," to prepare the reader for the details that follow.
FACTS and EVENTS	Details that describe what happened, what has been done, and what action needs to be taken.
OUTCOME	The results and effects.

Figure 4-1 Basic writing plan for short reports. The plan is modified slightly to suit each situation.

Writing Style

The reports described in this chapter are written in a direct, informative style that is crisp and to the point. The writers are usually describing events that have already occurred, so they write mostly in the past tense, which helps them to be consistent. They shift gear into the present or future tense only when they have to describe something that is presently occurring, outline what will happen in the future, or suggest what needs to be done. All three tenses occur in the report shown in Figure 4-2.

Dan Skinner has written the Background and most of the Facts paragraph mainly in the past tense because they deal with what has already been done. At the end of the Facts he has shifted into the present tense to report how the station manager feels now. Then for the Outcome he has jumped into the future tense to outline what he plans to do. This past-present-future arrangement is natural and logical; reader Don Gibbon will feel comfortable making the transitions from one tense to the next. Dan's Summary even follows the same pattern.

Incident Report

Anna King is working late in the H L Winman office in Calgary when the telephone rings. The caller is Bob Walton, a member of the electrical engineering staff who is on a field trip to Dryden, Ontario. He tells Anna he has been involved in a traffic accident near Brandon, Manitoba, his co-traveller has been injured, and some of his equipment has been damaged. He wants Jim Perchanski, his department head, to send out replacement equipment by air express.

To: Don Gibbon: dgibbon.ho@winman.ab.ca
From: Dan Skinner: dskinner.fld@winman.ab.ca
Date: 24:1:98
Ref: Carpet problem at KMON-TV

The indoor/outdoor carpet we installed in KMON-TV has corrected the noise problem but is "pilling" badly. I will examine the carpet with the manufacturer's representative to find the cause.

SUMMARY

The carpet was installed in the satellite studio control room during the night of January 8–9, to reduce the ambient noise level by 3.6 dB.

BACKGROUND
Past tense

The writing plan is followed closely in this short report

At the station manager's request, I returned to the control room today and checked the carpet's condition. After only two weeks it has tight little balls of carpet material adhering to its surface. I called the manufacturer's rep, who said that the condition is not unusual and does not mean that the carpet is wearing quickly. He suggested that it may be caused by improper cleaning techniques and probably can be easily corrected. However, our client is not pleased with the carpet's appearance.

FACTS
Mainly past tense

Present tense

The manufacturer's rep and I will return to the control room between midnight and 2 a.m. on January 31 to study the carpet-cleaning techniques used by maintenance staff. I will email our findings to you later in the day.

OUTCOME
Future tense

Figure 4-2 A short report transmitted by electronic mail.

Anna jots down notes while Bob talks. Because she will be out of the office the following day, she keys a report of the conversation into her computer and sends it by electronic mail to Jim Perchanski (see Figure 4-3), knowing he will access his email immediately after he arrives in the morning. She tells Jim what has happened to two members of his staff, where they are now, how soon they will be able to move on, and that one of them is injured. She also tells him that equipment is damaged and replacements are needed.

To: Jim Perchanski: prchnski.ho@winman.ab.ca
From: Anna King: aking.ho@winman.ab.ca
Date: 16:09:97
Ref: Accident report and request for spare parts ❶

Bob Walton and Pete Crandell have been involved in a high-way accident, which will delay their inspection of the Sledgers Control project at Dryden, Ontario. They need replacement parts shipped to them tomorrow (Wednesday, September 17). ❷

Bob telephoned from Brandon, Manitoba, at 19:35 to report the accident, which occurred at 17:15 some five kilometres west of Brandon. Pete has been hospitalized with a fractured left knee and a suspected concussion. Bob was unhurt. The panel van and some of their equipment were damaged. ❸

Bob wants you to ship the following items by air express on an Air Canada Wednesday evening flight to Winnipeg, and to mark the shipment "HOLD FOR PICK UP BY R WALTON SEP 18":

- 1 Spectrum analyser, HK7741
- 1 Calibrator, Vancourt model 23R ❹
- 24 Glass phials, 300 mm long × 50 mm dia.

He will rent a van and drive to Winnipeg to pick up the items Thursday morning. He will then drive on to Dryden and expects to arrive there about 16:00 hr. He has informed site RJ-17 at Dryden of the delay. ❺

Bob is preparing an accident report for you. He is staying at Hunter's Motel in Brandon (Tel: 204-453-6671). ❻

Figure 4-3 A third-person incident report.

Anna's message is an incident report, written pyramid style (see Figure 4-1), in which

- the **Summary Statement** is in the paragraph identified as (2),
- the **Background** is at the start of paragraph (3),
- the **Facts** are in the remainder of paragraph (3) and all of paragraph (4), and
- the **Outcome** is in paragraphs (5) and (6).

Because she will not be available to answer questions, Anna takes care to describe the situation clearly:

1 She knows that a subject line must be informative; it must tell what the message is about and stress its importance to the reader. If Anna had simply written "Transcript of Telephone Call from R Walton," she would not have captured Jim Perchanski's attention nearly as sharply.

2 This brief summary gets right to the point by immediately telling Jim Perchanski in general terms what he most needs to know:
> Why the message was written.
> What happened.
> What action has been taken.
> What action he has to take.

3 In this paragraph Anna tells what she knows about the accident and its effects. It serves as background to the important facts that follow.

4 Anna knows that Jim Perchanski must act quickly to ship the replacement equipment, so she uses a list as an attention-getter: if Anna had described the items in a paragraph, they would not have been nearly as noticeable:

> He will need replacements for an HK7741 Spectrum Analyser, a Vancourt 23R Calibrator, and 24 glass phials, each 300 mm long × 50 mm dia. He wants you to ship these items air express to the Air Canada terminal at Winnipeg, and to mark them...

5 Instructions and movement details must be explicit, otherwise the equipment and Bob Walton may not meet in Winnipeg. Anna has identified specific days, and once even the date, to make sure that no misunderstandings occur. To state "tomorrow" or "the day after tomorrow" would be simple but might cause Jim Perchanski to assume a wrong date, since he will be reading the memorandum one day later than it was written.

Accuracy of information is essential in report writing

6 In this brief closing paragraph Anna indicates what further action is being taken and where Jim can contact Bob Walton if he needs more information.

H L WINMAN AND ASSOCIATES

INTER-OFFICE MEMORANDUM

To: Jim Perchanski   From: Bob Walton
Date: September 17, 1997   Subject: Report of Traffic Accident
                                                  at Brandon, Manitoba

Pete Crandell and I were involved in a multiple-vehicle accident on
September 16, which resulted in injuries to Pete, damage to our panel van
and some equipment, and a two-day delay in our inspection of the Sledgers
Control project.   **A**

The accident occurred at 17:15 hr on Highway 1, about 5 km west of Brandon,
Manitoba. We were travelling east in company panel van TLA 711, on our   **B**
way to site RJ-17 at Dryden, Ontario. Pete was driving and we were
approaching the intersection with Highway 459.

Other vehicles involved in the accident were:

• Toyota Tercel, license 881 FLM, driven by D Varlick
• Ford truck, license TRB 851, driven by F Zabetts
• Pontiac Grand Am, license 372 HEK, driven by K Schmitt.

Positions of the vehicles and our panel van immediately before the accident
are shown on the attached sketch.

As the Toyota attempted a right turn into Highway 459 it skidded into the
Ford truck, which was standing at the intersection waiting to enter Highway
1. The impact caused the Toyota's rear end to swing into our lane, where   **C**
Pete could not prevent our van from colliding with it. This in turn caused the
van to slide broadside into the westbound lane, where the Pontiac
approaching from the opposite direction collided with its left side.

Pete was taken to Brandon general hospital with a broken left knee and a
suspected concussion; he will be there for several days. The panel van was
extensively damaged and was towed to Art's Autobody, 1330 Kirby Street,
Brandon. As some of our equipment also was damaged or shaken out of cal-   **D**
ibration, I telephoned Anna King on Tuesday evening and requested
replacements (she has prepared a list for you).

I have rented a replacement van from Budget, and have informed the duty
engineer at site RJ-17 that my inspection of the Sledgers Control project will
start on Friday, September 19, two days later than planned.

Bob

**Details of other people
involved, and their
vehicles, belong in the
Background, not the
Event**

Figure 4-4 A first-person incident report.

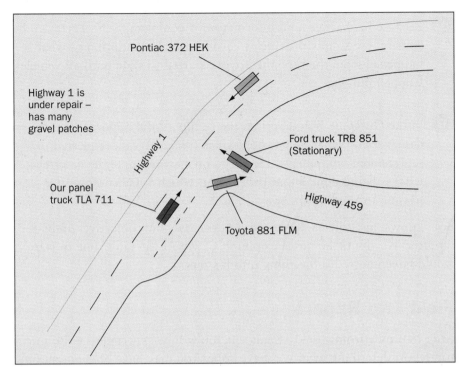

Figure 4-5 Attachment to Bob Walton's report (Figure 4-4).

When Jim walks in on Wednesday morning, he will know immediately what has happened and what action he has to take. He does not need to ask questions because he has been placed fully in the picture.

Bob Walton also used the report writer's pyramid when he subsequently wrote to his supervisor, from Hunter's Motel in Brandon, Manitoba, to describe the accident and its effect. His report is in Figure 4-4. Its focus and emphasis differ from those in Anna's earlier report, but it still is an incident report with the following parts:

Ⓐ This is his **Summary**: it takes a main piece of information from each compartment that follows.

Ⓑ This is the **Background**. By clearly describing the situation (*who? where? why? when?*) Bob helps Jim more easily understand what happened. Notice how he

- establishes where they were, how they happened to be there, in what direction they were travelling, and who else was involved;
- itemizes vehicles, licence numbers, and drivers' names in an easy-to-read list; and
- mentions that he is enclosing a sketch (Figure 4-5), so Jim can look at it *before* he reads on.

Because his background information is complete, Bob's **Facts** can be concise. He simply provides a chronological description of what happened from the time the Toyota started to slide until all vehicles stopped moving.

D In the **Outcome** Bob describes the results of the accident (injuries, damage) and what he has done since (rented a van, requested replacement equipment). He closes on a strong note: he describes what is being done about the project, which was his reason for passing through Brandon.

Bob knows his role is to be an informative but objective (unbiased) reporter. No doubt he has an opinion of who is at fault, but to state it would have injected subjectivity into his report.

Field Trip Report

After returning from a field assignment you will be expected to write a field trip report describing what you have done. You may have been absent only a few hours, inspecting cracks in a local water reservoir; you may have spent several days installing and testing a prototype pump at a power station in a nearby community; or you may have been far away for two months, overhauling communications equipment at a remote defence site. Regardless of the length and complexity of your assignment, you will have to remember and transcribe many details into a logical, coherent, and factual trip report. To help you, carry a pocket notebook for jotting down daily occurrences. Without such a record to rely on, you may write a disorganized report that omits many details and emphasizes the wrong parts of the project.

The simplest way to write a trip report is to answer the four questions shown in Figure 4-6, which, like all reports in this chapter, is a modification of the basic writing plan in Figure 4-1.

Short Trip Reports

Short trip reports do not need headings. A brief narrative following the Summary-Background-Facts-Outcome pattern carries the story:

Summary	A prototype automatic alarm has been installed at site RJ-17 for a one-month evaluation by the Roper Corporation.
Background	Dave Makepiece and I visited the site from January 15 to 17.
Facts	We completed the installation without difficulty,

following installation instruction W27 throughout, and encountered no major problems. However, we omitted step 33, which called for connections to the remote control panel, because the panel has been permanently disconnected.

Outcome   The alarm will be removed by M Tutanne on February 26, when he visits the site to discuss summer survey plans.

(In practice, the very short Background probably would be combined with either the Summary or the Facts to form a single paragraph.)

Longer Trip Reports

Long trip reports require headings to help their readers identify the compartments. Typical headings might be:

- **Summary.**
- **Assignment Details** (Background).
- **Work Accomplished** and **Problems Encountered** (Facts; best treated as two separate headings).
- **Suggested Follow-up** or **Follow-up Action Required** (Outcome).

Anna King's instructions to H L Winman and Associates' engineers (see Figure 4-7 on pages 94 and 95) tell them how to organize their longer trip reports, describes the information that normally would follow each heading, and includes excerpts and sample paragraphs.

With the exception of the Outcome section, trip reports should be written entirely in the past tense.

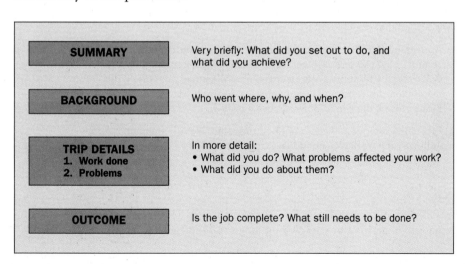

The writing plan is still based on the pyramid; we've just removed it from the Background

Figure 4-6 Writing plan for a field trip report.

H L WINMAN AND ASSOCIATES

475 Lethbridge Trail, Calgary AB T3M 5G1

Guidelines for Writing Long Trip Reports

A standard format is to be used for long trip reports. These instructions suggest how you can organize your information under five main headings: *Summary; Assignment Details; Work Accomplished; Problems Encountered;* and *Follow-up Action.* The headings may be omitted from very short reports.

This may look like a model report...

Summary
Make your summary a short opening statement that says what was and was not accomplished. Highlight any significant outcomes.

Assignment Details
State the purpose of the trip and include any other information the reader may want to know. If the information is lengthy, use subheadings such as

- Purpose of Trip
- Background
- Project No./Authority
- Personnel Involved
- Person(s) Contacted
- Date(s) of Field Trip

Work Accomplished
Describe the work you did. Normally present it in chronological order unless more than one project is involved, in which case describe each project separately. Keep it short: don't describe at great length routine work that ran smoothly. Whenever possible refer to your work instruction or specification, and attach a copy to your report:

...but really it's an instruction

> The manual control was disconnected as described in steps 6 to 13 of modification instruction MI1403, enclosed as attachment 1.

Go into more detail only if you encountered difficulty, or if work was necessary beyond that anticipated by the job specification:

> At the request of the site maintenance staff, we installed a manual control in the power house as a temporary replacement for a defective GG20 control. I left the parts removed from the panel, together with instructions for returning the panel to its original configuration, with Frank Mason, the senior power house engineer.

If parts of the assignment could not be completed, identify them and explain why the work was not done:

1

Figure 4-7 Anna King's instructions for writing long trip reports.

We had to omit Test No. 46 because the RamSort equipment had been removed.

Problems Encountered

Describe problems in detail. Knowledge of problems you encountered and how you overcame them can be invaluable to the engineering or operating departments, which may be able to prevent similar problems elsewhere.

Avoid statements that do not tell the reader what the problem was or how it was overcome. For example:

Considerable time was spent in trying to mount the miniature control panel. Only by fabricating extra parts were we able to complete step 17.

If this information is to be used by the engineering or operating department, it must be more specific:

We spent three hours trying to mount the miniature control panel according to the instructions in step 17. Because the main frame had additional equipment mounted on it, which prevented us from using most of the parts supplied, we had to fabricate a small sheetmetal extension to the main frame and mount it with the miniature panel, as shown in attachment 2.

Follow-up Action

Tie up any loose ends here. If any work has not been completed, draw attention to it even though you may already have mentioned it under "Work Accomplished." Identify what needs to be done, if possible indicate how and when it should be done, and say whose responsibility it now becomes:

The manual control mounted as a temporary replacement in the power house is to be removed when a new GG20 control panel is received on site. This will be done by Frank Mason, with whom I left instructions for doing the work.

The Outcome looks forward, says "who will do what"

In some cases you may direct follow-up action to someone else in your own or another department:

The manual control is to be removed from the power house by R Walton, who will visit the site on May 12.

If your report is very long, insert subheadings and use a paragraph numbering system to increase its readability.

Anna King
January 20, 1998

2

Occasional Progress Report

Progress reports keep management aware of what its project groups are doing. Even for a short-term project, management wants to hear how the project is progressing, especially if problems are affecting its schedule. Because delays can have a marked effect on costs, management needs to know about them early.

Jack Binscarth, one of Macro Engineering Inc's technologists in Toronto, has been assigned to Cantor Petroleums north of Edmonton to analyse oil samples. The job is expected to take five weeks, but problems have developed that have prevented Jack from completing the work on time. To let his chief know what is happening, he writes the brief progress report in Figure 4-8, adapting the standard Summary-Background-Facts-Outcome arrangement into the five-compartment past-present-future pattern shown in Figure 4-9 (see page 98).

Periodic Progress Report

If a project is to continue for several months, management normally will specify that progress reports be submitted at regular intervals.

A periodic progress report may be no more than a one-paragraph statement describing progress of a simple design task, or it may be a multipage document covering many facets of a large construction project. (There are also form-type progress reports, which call for simple entries of quantities consumed, amount of concrete poured, and so on, with cryptic comments.) Regardless of its size, the report should answer four main questions that the reader is likely to ask:

1. Will your project be completed on schedule?
2. What progress have you made?
3. Have you had any problems?
4. What are your plans/expectations?

To answer these questions, a periodic progress report can readily use the standard Summary-Background-Facts-Outcome arrangement:

Summary	A brief overview of the project schedule, progress made, and plans (*answers the first question*).
Background	The situation at the start of the report period.
Facts	Progress made (*answers the second question*) and problems encountered (*answers the third question*).
Outcome	Plans/expectations for the next period (*answers the last question*).

MACRO
ENGINEERING INC.
600 Deepdale Drive, Toronto ON M5W 4R9

FROM:	Jack Binscarth (at Cantor Petroleums)	DATE:	October 14, 1997
TO:	Fred Stokes Chief Engineer, Head Office	SUBJECT:	Delay in Analysis of Oil Samples

My analysis of oil samples for Cantor Petroleums has been delayed by problems at the refinery. I now expect to complete the project on October 25, nine days later than planned.

Summary

The first problem occurred on September 23, when a strike of refinery personnel set the project back four working days. I had hoped to recover all of this lost time by working a partial overtime schedule, but failure of the refinery's spectrophotometer on October 13 again stopped my work. To date, I have analysed 111 samples and have 21 more to do.

Progress

Progress reports follow a past-present-future arrangement

The spectrophotometer is being repaired by the manufacturer, who has promised to return it to the refinery on October 19. Today I informed the refinery manager of the delay, and he has agreed to an increase in the project price to offset the additional time. He will call you about this.

Situation Now

Providing there are no further delays I will analyse the remaining samples between October 20 and 24, and then submit my report to the client the following morning. This means I should be back in the office on October 26.

Plans

Jack

Figure 4-8 An occasional progress report. (Because the report is short, the Background component has been omitted.)

Figure 4-10 on page 99 shows how survey crew chief Pat Fraser used these four compartments to write an effective progress report (the numbers below are keyed to parts of the report):

1 The **Summary** tells civil engineering coordinator Karen Woodford how closely the survey project is adhering to schedule, and predicts future progress. This is the information she wants to read first.

2 The **Background** section reminds Karen of the situation at the end of the previous reporting period and predicts what Pat expected to accomplish during this period. Background should always be stated briefly.

The past-present-future structure is equally apparent here

3 The Facts (or Discussion) section is broken into two parts:
• Work done during the period (3a)
• Problems affecting the project (3b)

Pat Fraser opens each paragraph of this compartment with a topic sentence (a summary statement) that states the main point of the paragraph in general terms:

• *Dry, clear weather...enabled us to progress faster than anticipated.*
• *The electrical fault in the EDM equipment...recurred on May 23.*
• *I have had difficulty hiring reliable people to clear brush along the route.*

Now the writing plan has extended beyond the four basic compartments

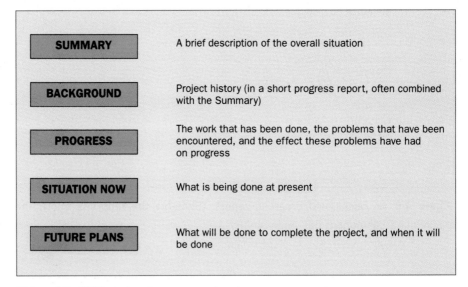

SUMMARY	A brief description of the overall situation
BACKGROUND	Project history (in a short progress report, often combined with the Summary)
PROGRESS	The work that has been done, the problems that have been encountered, and the effect these problems have had on progress
SITUATION NOW	What is being done at present
FUTURE PLANS	What will be done to complete the project, and when it will be done

Figure 4-9 Writing plan for an occasional progress report.

H L WINMAN AND ASSOCIATES

INTER-OFFICE MEMORANDUM

To: Karen Woodford, Coordinator
Civil Engineering

From: Pat Fraser, Survey Crew Chief

Subject: Progress Report No. 4—
Allardyce Survey Project

Date: May 31, 1997

The Allardyce Route survey has progressed well during the May 16 to 31 period. The survey crew has regained two days, and now is only four days behind schedule. We expect to be back on schedule by June 30.

The Summary sums up key features from the report's body

Project plan AR-51 shows we should have surveyed positions 30 to 34 during this period. But, as stated in my May 15 report, we were six days behind schedule at the end of the previous period, having surveyed only as far as position 28. Consequently, we expected to survey only to position 32 by May 31.

Dry, clear weather from May 18 to 23 enabled us to progress faster than anticipated. We reached position 31 on May 23, carried out a terrain analysis for the Catherine Lake diversion scheme on May 24 and 25, resumed surveying on May 26, and reached position 32 at 09:00 on May 29, two days earlier than expected. At end of work on May 31, we were just 300 metres short of position 33. Survey results are attached.

3a

Two problems affected the project during this period:

1. The electrical fault in the EDM equipment, which delayed us several times early in the project, recurred on May 23. I had the unit repaired at Fort Wilson on May 24 and 25, while we conducted the terrain analysis, and it has since worked satisfactorily.

2. I have had difficulty hiring reliable local people to clear brush along the route. Most remain with us for only a few days and then quit, and I have had to waste time hiring replacements. This problem will continue until mid-June, when the college students we interviewed in March will join the crew.

The "Present Work" compartment may be omitted from a progress report

We plan to advance to position 37 by June 15, which should place us only two days behind schedule. If we can maintain the same pace, I hope to make up the remaining two days during the June 16 to 30 period.

Pat Fraser

Figure 4-10 A periodic progress report.

Pat then describes in more detail what happened, using *facts* (exact dates and position numbers, for example) to support each topic sentence. To prevent the report from becoming too long, Pat attaches the survey results to it and simply refers to them in the narrative. (Because of their length, they have not been printed with Figure 4-10.)

❹ In the **Outcome** paragraph Pat tells Karen what the crew expects to accomplish during the forthcoming period, and even suggests when they may eventually get back on schedule. This final statement clearly supports the opening paragraph, and so brings the report to a logical close.

Other factors you should consider when writing periodic reports are:

- If a progress report is long, use headings such as these to help readers *see* your organization:

Adherence to Schedule (This is your Summary).
Progress During Period (These are your Facts; state the Background information at the front of the Progress section.)
Problems Encountered
Projection for Next Period (This is the Outcome.)

- For lengthy progress or problems sections, start with a summarizing statement describing general progress, then write several subparagraphs each giving details of a particular aspect of the project. For example:

4.   Interior construction work has progressed rapidly but exterior work has been hampered by heavy rain.

   4.1 In the east wing, we erected all partitions, laid 80% of the floor tiles, and installed 20% of the light fixtures.

   4.2 In the west wing, we laid all remaining floor tiles, installed all light fixtures, bolted down 16 of the 24 benches and connected them to the water supply and drains.

   4.3 We started landscaping on September 16, but had to abandon the work from September 18 to 23 when heavy rains turned the soil into a quagmire. By the end of the month we had completed only the outer areas of the parking lot.

- Be as brief as possible when describing routine work. Quote specifics rather than generalizations, and place lengthy details in an attachment. If, for example, you are reporting an extensive analysis, in your progress section you might write:

We analysed 142 samples, 88 (62%) of which met specifications. Results of our analyses are shown in attachment 1.

Heading titles parallel the pyramid's parts

Attachment 1 would contain several pages of tabular data (numbers, quantities, measurements), which if included as part of the report narrative would inhibit reading continuity.

- Describe problems, difficulties, and unusual circumstances in depth. State clearly what the problem was, how it affected your project, what measures you took to overcome it, and whether the remedial measures were successful. For example:

Topic Sentence	Juvenile vandalism has proved to be a petty but time-consuming problem. On September 3 (Labor Day) youths scaled the fence around the materials
Facts	compound and stole about $300 worth of building supplies. On September 16 they started up a front end loader, drove it into the excavation, then got it stuck in the mud and burned out the clutch. To
Outcome	prevent a recurrence, from September 18 I have doubled the night watch and have had the site policed by a patrol dog. There have been no further attempts at vandalism.

Each problem description is shaped like a miniature pyramid

- Forewarn management of any situation that, although it may not yet affect your project, may become a future problem. With such knowledge, management may be able to avert a costly work stoppage or equipment breakdown. Here is a typical situation:

    7.1 Unless the strike at Vulcan Steel Works ends shortly, it will soon curtail our construction program. Our present supply of reinforcing barmats will last until mid-October, after which we must find an alternative source of supply. I have researched other suppliers, but have been warned by union representatives that any attempt to obtain steel elsewhere may result in a walk-out at other plants.

Predict potential developments...

Where should such an entry appear in your progress report? The best position would be at the end of the Facts (Problems) section, immediately before the Outcome.

- If your report is lengthy or comprehensive, number your paragraphs and subparagraphs (see the examples above). The paragraph numbers can help you refer to a specific part of a previous report, like this:

    The possibility of a shortage of steel mentioned in para 7.1 of my September report was averted when the strike at Vulcan Steel Works ended on October 6.

...and then in a subsequent report describe the outcome

- Maintain continuity between reports. If you introduce a problem that has not been resolved in one report, then refer to it again in your next report, even though no change may have occurred or it was solved only a day later (see the example in the previous paragraph). You must never simply drop a problem because it no longer applies.

- If management expects you to include project cost information in your progress report, insert it in three places:

  In the **Summary** (comment briefly on how closely you are adhering to projected costs).

  In the **Progress** section (give more details of costs, and particularly cost implications of problems).

  In the **Outcome** section (indicate future cost trends).

  Costs are usually closely linked with your adherence to schedule: the more you drop behind schedule, the more likely you will have to report a cost overrun.

Project Completion Report

It's mostly the Facts compartment that gets expanded and relabelled

A project completion report may be the only report evolving from a short project, or the last in a series of progress reports concerning a lengthy project. Thus the Summary-Background-Facts-Outline arrangement shown in Figure 4-1 can be adhered to fairly closely, with the Facts compartment being separated into two compartments labelled **Project Highlights** and **Exceptions** (see Figure 4-11). The Exceptions section draws attention to deviations from the original project plan.

The project completion report written by Jack Binscarth at the end of his analysis of oil samples for Cantor Petroleums has the five writing compartments identified beside each part of the report. (See Figure 4-12; Jack's progress report for this project is in Figure 4-8.) Note particularly that in

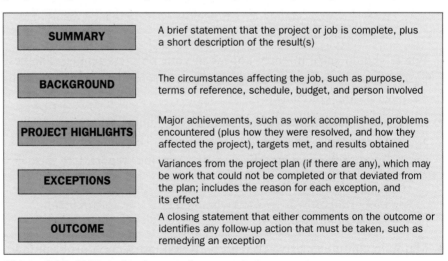

SUMMARY	A brief statement that the project or job is complete, plus a short description of the result(s)
BACKGROUND	The circumstances affecting the job, such as purpose, terms of reference, schedule, budget, and person involved
PROJECT HIGHLIGHTS	Major achievements, such as work accomplished, problems encountered (plus how they were resolved, and how they affected the project), targets met, and results obtained
EXCEPTIONS	Variances from the project plan (if there are any), which may be work that could not be completed or that deviated from the plan; includes the reason for each exception, and its effect
OUTCOME	A closing statement that either comments on the outcome or identifies any follow-up action that must be taken, such as remedying an exception

Figure 4-11 Writing plan for a project completion report.

MACRO
ENGINEERING INC.
600 Deepdale Drive, Toronto ON M5W 4R9

FROM: Jack Binscarth   DATE: October 25, 1997
TO: Fred Stokes   SUBJECT: Finalizing Cantor
                                     Petroleums' Project

Summary Statement

I completed the analysis of oil samples for Cantor Petroleums on October 24, eight days later than planned. The work was done at the refinery, as requested in Cantor Petroleums' pur-

Background

chase order No. 376188 dated September 4, 1997, and was scheduled to start on September 11 and end on October 16. I was assigned to the project under work order No. 2716.

Project Highlights

The work plan called for me to analyse 132 oil samples within the five-week period, but three problems caused me to over-run the schedule and complete four fewer analyses than spec-ified. The delay was caused by a strike of refinery personnel

Exceptions

and a faulty spectrophotometer that had to be sent out for repair and recalibration. The incomplete analyses were caused by four contaminated samples that could not be replaced in less than six weeks.

Outcome

Russ Dienstadt, the refinery manager, agreed to a cost over-run and has corresponded with you separately on this subject. He also agreed that it would be uneconomical for me to return to analyse replacements for the four contaminated samples. When I delivered the 128 analyses to him on October 24, he accepted the project as being complete.

Jack

Good Background infor-mation contributes to leaner, better focused Facts

Figure 4-12 A project completion report.

a short report like this it's acceptable to combine two, or sometimes more, writing compartments into a single paragraph. In Jack's project completion report, paragraph 1 contains both the **Summary** and the **Background**, and paragraph 2 contains both the **Project Highlights** and the **Exceptions**.

Inspection Report

An inspection can range from a quick check of a small building to assess its suitability as a temporary storage centre, to a full-scale examination of an airline's aircraft, avionics equipment, repair facilities, and maintenance methods. In both cases the inspectors will report their findings in an inspection report. The building inspector's report will be brief: it will state that the building either is or is not suitable, and give reasons why. The airline inspector's report will be lengthy: it will describe in detail the condition of every aspect of the airline's operations and list every deficiency (condition that must be corrected). In both cases the inspectors' reports can follow the Summary-Background-Facts-Outcome arrangement, as shown in Figure 4-13.

For an inspection report, these four compartments are:

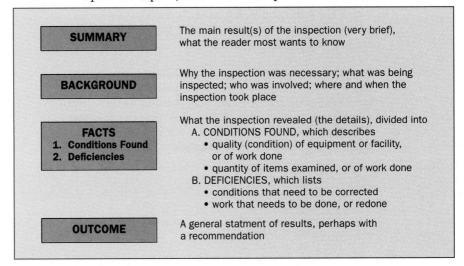

It's better to describe conditions and deficiencies in two separate compartments

Figure 4-13 Writing plan for an inspection report.

Kevin Doherty's building inspection report in Figure 4-14 shows how these compartments helped him shape his report into a logical, easy-to-follow document. Note particularly how:

- His **Summary** (1) tells the Production Manager the one thing he most wants to know: can they use the building?
- The **Background** (2) describes who went where, why, and when.
- Kevin has opened the **Conditions** section (3) with a summarizing general statement, and then supported it with facts (3A).

- He has presented the **Deficiencies** (3B) as a briefly stated list, which makes it easy to identify what has to be done, and has used active verbs to demonstrate that the actions *must* be performed.
- The recommendation in his **Outcome** (4) supports his summary.

For a short inspection report like this, Kevin was correct in presenting all the Conditions first and then listing all the Deficiencies. But such an arrangement could become cumbersome for a long report that covers many items. For example, if Fran Hartley followed this sequence for an inspection at Remick Airlines, the organization of the Facts section would be like this:

A. CONDITIONS FOUND:
　　1. Electrical Shop
　　2. Avionics Calibration Centre
　　3. Flammable Materials Storage
　　　　(etc...)
B. DEFICIENCIES:
　　1. Electrical Shop
　　2. Avionics Calibration Centre
　　3. Flammable Materials Storage
　　　　(etc...)

A plan for a short inspection report

The more departments Fran inspects, the longer the report becomes and the further apart each department's Conditions and Deficiencies sections grow.

To overcome this difficulty, Fran should treat each department as a *separate* inspection and reorganize the report so that for each department the Deficiencies section immediately follows the Conditions section. The organization of the whole report would then become:

SUMMARY
BACKGROUND
FACTS:
　　1. Electrical Shop
　　　　A. Conditions Found
　　　　B. Deficiencies
　　2. Avionics Calibration Centre
　　　　A. Conditions Found
　　　　B. Deficiencies
　　3. Flammable Materials Storage
　　　　A. Conditions Found
　　　　B. Deficiencies
　　　　　　(etc...)
OUTCOME:
Conclusions
Recommendations

The plan for a longer, more detailed inspection report

MACRO
ENGINEERING INC.

600 Deepdale Drive, Toronto ON M5W 4R9

FROM:	Kevin Doherty	DATE:	January 8, 1998
TO:	Hugh Smithson	SUBJECT:	Inspection of Carter
	Production Manager		Building

① The Carter Building at the corner of River Avenue and 39th Street will make a suitable storage and assembly centre for the Dennison contract.

② Christine Lamont and I inspected the Carter Building on January 6 to assess its suitability both for storage and as a work area for 20 assemblers for 15 months. We were accompanied by Ken Wiens of Wilshire Properties.

③ We found the interior of the building to be spacious and to have good facilities, but to be unsightly. Our inspection showed that:

③a
- There are 460 m² of usable floor space (see attached building plan, supplied by Mr Wiens); we need 350 m² for the project.
- There are two offices, each 16 m², and a large unimpeded space ideal for partitioning into a storage area and four work stations.
- The building is structurally sound and dry, but it is very dirty and smells strongly (the previous tenant was a fertilizer distributor).
- There are numerous power outlets, newly installed with heavy-duty circuits, and the building has excellent overhead lighting.
- Several walls are damaged and many contain obnoxious graffiti.
- There is a new loading ramp on the north side of the building, suitable for semitrailers.
- Washroom facilities are adequate for up to 30 people, but one toilet and two washbasins are broken.

Before we rent the building, the rental agency will have to

③b
1. clean it thoroughly,
2. repair damaged walls, partitions, and toilet facilities, and
3. redecorate the interior.

Ken Wiens said his firm would be willing to do this.

④ I recommend we rent the Carter Building from Wilshire Properties, with the provision that the deficiencies listed above must be corrected before we move in.

Figure 4-14 A short informal inspection report.

Fran's inspection report would now have much more tightly knit, logical, coherent organization.

Laboratory Report

There are two kinds of laboratory reports: those written in industry to document laboratory research or tests on materials or equipment, and those written in academic institutions to record laboratory tests performed by students. The former are generally known as "Test Reports" or "Laboratory Reports"; those written by students are simply called "Lab Reports."

Lab reports are written frequently in colleges, less often in industry

Industrial laboratory reports can describe a wide range of topics, from tests of a piece of metal to determine its tensile strength, through analysis of a sample of soil (a drill core) to identify its composition, to checks of a microwave oven to assess whether it emits radiation. Academic lab reports can also describe many topics, but their purpose is different since they describe tests that usually are intended to help students learn something or prove a theory rather than produce a result for a client.

Laboratory reports generally conform to a standard pattern, although emphasis differs depending on the purpose of the report and how its results will be used. Readers of industrial laboratory or test reports are usually more interested in results ("Is the enclosed sample of steel safe to use for construction of microwave towers that will be exposed to temperatures as low as $-40°C$ in the Canadian winter?" a client may ask), than in how a test was carried out. Readers of academic lab reports are usually professors and instructors who are more likely to be interested in thoroughly documented details, from which they can assess the student report writer's understanding of the subject and what the test proved.

A laboratory report comprises several readily identifiable compartments, each usually preceded by a heading. These compartments are described briefly below.

Part	Section Title	Contents
SUMMARY	Summary	A very brief statement of the purpose of the tests, the main findings, and what can be interpreted from them. (In short laboratory reports, the summary can be combined with the next compartment.)

Part	Section Title	Contents
BACKGROUND	Objective	A more detailed description of why tests were performed, on whose authority they were conducted, and what they were expected to achieve or prove.
FACTS		*There are four parts here:*
	Equipment Setup	A description of the test setup, plus a list of equipment and materials used. A drawing of the test hook-up may be inserted here. (If a series of tests is being performed, with a different equipment setup for each test, then a separate equipment description, materials list, and illustration should be inserted immediately before each test description.)
	Test Method	A detailed, step-by-step explanation of the tests. In industrial laboratory reports the depth of explanation depends on the reader's needs: if a reader is nontechnical and likely to be interested only in results, then the test description can be condensed. For lab reports written at a college or university, however, students are expected to provide a thorough description of their method.
	Test Results	Usually a brief statement of the test results or the findings evolving from the tests.
	Analysis (or Interpretation)	A detailed discussion of the results or findings, their implications, and what can be interpreted from them. (The analysis section is particularly important in academic lab reports.)

A generic writing plan for a lab report

Part	*Section Title*	*Contents*
OUTCOME	Conclusions	A brief summing-up, which shows how the test results, findings, and analysis meet the objective(s) established at the start of the report.
BACKUP	Attachments	These are pages of supporting data such as test measurements derived during the tests, or documentation such as specifications, procedures, instructions, and drawings, which would interrupt reading continuity if placed in the report narrative (i.e. in the Test Method section).

In practice, the writing plan is adapted to suit the industry and the circumstances

The compartments described here are those most likely to be used for either an industrial laboratory report or a college/university lab report. In practice, however, emphasis and labelling of the compartments will differ, depending on the requirements of the organization employing the report writer or, in an academic setting, the professor or instructor who will evaluate the report.

ASSIGNMENTS

Project 4.1: Interrupted Project Work

You are an independent consultant working under contract to H L Winman and Associates. The contract calls for you to take all the water-level data recorded over the past 20 years on lakes and rivers in the "Power Bay Territory" of northwestern Ontario. You have to enter this data into a new database, from which H L Winman and Associates will create spreadsheets to support a water-level study they are conducting for Ontario Hydro. For the years up to and including 1992, you are key-stroking the data manually, extracting the information from hard copy printouts. For the years from 1993 up to last year, you are transferring the data from computer-generated records supplied to you on CD-ROM disks. The records in both cases have been provided by the Environmental Systems Division of the Province of Ontario.

You started the project on March 15 and are due to complete it on June 30. Today's date is Friday, June 27.

Two nights ago you were wakened by a violent electrical storm. The next morning (Thursday), when you unlocked your office door you noticed an acrid, pungent smell. You checked the surge protector and saw that it had tripped "off" during the night, which meant there must have been a major surge on the line. You reset the surge protector and switched on your computer. There was no response and the computer screen remained blank.

You checked the surge protector again: the "on" light was glowing. *Then* you noticed that the computer was plugged into *an ordinary socket*, not the surge protector! (Only then did you remember the note the office cleaner had left on your desk a week earlier: "Sorry. The vacuum cleaner head got caught in some of the wires under your desk and I had to plug them in again." You hadn't checked the connections at the time; if you had, you would have noticed the fault.)

You took the computer to Westside Computer Centre and borrowed a "loaner." In your office, you loaded the conversion programs into the loaner's hard disk, plus the work you had copied onto a backup disk, and checked that the system worked (it did). But then you realized you had been careless on June 23 to 25 and had not made backup copies of your work for those days.

On Friday morning Pam Withrow telephoned from Westside Computer Centre. "I'm afraid the damage to your computer is extensive," she reported. "The hard disk and the CD-ROM drive have been destroyed, and some other circuitry seems to have been affected. You're looking at a $1800 repair bill. Frankly, I'd write the unit off and buy a new one." Pam also mentioned that a CD in the CD-ROM drive had been "fried" and was welded to the drive.

You groan inwardly. Not only have you to replace your computer, but you also have to redo at least three days' work *and* get hold of a replacement CD.

Part 1

Write a letter-form incident report to the Environmental Systems Division, Province of Ontario (address: Suite 1701, 330 Oswald Place, Toronto, Ontario, M5W 2R6) and fax it to them. Tell them their CD has been destroyed and ask for a replacement. It's disk No. 3 of a three-disk series titled "Water Levels, Power Bay Territory." You notice that disks 1 and 2 carry the identification numbers OES3301 and OES3302. Ask for a fast response because you need the disk to continue with the data conversion.

Part 2

Write an email progress report to Vern Rogers, H L Winman and Associates's branch manager in the capital city of the province where your office is located. Inform him that you will not be sending a progress report on disk this week (every Friday afternoon you copy the results of that week's work onto a disk and mail it to him by Priority Post, for delivery on Monday morning). Inform him you are four days' behind schedule and suggest how you propose to make up the lost time. Remember that to some extent you are depending on the speed with which the Environmental Systems Division responds to your request for a replacement CD.

Here are some additional details you will need to write these reports:

- Your company's name is Pro-Active Consultants Limited, and you run the business from your home (for more information, see project 3.6 on page 76).
- · The address of the H L Winman and Associates's branch office is 300 Broad Street of the capital city in your province (create a suitable postal code).
- The contract number you are working on is HLW2230, and it's dated March 8.
- Your fax number is the same as your telephone number.
- The Environmental Systems Division's fax number is (416) 293-4601.
- Vern Rogers's email address is: v_rogers@winman.br3.ca.

Project 4.2: Accident at Cormorant Dam

You are an engineering technologist employed in the local branch of H L Winman and Associates. Currently you are supervising installation work at a remote construction project at Cormorant Dam.

The day before you left for the construction site, your branch manager (Vern Rogers) called you into his office. "I'd like you to meet Harry Vincent," he said, and introduced you to a tall, grey-haired man. "Harry is with the Department of the Environment and he wants you to take some air pollution readings while you're at Cormorant Dam."

Mr Vincent opened a wooden box about $35 \times 25 \times 25$ centimetres, with a leather shoulder strap attached to it. In the box, embedded in foam rubber, you could see a battery-powered instrument. "It's a Vancourt MK 7 Air Sampler," he explained, "and it's very delicate. Don't check it with your luggage when you fly to Cormorant Dam. Always carry it with you."

For the next hour Mr Vincent demonstrated how to use the air sampler, and made you practise with it until he was confident you could take the twice-daily measurements he wanted.

Now it is 10 days later and you have just finished taking the late-afternoon air sample measurements. You are standing on a small platform halfway up some construction framework at Cormorant Dam, and are replacing the air sampler in its box.

Suddenly there is a shout from above, followed immediately by two sharp blows, one on your hardhat and the other on your shoulder. You glimpse a 3-metre length of 80 millimetre square construction lumber

A painful cause for writing an incident report

tumble past you followed by the air sampler box, which has been knocked out of your hand. The box turns end over end until it crashes to the ground. When you retrieve it the box is misshapen and splintered and the air sampler inside it is twisted. Your arm also is throbbing badly and you cannot grip anything. An examination at the medical centre shows you have a dislocated shoulder, and now your arm is supported by a sling. (Fortunately, it is not your writing hand.)

Part 1

Write an incident report to Harry Vincent of the Department of Environment. Tell him

- what has happened,
- that you have shipped the damaged air sampler to him on Remick Airlines Flight 751, for him to pick up at your city's airport (you enclose the airline's receipt with your report), and
- that if he wants you to continue taking air pollution measurements, he will have to send you another air sampler.

Harry Vincent's title is Regional Inspector and his address is Department of the Environment, Suite 306, 444 Waltham Avenue of your city.

Part 2

Write a memorandum-form incident report to Vern Rogers. You can mention that you were absent from the construction site for 24 hours, but that otherwise the incident has not affected your supervision work.

Project 4.3: Installing an Automatic Car Wash

(Note: There are four assignments in this project. Your instructor will tell you whether you are to write all four reports or only selected reports.)
You are an independent consultant managing a turnkey project for High Gear Truck and Car Rentals (HGTCR). The contract calls for the installation of three drive-through automatic car washes, at one of the HGTCR branches in Regina, Calgary, and Edmonton. In each city the car wash is to be installed at the most central and easily accessible rental outlet, and will be used to wash cars from all the HGTCR outlets in that city.

When Frank Moroni awarded the contract to you, he said: "This is an experiment. If having a central car wash works well, we'll be putting car washes in 12 other cities in Canada. And if we're happy with how you handle the installation phase, we'll be calling on you to manage the remaining 12 installations." That was on March 20.

The car wash manufacturer is AutoWash Limited of Thunder Bay, Ontario. AutoWash will provide a team who will travel to the three sites sequentially to install the hardware (the overhead ramp, the rotation

In a turnkey project, you handle the details; the client depends on you for the finished product

brushes, the spray jets, the motors and gears, and so on): Regina in week 1, Calgary in week 2, and Edmonton in week 3. Your role is to hire and coordinate the work of local contractors, primarily a construction company which will excavate the pit and pour the concrete footings, an electrician, a pipe fitter, and a computer consultant who will perform tests on and troubleshoot the microprocessor control equipment. For each contractor, AutoWash will provide detailed instructions and a checklist.

The AutoWash installation team will spend one week, Monday to Friday, in each city. The schedule is as follows:

City	Installation Dates	Date Car Wash to Start Operating
Regina	May 8 to 12	May 13
Calgary	May 15 to 19	May 20
Edmonton	May 22 to 26	May 27

To ensure a smooth operation, you arrange with AutoWash for the equipment to be shipped to and in place at each site one week in advance of the installation start date. In other words:

At Regina:	May 1
At Calgary:	May 8
At Edmonton:	May 15

During the week of April 3 to 7, you visit the three cities and interview and hire the building contractors. In the contract you issue to each, you specify that the base construction must be completed *before* May 1 (Regina), May 8 (Calgary), and May 15 (Edmonton), and that they will incur a penalty of $500 for *each* day they are late.

On May 2 you visit the Regina HGTCR location and inspect both the construction work (for quality and completeness) and the delivered car wash equipment (for quantity and completeness, using a checklist provided by AutoWash). All is satisfactory. You send a fax to AutoWash to tell them that everything is ready for work to start on Monday May 8.

The project gets off to a good start...

On May 8 you return to Regina and check that the installation is progressing satisfactorily and that the contractors are ready to come in at the appropriate moment. Then on May 9 you take an early morning flight to Calgary to inspect the construction work and the equipment there.

Part 1. Inspection Report
You realize immediately that the inspection in Calgary is not going to be as straightforward as the inspection in Regina. For one thing, there is debris scattered around the concrete shell: two piles of gravel and soil; broken pieces of wooden formers used for the concrete pours; scraps of

metal; and lunch bags and drink containers. You look up the contract: sure enough in para 17[c] it says "The contractor shall remove all construction debris from the site."

With a tape measure you measure the length, width, and depth of parts of the structure. The main structure is satisfactory but you are dissatisfied with the concrete's rough edges: in some places they should have been sanded down to remove furring and spikes. This has not been done along about 20% of the structure's edges. Paragraph 12(b) says "The contractor shall remove all evidence of rough construction by smoothing the joints and edges of all new concrete."

...then you encounter problems

You are dismayed, too, to find that the groove for the cables and wires cut in the concrete between the locations of the entry control box and the computer control unit is too shallow. You measure it: it varies in depth between 31 mm and 36 mm, and in width between 42 mm and 51 mm. You check construction specification No. AW-2121, which says in step 24[c]: "The cable troughs must be a minimum 34 mm deep and 50 mm wide." This means the contractor must return and saw the edges and the base of the trough until it meets the specified depth and width over its whole length.

You decide to write in the list of deficiencies that the work must be completed by noon on Friday May 12, after which the $500 per day penalty clause will come into effect.

Your check of the AutoWash equipment is more straightforward: everything is satisfactory except for three items: a 100 metre roll of No. 22 AWG cable is missing; the seal over a 50 metre roll of No. 10 hookup wire has been removed and some of the wire has been cut off; and the lenses of the red and green "stop/go" lamps are cracked and the bulbs are missing. (You wonder if the missing items were not shipped or have been stolen since the equipment was delivered.)

Write an inspection report and prepare a fax cover sheet addressed to both AutoWash and the construction company. AutoWash's fax number in Thunder Bay is 807-224-2164, where your contact is Anja Wiederhausen. The construction company in Calgary is Westan Builders Ltd; your contact is the owner, Stan West, and his fax number is 947-2233.

Part 2. Progress Report

Work stops, for no reason you can discern

You fly back to Regina on Wednesday May 10, arriving at 7:20 p.m. You check in at the Ramada Inn, then walk over to Albert Street to see how the installation has progressed. To your amazement, only the basic framework has been installed and there are four picketers with placards tied across their chests marching back and forth in front of the HGTCR rental outlet.

Inside, branch manager Janice Phelps is *not* impressed. "Can you imagine what this is doing to our business?" she growls. "Customers are staying away in droves, and none of it's my doing!"

You ask what happened.

"Something to do with using nonunion labor, I think. I didn't get the full story. You'd better talk to the installers from AutoWash."

You call the Travelodge, where the three members of the AutoWash crew are staying, but they are not there. You keep calling until 11 p.m., then leave a message that you will meet them for breakfast in the Ramada Inn coffee shop at 8 a.m. Thursday morning.

"What's the problem?" you ask as they sit down.

"It's simple," crew leader Bev Shallenberg replies. "The electricians and pipefitters are union labor. We're not, and they won't work with us."

(You had hired unionized contractors intentionally, to forestall any such problems.)

"When did this happen?"

"Tuesday morning, right before lunch, when the electrician showed up on site. He asked to see our union cards, and that was it!"

You phone the Department of Labor and ask if the strike is legal. It is. You phone the union boss, William Persimmon.

"I've been expecting you to call," he says, and he explains that no union member will touch the job. "And if you attempt to hire scab labor, my members will be down there in a flash, in full force. Nobody will want to drive onto the lot. High Gear's not going to like that!"

You try all day to get one side or the other to back down, but everyone is intransigent. In desperation, you call Freja Arundsen of the Department of Labor and ask her to arrange a meeting between all four parties: yourself, William Persimmon, Bev Shallenberg, and herself. She agrees: 9 a.m. on Friday, in her office.

You acquire a human relations problem, and both sides are intransigent

The meeting lasts one-and-one-half hours, after which Freja suggest that perhaps the installer could hire a local *union* construction worker to become part of their team. Each representative then confers by telephone with their organization. The union reluctantly says yes (they had wanted the whole crew to be unionized). AutoWash management also reluctantly says yes (reluctant because doing so will add $105 a day to their installation costs, and they will not be able to recover the extra cost from HGTCR).

At 11:15 William Persimmon calls off the picketers and assigns a union construction worker to join the crew.

You discuss progress with Bev Shallenberg, who says: "Management at AutoWash won't let us work weekends—they'd have to pay us time and a half if we do—so we're effectively three days behind schedule. There's no way we can make up the lost time.

You readjust your installation plan and say you will arrange to hire a unionized construction worker to work with the installation crew in Calgary and Edmonton.

"It's a total waste!" Bev says. "They won't know enough to be really useful, or be able to speed up the work."

Write a letter-form progress report to Frank Moroni at HGTCR. Inform him of the problem and the delay it has caused, and predict a new completion date for each location. Prepare a fax cover sheet. The address of HGTCR's head office is 2130 Malton Road, Toronto, Ontario, M3J 2P6. Their fax number is 905-963-2241.

Problem resolved (more or less!); now you have to report the results

Part 3. Incident Report

It's now Friday May 19 and it's day two of the installation at Calgary. The four sides and the roof of the building have been erected and the crew is now installing the major components of the overhead travelling arm, from which the brushes are suspended. It's 3:15 and you have just returned from inspecting the work at the site of the Edmonton installation (everything was okay).

At 3:22 there is a shout from a team member assembling parts on the overhead structure. One end of the heavy transverse crossbeam, which has been hoisted up by two pulleys with a rope tied to each end of the beam, slips out of his hands, drops, and swings across the car wash, giving a glancing blow to the union construction worker holding the rope at the other end. He is knocked sideways, lets go of the rope, and the whole beam crashes downward, landing on his feet. In its fall, one end scrapes against a side window and scatters glass all over him.

You call an ambulance, because the injured worker (whose name is Steve Hallohan) cannot stand or use his left hand and is bleeding from multiple cuts to his face and arms.

An injury stops work and sends a crew member to hospital

The ambulance arrives at 3:41 and whisks Steve to the nearest hospital. No further work is done that day. (No damage is done to the beam, but a replacement window has to be ordered from AutoWash.)

The following morning (Saturday May 20) you write an incident report which you fax to Anja Wiederhausen at AutoWash in Thunder Bay, with a copy to Frank Moroni at HGTCR in Toronto and another to Carla Strothers in Calgary (she is manager of the affected car rental outlet, and you hand the report to her personally, rather than mail it). You also send a copy to Phil Evershed, president of the union office in Calgary (fax: 361-2255).

Part 4. Project Completion Report

It's now Friday morning, June 2, and you are at Edmonton preparing to write your project completion report to Frank Moroni (as a letter, which you will mail to him). An hour ago you telephoned him to say the project is complete (but four working days or six calendar days behind schedule). All three car washes are operating as planned.

"I'm getting good feedback from Regina," Frank says. "They like the system. Of course, distances between outlets are less there than they are in Calgary and, particularly, Edmonton."

You tell him that you're still waiting for the red and green lenses. "I took the good ones from Edmonton and installed them into the Calgary wash, expecting replacements to arrive before we finished at Edmonton. They haven't, so I've taped the cracked lenses in place as a temporary measure and have left instructions with Kevin Hees, the HGTCR manager in Edmonton, on how to install them."

The project is complete, with only two outstanding items

The only other delay you experienced was on the second day in Edmonton, when the electrical contractor had double-booked and failed to show up. A phone call to his home and then where he was working brought him in half a day late. (You have also heard the Regina construction worker is out of hospital and his broken left foot is recovering.)

Now write your project completion report, drawing attention to the main highlights experienced during the project, and mentioning any exceptions to the project plan.

Project 4.4: Theft at Whiteshell Lake

You are the team leader of a four-person inspection crew en route to a remote site 815 kilometres from your office, where construction of a nuclear power generating station is in progress. You are travelling in a panel van and after 580 kilometres you and the crew agree to stop for the night. At 8:05 p.m. you pull into the Clock Inn, a small motel beside the road that skirts around Whiteshell Lake.

The following morning you are having breakfast in the motel's tiny dining room when Fran Pedersen, one of the crew, goes out to the van to fetch the road map. She returns almost immediately and gasps, "The van has been broken into!"

The four of you scramble out to the parking lot and can see right away that the window on the front passenger's door has been smashed.

"They were after the radio," Shawn Mahler observes, pointing to a gaping hole in the dash.

"Check if anything else is missing," you suggest. Already you are expecting the worst, but to your surprise find that only two other items have been taken, one inconsequential and one important: about $6.00 from a tray in the dash (parking meter loonies and quarters), and a video camera and videotapes from a storage box in the rear of the van.

Expensive equipment, for which you need replacements

You try telephoning your office, but it is too early and no one answers. The motel has a fax machine, so you write a memo to your manager and send it by fax. In it you describe what has happened and ask for a replacement video camera to be sent to you. Here is some additional information you draw on to write your report:

- You are driving company panel van licence number HLW 279; it is a Ford.

- Your trip was authorized by Travel Order N-704, dated one week ago, and was signed by your manager.
- The power generating station is being constructed beside the Mooswa River, 27 kilometres north of the small town of Freehampton.
- The Clock Inn is 3.5 kilometres west of Clearwater Village, on highway A1136.
- The third member of your crew is Servi Dashi.
- The video camera is a Nabuchi TX200 "Portacam." You rented it from Meadows Electronics at 2120 Grassmere Road of your city. Its serial number is 21784B.
- Your manager's name is M B Corrigan.
- The purpose of the video camera is to record construction progress visually. The videotapes will be edited and then shown at the Power Authority Directors' Meeting scheduled for the 22nd of next month.
- You telephoned the RCMP detachment at Clearwater Village to report the break-in and theft. They ask you to drop in and make your report in person. You plan to do this at the start of your drive to the construction site (which will be *after* you have sent your fax).
- In your report you ask your manager to ship you a replacement video camera by Greyhound bus the day after tomorrow. One bus a day passes through Freehampton, but it stops only on request. (You will drive to Freehampton to meet the bus, and will telephone your manager tomorrow to check that the video camera *will* be on that particular bus.)
- You use today's date as the date of your report.

A single incident can evolve into several reports or letters

Part 1.

Write the incident report to M B Corrigan. Prepare it as a memorandum with a fax cover sheet.

Part 2.

Write a letter to Meadows Electronics, to explain the loss of their video camera. You may mention that M B Corrigan will be contacting them re insurance coverage.

Project 4.5: Problem on Top of a Mountain

From February 14 to 21 you have been enjoying a 10-day skiing vacation at Whistler Mountain in British Columbia, a two-hour scenic drive north from Vancouver along provincial highway 99. However, when you return to Holiday Guest House at the end of your fourth day (February 17), proprietor Rita Corbin hands you a fax from Jim Perchanski at H L Winman and Associates in Calgary (your employer). His message reads:

> The cafeteria at the top of the gondola lift at Blackcomb mountain has a problem with their electromechanical compacting and waste disposal system. Please investigate and obtain a fix. Charge your time to work order 2730. Thanks. Jim.

You groan inwardly: the skiing has been excellent and the weather forecast promises ideal conditions for tomorrow. You're going to take the gondola lift to the top of the mountain all right, but to work, not to ski!

At the top of Blackcomb you discover two faults: a burnt out automatic ALR switch type 261058, and a defective microprocessor (Tolstar model 66A). You suspect a power surge caused the switch to burn out and the microprocessor to overload. So you place two telephone orders for urgent delivery of three parts:

Parts are needed: there will be a delay

- From Fraser Electronic Supply House in Vancouver,
  1. a replacement ALR switch, and
  2. a power surge protector (which you will install ahead of the switch).
- From Mercier Distributors in Montreal,
  3. a replacement Tolstar 66A microprocessor.

You make a note in your diary: "Feb 18 – Time spent travelling to and from cafeteria and investigating problem: 5.5 hours."

On the sixth day you ski. When you return to Holiday Guest House at 4:30 p.m., the shipment from Vancouver has already arrived. The shipment from Montreal arrives an hour later.

On the seventh day (February 20) you install the parts and test the system. It works perfectly. Time spent: 3.5 hours. It's already 1:30 p.m., so you decide to sit in the cafeteria, admire the brilliant scenery, and write a trip report to fax to Jim Perchanski from the Guest House. In your report you also plan to ask for your vacation to be extended for two more days, and for the company to reimburse you for two nights accommodation ($125 per night), two "observer" (non-ski) lift tickets ($44 each), a per diem rate for two days at $30 per day, and the parts from Fraser Electronic Supply House ($77 plus $15 shipping fee), which you had to charge to your personal VISA card. The cost of the Tolstar microprocessor ($240 plus $30 shipping) was charged directly to the H L Winman account. You should, however, inform Jim Perchanski of that cost, so he will know to include it in the invoice to the cafeteria.

Now you try to recover your personal lost time (and expenses!)

Write your report.

Project 4.6: Effect of a Power Outage

H L Winman and Associates has been carrying out a series of extreme cold and heat tests on electronic and mechanical switches for Terrapin Control Systems of Coquitlam, BC. The tests have been running for four months and will last another two months. The schedule is tight because of initial problems with measuring equipment, which delayed the start by nine days and used up any spare time the project had available.

Currently, you are testing the switches for continuous periods of from 8 to 14 hours. The tests have two parts:

1. For the first 6 hours each day you increase or decrease temperature in 2°C increments until a predetermined high or low temperature is reached. At each 2° increment you test the switches and record how they perform.
2. For the remaining 2 to 8 hours you bake or deep-freeze the switches at the preselected temperature. No monitoring is necessary during this period (although the switches are tested at room temperature the following day).

To avoid having a technician stay throughout part 2, which on some evenings runs as late as 12:30 a.m., you have installed electrical timers in the circuits of the oven and freezer chamber. The timers are set to switch off at the end of the prescribed bake and deep-freeze periods.

This morning when you remove batches 92H and 98C from the oven and freezer chambers you notice that, instead of being close to room temperature, the oven is still hot and the freezer is still cold. You check the electrical timers, and both are "off." Then you notice that the electric clock on the lab wall reads only 3:39; your wristwatch reads 9:03—a different of 5 hours and 24 minutes. You telephone the local power company.

"Was there a power cut last night?" you ask.

"Yes, there was," the voice answers. "We had a transformer blowout at Penns Vale. It affected everyone in your area."

You ask when the power cut started and ended.

An electrical fault renders a day's work useless...

"The transformer blew out at 9:23 last night," the voice announces. "And we restored power to your area at 2:47 a.m."

You thank the voice, and consult your log for the previous day's tests:

- You started part 1 at 9:55 a.m.

- You started part 2 at 3:55 p.m., and set the timers to run for 8 hours (they were to switch off at 11:55 p.m.).

You consider what has happened:

- The continuous bake and deep-freeze periods were interrupted part way through.

- The oven temperature dropped, and the freezer temperature rose, for 5 hours and 24 minutes (but to what temperature?).

- The power was restored and the oven temperature again increased, and the freezer temperature decreased (but to what temperature?).

- The electric timers switched off at 5:19 a.m. (after their eight hours *total* running time).

You consider the implications of the power cut:

- The batches have had uncontrolled, nonstandard testing and will have to be discarded.

- Yesterday's tests will have to be run again (on two new batches). The cost:

Labor:	14 hours (7 hours per batch) = $336.
Materials:	Two complete batches at $92 each.
Time:	One day extra to be added to the program schedule.

Write an incident report to your project coordinator (J H Grayson). Describe what has happened and the implications, and suggest what might possibly be done to prevent a recurrence.

...so you reassess the situation and report to management

Project 4.7: A Computer Goes Missing

It's 9 p.m. and you are flying from Chicago O'Hare airport to your home city, aboard Remick Airlines flight 717. Your journey started in Washington, DC, at 5:30 p.m., when you boarded United Airlines flight 1216 at Washington National airport and flew to Chicago, arriving there at 6:25 p.m. Flight 717 left Chicago at 8:05 p.m. (Actually, you were to have flown a direct route to your home city, on a flight that left Washington at 4:30 p.m., but the meeting you were attending ran later than you expected, you got caught in rush-hour traffic, and you missed the flight by eight minutes.) The date is today.

You are a member of the Society of Engineering Technologists (SET), and you serve as a member of a team that plans conferences and educational seminars for SET members. The particular meeting you attended is an annual event during which 12 delegates make up a long-range educational plan to submit to SET's executive committee. Your involvement is supported by your employer (the local branch of H L Winman and Associates), who donates your time. Your travel expenses are covered by SET.

You are this year's secretary to the planning committee and, as you have over an hour's flying time ahead of you, you decide to start typing the minutes of the meeting into your portable computer. You reach under the seat ahead of you, and lift up and open your computer-size black carrying case. But instead of your computer, you pull out an InFocus 550 LCD panel! Clearly, this is not your case, although it looks exactly like it.

Every business traveller's nightmare: a lost computer

But where and when did you lose your case? (Or, rather, when did you pick up the wrong one?) There are three possibilities: (1) when you were checking in at Washington National and you put it on the floor beside you while you lifted your clothing case onto the scales; (2) when you disembarked from United flight 1216 and lifted it down from the overhead rack; and (3) when you stopped for a slice of pizza and a soft drink at Camille's crowded café in Chicago airport, and shared a table with a woman to whom you hardly spoke. There were three chairs and you both placed your coats and hand luggage onto the third chair.

You search the LCD panel and the bag for identification of the owner, but there are no clues. (You realize that your computer has no identification on it, either.) You doubt that your computer was intentionally stolen, because from your knowledge of LCD panels you realize that the one you have would cost at least $1000 more than your computer!

Unfortunately, it isn't *your* computer that is missing, but a company computer you have on loan from H L Winman and Associates. It's a Nabuchi CD801, serial number 2106A711.

There is a pad of lined paper in the LCD case, and you pull it out to write an incident report to Vern Rogers, your branch manager.

Chances of tracing your computer seem unlikely

At your home airport you report the loss to Remick Airlines Customer Service Representative Ian Coulson, who takes down details and tells you he'll put a tracer on it for you. "But, really," he says, "hand luggage is hard to trace and it's hardly our responsibility. Now, if it had been checked luggage, I could have it here within 12 hours."

You take a taxi home and complete your incident report, ready to give to Vern Rogers when you go to work in the morning.

Project 4.8: Problem Connectors at Site 14

You are an independent consultant running a business you call Pro-Active Consultants Limited from your home office. One of your clients is H L Winman and Associates. One week ago you received a telephone call from HLW's electrical engineering project coordinator Don Gibbon, who assigned you to conduct an investigation report. He told you that H L Winman and Associates is management consultant to Interprovincial Power Company, and currently is supervising the installation of parallel HV DC power transmission lines and a microwave transmitting system along a corridor between Weekaskasing Lake and Flint Narrows. The microwave transmission towers are located approximately 65 kilometres apart, and are numbered consecutively from No. 1 at Weekaskasing Lake to No. 17 at Flint Narrows. Each tower site has a small residential community and a maintenance crew.

A field trip to investigate a problem

The maintenance crew supervisor at tower site No. 11 is Karen Wasalyshyn, and she telephoned Don Gibbon yesterday to say she found nine faulty connectors type MT-27 at her site and had to replace them. Don wanted to know if the problem is purely local or is prevalent elsewhere, so he instructed you to fly to site No. 14, the nearest site to your office, to investigate whether there are any other faulty MT-27 connectors. "I'll fax you a test procedure you can use to test the connectors," he added.

You flew to Site No. 14 two days ago and stayed there until this morning, when you flew back to your home city. The site maintenance crew supervisor was Don Sanderson, who asked you to send him a copy of your report. Here are the details you discovered:

1. There are 317 type MT-27 connectors on site, with 92 in stock and 225 installed along the lines and up the tower.
2. You tested 278 of the connectors.
3. You could not test the remaining 39 because they were along part of the transmission line that was powered-up throughout your visit.
4. You placed each connector under tension using test procedure TP-33.
5. 241 of the connectors were OK.
6. 37 of the connectors proved to be faulty.
7. You identified the fault as a hairline crack, which became visible when a faulty connector was placed under tension.
8. You also noticed that, although the connectors looked similar, there seemed to be two kinds of connectors on site. One batch of connectors had the letters GLA on the base. The other had the letters MVK on the base.
9. Of the 278 connectors you checked, 201 were stamped MVK, and 77 were stamped GLA.
10. All the faulty connectors had the letters GLA stamped on the base. There were no faulty connectors with the letters MVK on the base.
11. You figured that the letters must identify either different manufacturers or different batches made by the same manufacturer.
12. You recommended to Don Sanderson that he replace all installed GLA connectors with MVK connectors, and to place all the GLA connectors in a separate box marked NOT TO BE INSTALLED, until he receives instructions from Don Gibbon.

A trip report to describe your findings

Part 1
You have returned to your home office. Write a trip report to Don Gibbon, as a letter with a fax cover page.

Part 2
Don Gibbon telephones the next day. "Will you email the maintenance crew supervisor at each of the 17 sites," he asks. "Tell them to test the connectors the same way you did, and to replace *all* GLA connectors—not just the faulty ones—with MVK connectors. They are to ship the GLA connectors to me: you know the Calgary address. I'll email you a group address alias you can use. It will distribute your message to all 17 sites." Finally, he says to address a copy of the message to him.

Here are some email addresses you will need to complete this assignment:

- IPC Supervisors' group alias: ipc@winman.grp17.ca
- Don Gibbon's address: dgibbon@winman.ho.ca
- Your email address: pro_active@mbupline.net

WEBLINKS

Lab Reports
www.rpi.edu/dept/llc/writecenter/web/text/labreport.html

This handout from the Rensselaer Polytechnic Institute's Writing Center describes each part of a lab report, including title page, abstract, introduction, methods and materials, experimental procedure, results, discussion, conclusion, references, and appendices.

Engineering Lab Reports
www.iit.edu/~writer/mmae_out.htm engineering

The Illinois Institute of Technology's Writing Center maintains this detailed site about writing engineering lab reports.

Progress Reports
www.io.com/~hcexres/tcm1603/acchtml/progrep.html

This document is one chapter from the online textbook used in Austin Community College's online course, Online Technical Writing (www.io.com/~hcexres/tcm1603/acchtml/acctoc.html). It deals with the purpose, timing, format, and organization of progress reports. A revision checklist alerts the reader to specific problems to avoid.

Chapter 5
Longer Informal and Semiformal Reports

The previous chapter discussed short reports that deal primarily with facts, reports in which the writer identifies the relevant details and presents them briefly and directly. This chapter describes longer reports that often deal with less tangible evidence, reports in which the writer analyses a situation in depth before drawing a conclusion and, sometimes, making a recommendation. They may describe an investigation of a problem or unsatisfactory condition, an evaluation of alternatives to improve a situation, a study to determine the feasibility of taking certain action, or a proposal for making a change in methods or procedures. All are written in a fluent narrative style that is both persuasive and convincing; their writers have concepts or new ideas to present and they want their readers to understand their line of reasoning.

Investigation Report

The term "Investigation Report" covers any report in which you describe how you performed tests, examined data, or conducted an investigation using tangible evidence. You start with known data and then analyse and examine it so that the reader can see how the investigation was conducted and the final results were reached. The report may be issued as a letter, as an interoffice memorandum, or as a semiformal report.

Although they are not always readily identifiable, there are standard parts to a well-written investigation report that help shape the narrative and guide the reader to a full understanding of its topic. They are shown in Figure 5-1, which is an expanded version of the basic writing plan described at the start of Chapter 4. These parts are easy to recognize in long investigation reports, where headings act as signposts introducing each parcel of information. They are more difficult to identify in short reports that use a continuous narrative. In the three-page investigation report in Figure 5-2, the parts are identified by circled numbers:

1 This is the **Summary**.

2 The **Background** is only one sentence, which refers to the memo that instigated the investigation. Because the reader already knows the circumstances, the report writer can omit details.

3 The **Discussion** starts here, with a very brief reference to the **Method**.

4 These are the **Findings**.

5 This is the first of three **Ideas** for resolving the problem.

Comparing each plan, idea, method, or product against the criteria helps a writer be objective

6 These are the **Criteria**: the requirements against which each idea will be measured.

7 The **Analysis** starts here. The table provides a convenient, easy-to-access summary of what each idea will achieve and cost.

8 In the Analysis, each idea is compared to the Criteria. (Note that the ideas are *not* compared one against another.)

An expanded writing plan is still based on the pyramid seen in earlier chapters

SUMMARY	A brief statement of the situation or problem and what should be done about it.
INTRODUCTION	**Background** to the situation or problem.
DISCUSSION	The **Facts** or **Investigation Details** comprising:
METHOD	How the investigation was tackled.
FINDINGS	What the investigation revealed.
IDEAS	Different ways the situation can be improved or the problem resolved.
– CRITERIA –	Factors that influence the analysis.
ANALYSIS	Evaluation of each idea.
CONCLUSIONS	The **Outcome**, or result of the investigation; a summing-up.
RECOMMENDATION	A positive statement advocating action.*
ATTACHMENTS	Evidence: detailed facts, figures, and statistics that support the Discussion.*

Included only when appropriate.

Figure 5-1 Writing plan for an investigation report.

⑨ The **Outcome** starts by drawing **Conclusions**...

⑩ ...and continues with a **Recommendation**.

⑪ The **Attachments** contain drawings, specifications, and detailed cost estimates for each Idea.

This is known as the *objective* approach for comparing ideas, because the writer's opinions do not become apparent until the Conclusions and Recommendations. In the investigation report in Figure 5-2, for example, Phyllis van der Wyck does not let her opinions become known right until the end of the report.

Sometimes a report writer will allow his or her comments and opinions to become apparent much earlier in the report, during the presentation of the ideas. This is known as the *subjective* approach. If Phyllis had used this approach, she would have established her Criteria right up front, immediately after or as part of the Introduction, and would have analysed each Idea in the same paragraph that she presented it. For idea 1, for example, she might have written:

> 1. We could move the air-conditioning equipment to a storage room at the other end of the building for a projected sound level reduction of 10 to 12 dB. This would achieve the greatest sound level reduction among the three methods—in fact the only one to fully meet the requirements—but at $20 000 to $24 000 it would be the most costly.

Letting your presence be evident shows you are being subjective

Because each idea is analysed when it is presented, in a subjective comparison the Ideas are followed immediately by the Outcome. There is more information on writing comparative analyses in Chapter 6, on pages 170 to 171, and in Figure 6-2.

The Parts of an Investigation Report

Every investigation report will differ, depending on the topic you are investigating and the results you obtain. Sometimes you will use all of the standard parts listed below, sometimes you will use only some of them, and occasionally you will need to devise and insert additional parts of your own. Typical standard parts are described here.

Summary
Start with a brief description of the whole report, stated in as few words as possible. This will give busy readers a quick understanding of the inves-

MEMORANDUM   KCMO-TV

TO: Dennis Carlisle   FROM: Phyllis van der Wyck
           Operations Manager   Engineering Department

DATE: October 21, 1997   REF: Investigation of High
                                                Ambient Sound Level:
                                                Satellite Studio Control Room

I have investigated the high ambient sound level reported in the control room of our satellite studio at 21 Union Road, and have traced it to the building's air-conditioning equipment. The sound level can be reduced to an acceptable level by replacing the blower motor and soundproofing the air-conditioning ducts and blowers. The cost will be $9200.

My investigation was authorized by your memo of August 28, 1997, in which you described the audio difficulties your production crews are experiencing when programming from the satellite studio.

Tests conducted with a sound level meter at various locations in the control room established that the average ambient sound level is 36.8 dB, with peaks of 38.7 dB near the west wall. This is approximately 7 to 9 dB higher than the sound levels measured in the control room for No. 1 studio on Westover Road, where the ambient sound level is 29.5 dB with peaks of 30.2 dB near the south wall.

The unusually high sound level is caused by the air-conditioning equipment, which is in an annex adjacent to the west wall of the control room. Air-conditioner rumble and blower fan noise are carried easily into the control room because the short air ducts permit little noise dissipation between the equipment and the work area. The flat hardboard surface of the west wall also acts as a sounding board and bounces the noise back into the room.

I have considered three methods we could use to reduce the ambient sound level:

1. Move the air-conditioning equipment to a storage room at the other end of the building, for an estimated 10–12 dB reduction in sound level. This would, however, require major structural alterations that will cost between $20 000 and $24 000.

This two-and-one-half page memo-report would benefit from having headings inserted at appropriate places

Ideas are numbered consecutively for ease of reference

Figure 5-2 A memorandum investigation report with primarily objective development.

2. Replace the existing blower fan assembly with a model TL-1 blower manufactured by the Quietaire Corporation of Hamilton, Ontario, and line the ducts with Agrafoam, a new soundproofing product developed by the automobile industry in Germany. Together, these methods would reduce the ambient sound level by about 6.5–7.5 dB. The cost will be $9200.

3. Cover the vinyl floor tiles with Monroe 200 indoor/outdoor carpet, a practice that has proved successful in air traffic control centres, and mount carpet on the control room's west wall, for a sound level reduction of about 4.0–4.5 dB. The cost will be $2100.

The remedy we select must

*   reduce the ambient sound level by at least 7.3 dB, to provide conditions similar to those at the Westover Road control room, **6**

*   be implemented quickly (ideally by November 15, when the Christmas Pageant programs will be recorded), and

*   cost no more than $10 000, if the modifications are to be completed within the 1997–98 budget year.

As the table shows, none of the three methods meets all of the above criteria, although one comes close to doing so. **7**

	Required	Method 1 Relocation	Method 2 Blower/Ducts	Method 3 Carpet
Projected sound level reduction (min)	7.3 dB	10–12 dB	6.5–7.5 dB	4.0–4.5 dB
Time to implement (max)	3 weeks	12 weeks	3 weeks	1.5 weeks
Approximate cost (max)	$10 000	$20–24 000	$9200	$2100

A table simplifies a comparison, makes it easier to analyse

*   Method 1—relocating the air-conditioning equipment—would reduce the sound level more than the required minimum but cannot be implemented quickly or within budget.

*   Method 2—replacing the blower motor and lining the ducts—probably would reduce the sound level to an acceptable level, but only just. It could be implemented quickly and within budget.

- Method 3—covering the floor and one wall with carpet—would reduce the sound level by only one-third of the desired reduction. It could be implemented quickly and within budget.

(9)

The only method that comes close to meeting our immediate requirements is method 2. If we were to combine it with Method 3, we could achieve a probable total sound reduction of 8.3–9.8 dB, which would meet the required reduction but would exceed the budget by $1200. (Note that, when combining the methods, the total sound level reduction that will be achieved will be *less* than the summation of the two individual sound level reductions.)

A major recommendation and a minor recommendation

(10)

Because method 2 comes close to the required minimum reduction in sound level, I recommend we replace the blower motor and line the ducts with Agrafoam for a total cost of $9200. However, because actual sound level reductions can differ from those projected, I suggest we retest the sound levels following installation. If a further reduction in sound level proves necessary, then I recommend we install Monroe 200 carpet on the floor and west wall in March 1998 for a total cost of $2100, using $900 from the 1997–98 budget year and $1200 from the 1998–99 budget year.

These modifications should provide the quieter working environment needed by your production crews.

P. Van der Wyck

(11) Att: Specifications and cost estimates

tigation from which they can learn the results and assess whether they should read the whole report.

Background

Describe events that led up to the investigation, knowledge of which will help the reader place the report in the proper perspective. This part is often referred to as the **Introduction**.

Investigation Details

Describe the investigation fully, carefully organizing the details so that readers can easily follow your line of reasoning. This is the **Discussion**. In a long report, the Discussion will contain all or some of the information described here, and the parts should appear roughly in this order:

Introduce Guiding Factors. These are the requirements or limitations that controlled the direction of the investigation. They may be as diverse as a major specification stipulating definite results that must be attained, or a minor limitation such as price, size, weight, complexity, or operating speed of a recommended prototype or modification.

Outline Investigation Method. Tell readers the planned approach for the investigation so that they can understand why certain steps were taken.

Describe Equipment Used. If tests were performed that called for special instruments, describe the test setup and, if possible, include a sketch of it. In long reports this may have to be done several times, to keep details of each test setup close to its description.

> In its expanded form, an investigation report can seem similar to a research paper (see page 248)

Narrate Investigation Steps. Describe the course of your investigation as a series of steps, so that readers can visualize what you did. Sometimes you may be able to use a chronological description, while at other times you may want to vary the sequence to help readers better understand what was done and what was found out.

Discuss Test Results (or **Investigation Results**). Present the results of your investigation and discuss what effect they have had or may yet have. (Simply tabulating the results and assuming that readers will infer their implications does not suffice.)

Develop Ideas and Concepts. At some point in your narrative you may want to introduce ideas and concepts that evolve as a result of your investigation. You may introduce them periodically throughout the report to demonstrate what you had in mind before taking the next investigative step, or group them together near the end of the Discussion.

Establish Evaluation Criteria. If you have developed alternative ideas (methods) for resolving a problem, you will have to evaluate them to

> Each criterion must seem necessary, logical, and relevant

determine which is best. Rather than compare one method against another, you should first establish the criteria that will result in the best or optimum resolution. The criteria can include maximum cost, a desired level of performance, a modification time frame, and so on. In most cases you will also need to prove the criteria, to show why they are valid. For example, Phyllis van der Wyck could have written (see point 6 on page 129):

> The selected remedy must...reduce the ambient sound level by at least 7.3 dB.

But then readers might ask: "Why?" So she added:

> ...to provide conditions similar to those at the Westover Road control room.

Analyse Ideas. Now you compare each idea against the criteria you have established, to show how effective the Idea will be. You may do this in two ways:

1. Present all the ideas, then establish the criteria, then evaluate each idea against the criteria (as Phyllis van der Wyck has done in Figure 5-2).
2. Establish the criteria first, then present each idea in turn and evaluate it immediately in light of the criteria, before presenting the next idea.

The former helps you to appear objective, since you present only facts. The latter allows you to be subjective, since you can offer comments and opinions.

Conclusions

In a brief summing-up, draw the main conclusions that have evolved from your investigation.

Conclusions must summarize only key outcomes, never introduce new information

Recommendation(s)

State what steps need to be taken next. Recommendations are optional and must develop naturally from your Discussion and Conclusions.

Evidence

Attach detailed data such as calculations, cost analyses, specifications, drawings, and photographs that support the facts presented in the **Investigation Details** section but would interrupt reading continuity if included with it. These pages of evidence are numbered consecutively and named **Attachments** (or, in a formal report, Appendices), and placed at the end of the report.

Some companies preface their investigation reports with a standard title and summary page similar to the H L Winman and Associates' design illustrated in Figure 5-3. This page saves a reader the trouble of searching

H L WINMAN AND ASSOCIATES

INVESTIGATION REPORT

REPORT NO: 70/26   FILE REF: 53-Civ-26
DATE: March 20, 1997
PREPARED FOR: City of Montrose, Alberta
AUTHORITY: City of Montrose letter Hwy/69/38, Nov 7, 1996

REPORT PREPARED BY: *G G Waterston*

APPROVED BY: *M. Warner*

Authorization and other details are grouped together in an easy-to-find arrangement

SUBJECT OR TITLE

Investigation of Stormwater Drainage Problem
Proposed Interchange at Intersection of
Highways 6 and 54

SUMMARY OF INVESTIGATION

The proposed interchange to be constructed at the intersection of Highways 6 and 54, on the northern perimeter of Montrose, incorporates an underpass that will depress part of Highway 54 and some of its approach roads below the average surface level of the surrounding area. A special method for draining the stormwater from the depressed roads will have to be developed.

The summary is a miniature pyramid: para 1 = background...

Two methods were investigated that could contend with the anticipated peak runoff. The standard method of direct pumping would be feasible but would demand installation of four heavy-duty pumps, plus enlargement of the 1.21 km drainage ditch between the interchange and Lake McKing. An alternative method of storage-pumping would allow the runoff to collect quickly in a deep storage pond that would be excavated beside the interchange; after each storm is over, the pond would be pumped slowly into the existing drainage ditch to Lake McKing.

Although both methods would be equally effective, the storage-pumping method is recommended because it would be the most economical to construct. Construction cost of a storage-pumping stormwater drainage system would be $984 000, whereas that of a direct pumping system would be $1 116 000.

...para 2 = investigation details, and para 3 = outcome

Figure 5-3 Title and summary page for a semiformal investigation report.

for the summary and the report's identification details. Subsequent pages (which have not been included with the example) contain the report narrative, starting with the Background (Introduction). Reports written in this way tend to adopt a slightly more formal tone than memorandum and letter reports and are less likely to be written in the first person. Sometimes they are called a form report, although the preferred name is semiformal investigation report.

Evaluation Report/Feasibility Study

Evaluation reports are similar to investigation reports, and the two names frequently are used interchangeably. Evaluation reports often start with an idea or concept their authors want to develop, prove, or disprove. Their authors first establish guidelines to keep their report within prescribed bounds, and then research data, conduct tests, and analyse the results to determine the concept's viability. At the end of the evaluation they draw a conclusion that the concept either is or is not feasible, or perhaps is feasible in a modified form.

The writing plan shown in Figure 5-1 also can be applied to an evaluation report. Morley Wozniak's evaluation of landfill sites in Figure 5-4 follows this plan.

Morley's report is significant in that it is preceded by a one-page cover letter, thus adopting the technique suggested on page 48 and illustrated in Figure 3-7. His cover letter is similar to the Executive Summary that often precedes a formal report (Executive Summaries are described in Chapter 6), since it *describes and comments on* key implications drawn from the report. Its addressee (Quillicom's town engineer) has the option of distributing it to the town councillors with the report, or detaching it and replacing it with a cover letter of his own.

The parts of the writing plan shown in Figure 5-1 are identified in Morley's report by circled numbers beside the narrative; they are keyed to the additional comments provided here.

1 Although several factors affect site selection, in his **Summary** Morley focuses primarily on environmental impact because he believes it is of overriding importance.

2 The **Introduction** provides Background details leading up to the study assigned to H L Winman and Associates, and then to Morley.

H L WINMAN AND ASSOCIATES

475 Lethbridge Trail, Calgary AB T3M 5G1

May 23, 1997

Mr Robert D Delorme, P.Eng
Town Engineer
Municipal Offices
Quillicom ON P8R 2A2

Dear Mr Delorme

Our assessment of the three sites selected as potential landfills for the Town of Quillicom shows that each has a disadvantage or limitation. The most serious exists at Lot 18, Subdivision 5N, which is the site preferred by the Town Council. A distinct possibility exists that a landfill located here could contaminate the town's water supply.

The disadvantages of the two other sites affect only cost and convenience. Lot 47, Subdivision 6E, will be considerably more expensive to operate, while Lot 23, Subdivision 3S, will have a much lower capacity and so will have to be replaced much sooner than either of the other sites.

If the Town Council still prefers to use Lot 18, a drilling program must first be conducted to identify the soil and bedrock structure between the lot and Quillicom. Providing the boreholes show no evidence that contamination will occur, then the site would be a sound choice.

The enclosed report describes our study in detail. I will be glad to discuss it and its implications with you.

Regards

Morley Waymark

for Vincent Hrabi
Branch Manager
H L Winman and Associates
Thunder Bay, Ontario
enc

City, province, and postal code, correctly placed all on one line

A cover letter accompanying an in-depth report or proposal may be signed by the author for the department manager

Figure 5-4 The cover letter preceding an evaluation report. This letter also is an executive summary.

3 The **Evaluation Details** start here, with a single paragraph in which Morley outlines his **Investigation Method** (i.e. how he tackled the study). Note that he mentions the three components *in the same sequence* that he will describe them further on in the report.

4 These are Morley's **Findings**—the results of his research. He describes the findings in detail because his readers must fully understand the geology of the area if they are to accept the conclusions he will draw later in his report. Note that he is totally objective here, reporting only facts without letting his opinions intrude.

Morley's rationale for organizing and writing his report

5 Morley presents the three possible landfill sites as his **Ideas** (even though they were originally presented to him by the client—the Town of Quillicom). In effect he is saying to his readers: "Now that I have described the geology of the land to you, here are three locations within the area for you to choose from." Note that he still is totally objective.

6 In his **Analysis** Morley evaluates each site (each Idea) to determine its suitability. The initial paragraph outlines three general criteria he will use to evaluate the sites. He does not identify specific criteria because they have not been defined.

7 Morley's evaluation must clearly establish the factors on which he will base his conclusions. Now he allows some subjectivity to appear in his writing (we can hear his voice behind his words).

8 Morley's **Conclusions** identify the main features affecting each site. Note that he simply offers the alternatives without saying or even implying which is preferable. This part of the report, together with the Recommendations, is the Outcome (sometimes referred to as the *terminal summary*).

9 In the **Recommendations** Morley states specifically what he believes the Town Council must do. He must sound definite and convincing, so he starts with "We recommend…" rather than the passive "It is recommended that…"

10 The **Attachment** brings together all the site details in an easy-to-read form, and simultaneously provides readers with Evidence to support what Morley says about the landfill sites in the report narrative.

H L WINMAN AND ASSOCIATES

475 Lethbridge Trail, Calgary AB T3M 5G1

EVALUATION OF THE PROPOSED LANDFILL SITES FOR THE TOWN OF QUILLICOM, ONTARIO

SUMMARY

Two of the three locations selected as potential landfill sites for the Town of Quillicom, both southeast of the town, are environmentally safe. There is insufficient data to determine whether the third site, to the north of Quillicom, poses an environmental risk. All three sites are financially viable although one, because of its greater distance from Quillicom, would be more costly to operate.

The Background traces the history leading up to the present study

INTRODUCTION

The Town of Quillicom in Northwestern Ontario currently operates a landfill 3.7 km southeast of the town. The landfill was constructed in 1952, and since 1968 has also served the mining community of Melody Lake, 2.8 km to the southwest of the landfill. In a report dated February 27, 1997, Quillicom town engineer Mr Robert Delorme identified that the existing landfill was nearing capacity and that a new landfill must be found and operational by April 30, 1999.

Previously, in 1994, the town had identified two sites as potential replacement landfills: Lot 18, Subdivision 5N, 3.4 km north of Quillicom; and Lot 47, Subdivision 6E, 14.6 km to the southeast. The costs to set up and operate both sites were determined, and Lot 18 proved to be more economical ($2000 more to purchase and develop, but $17 000 a year less to operate), and was favored by the Town Council. However, in a letter to the Council dated November 15, 1995, Mr Delorme expressed his concern that leachate from the site could possibly contaminate the town's source of potable ground water, and recommended that the town first carry out an environmental study.

The town subsequently engaged H L Winman and Associates to examine the sites and determine both their financial viability and their environmental safety. In a letter dated March 15, 1997, Mr Delorme commissioned us to carry out the study, and to include a third potential landfill site at Lot 23, Subdivision 3S, immediately adjacent to the existing landfill, in our assessment.

Figure 5-4 The evaluation report (6 pages).

STUDY PLAN

We divided our study into three components: (1) an examination of the area geology and its ability to constrain leachate movement; (2) an examination of the physical properties of the proposed landfill sites; and (3) an evaluation of the financial and environmental suitability of the sites.

AREA GEOLOGY AND HYDROGEOLOGY

Bedrock at Quillicom and in the area of all three proposed landfill sites is chiefly granite and gneiss lying 15 to 30 metres below the surface. A layer of till varying in thickness from 10 to 20 metres covers the bedrock, and is itself covered by 1 to 15 metres of lacustrine silts and clays.

The whole area has experienced repeated glaciation, with the most recent occurring about 20 000 years ago with the advance of the Late Wisconsonian Ice Field. The advancing ice severely scarred this granite and gneiss. When the ice began to retreat 10 000 years later, till was deposited over the region. In addition, meltwater streams below the ice field deposited vast quantities of alluvial material, which today exist as eskers.

The melting ice also created Lake Agassiz and caused silts and clays to be deposited to a depth of up to 30 metres over the entire lake bed. (In the Quillicom area these lacustrine deposits range from 5 to 15 metres deep.) Then, as the lake drained and water levels receded, streams cut into the lacustrine and till deposits. These streams eventually dried up and their channels were filled with windblown silts and sands. Today the channels are known as buried stringers and, if they are water bearing, as stringer aquifers in the weathered bedrock and eskers.

The Town of Quillicom obtains its potable water from a stringer aquifer on the surface of the weathered bedrock. Other stringer aquifers are known to exist in the area to the east and south of Quillicom, and likely exist to the west and north.

The Ontario Water Resources Branch provided us with logs obtained during drillings for ground water wells in the late 1970s, all to the south and east of Quillicom. We have plotted the locations and types of materials on a topographic map, which shows that

- the bedrock in the area slopes downward, toward the south, from the Town of Quillicom, and
- a major 350-metre wide glacial esker starts 1 kilometre southeast of Quillicom and continues for several kilometres southeast under Highway A806, to beyond the proposed landfill at Lot 47, Subdivision 6E.

As there are very few borehole records for the area north of Quillicom, we could not plot a similar map for that area.

2

Presenting technical details so they will be understood by all readers takes skill!

THE PROPOSED LANDFILL SITES

The attributes of the three proposed landfill sites are discussed briefly below and itemized in detail in the attachment. The anticipated life of each site is based on the 1997 population of Quillicom. Similarly, projected operating costs are based on 1997 prices.

Lot 23, Subdivision 3S

This narrow, 7.81-hectare strip of land is immediately east of and adjacent to the existing landfill, 3.7 km southeast of Quillicom on Highway A806. As it is the smallest site, it will cost only $9000 to purchase and develop. Its annual operating cost will be $47 000, the same as at the present landfill, and it will have an operational life of 12 to 14 years.

Lot 47, Subdivision 6E

The largest of the three proposed sites at 35.91 hectares, but also the most distant, Lot 47 is a rectangular parcel of land 14.6 km southeast of Quillicom on Highway A806. Its combined purchase and development price will be $20 000, and its annual operating cost will be $64 000. (The high operating cost is caused primarily by the much greater distance the garbage collection vehicles will have to travel.) It will have a lifespan of almost 60 years.

Judicious use of white space makes technical details more accessible...

Lot 18, Subdivision 5N

A roughly square, 22.75-hectare parcel of land, this lot is 3.4 km directly north of Quillicom on Highway B1017. It will cost $22 000 to purchase and develop, and $47 000 a year to operate (the same as at present). At the current fill rate, it will last for 36 to 40 years.

SITE COMPARISONS

We considered three factors when comparing the three proposed landfill sites: cost, environmental impact, and convenience.

Cost. We examined cost from two points of view: the immediate expense to purchase and develop the site, and the annual cost to operate it.

...as does the author's choice not to justify the right margin

- Lot 23, adjacent to the existing landfill, offers the lowest purchase and development cost at $9000, compared with $20 000 and $22 000 for the two alternative sites.
- Lot 23 and Lot 18 (the site north of Quillicom) offer comparable operating costs at $47 000 per year, whereas Lot 47 (14 km southeast of Quillicom) would have the highest annual operating cost of $64 000.

3

However, if the purchase and development costs are spread over 10 years and added to the operating costs, Lots 18 and 23 show a more comparable cost structure:

Site:	Lot 18	Lot 23	Lot 47
Annual Cost:	$49 200	$47 900	$66 000

Environmental Risk. The primary environmental risk is the effect that leachate from the landfill could have on Quillicom's source of potable water. If a landfill lies on a glacial esker, leachate from the landfill will probably contaminate ground water aquifers in the esker. If these aquifers are connected hydraulically to stringer aquifers, the stringer aquifers also probably will become contaminated.

- Lots 23 and 47 (and the existing landfill) lie on a major esker south of Quillicom but offer no environmental risk because the slope of the bedrock in the area is to the south, away from the town. Consequently, even if leachate from the landfill contaminates the ground water, it will flow away from Quillicom and will not contaminate the town's water supply.
- Lot 18, however, is in the uncharted area north of Quillicom, where neither the presence of eskers nor the slope of the bedrock has been determined. Consequently it offers a potential risk that leachate from a landfill located here could contaminate the town's water supply. This will be particularly true if the slope of the bedrock south of Quillicom is the same north of the town, since then leachate will flow south, *toward* Quillicom.

Convenience. To establish convenience we considered the size of each landfill site (measured as the number of years it can be used before another site must be found) and its proximity to Quillicom.

- Lot 47 is the largest site, offering close to 60 years of use, but is four times farther from Quillicom than either of the two other sites.
- Lot 18, to the north, is next largest and can provide between 36 and 40 years of use. It is a comfortable 3.4 km distant from Quillicom.
- Lot 23, although the same distance as Lot 18, has a life span of only 12 to 14 years.

CONCLUSIONS

The possibility of leachate contamination of the Town of Quillicom's water supply makes Lot 18, Subdivision 5N, a doubtful choice until sufficient drilling has been done to create a profile of the strata between the lot and Quillicom.

8

4

This analysis sets the scene for the conclusions the report author will draw

The remaining two sites are environmentally sound but have different advantages:

- Lot 47, Subdivision 6E, provides the greatest space but will be costly to operate.
- Lot 23, Subdivision 3S, offers the lowest cost but will have only a limited life span.

RECOMMENDATIONS

We recommend that the Town of Quillicom purchases Lot 23, Subdivision 3S, and operates it as a temporary landfill from 1999 to 2009. We also recommend that the town concurrently conducts a drilling program to the north of Quillicom to determine whether Lot 18, Subdivision 5N, will be an environmentally sound site to use after the year 2009.

Morley Wozniak, P.Eng

5

ATTACHMENT

⑩

**COMPARISON OF PROPOSED LANDFILL SITES
FOR THE TOWN OF QUILLICOM**

An "open" table (no lines drawn around it) is cleaner for a simple presentation

Comparison Factor	Lot 18 Sub 5N	Lot 23 Sub 3S	Lot 47 Sub 6E
Distance from Quillicom (driving dist in km)	3.4 N	3.8 SE	14.6 SE
Size (hectares)	22.75	7.81	35.91
Life (years)	36–40	12–14	58–60
Environmental risk (for site's potential for contaminating the Quillicom water supply)	Unknown	None	None
Development costs:			
Purchase price ($)	10 000	2 000	5 000
Construction cost ($)	12 000	7 000	15 000
Ten-year cost ($/yr)	2 200	900	2 000
Operating costs ($/yr)	47 000	47 000	64 000
Combined development and operating costs:			
Year 1, w/o amortization ($)	69 000	56 000	84 000
Per year, w amortization ($)	49 200	47 900	66 000

6

Compare how Phyllis van der Wyck and Morley Wozniak each use the first person (see Phyllis's investigation report in Figure 5-2). Phyllis uses the informal "I" because she is writing a memorandum report to another member of the television station where she works. Morley uses the slightly more formal "we" because he knows his semiformal report will be distributed to the Quillicom town councillors. (Note, however, that he uses "I" in the personal cover letter to Robert Delorme that accompanies the report.)

We encourage using the first person in letter and report writing

Like an evaluation report, a feasibility study starts by introducing an idea or concept, and then develops and analyses the idea to assess whether it is technically or economically feasible. The chief difference lies in the name and application of each document. An evaluation report is generally based on an idea that is originated and evaluated within the same company; hence it is nearly always informal. A feasibility study is normally prepared at a slightly higher level: the management of company *A* asks company *B* to conduct a feasibility study for it, because company *A* does not have staff experienced in a specific technical field. For example, if a national wholesaler engaged solely in the distribution of dry goods were to consider purchasing an executive jet, it would seek advice from a firm of management consultants. The consultants would examine the advantages and disadvantages, and publish their results in a feasibility study they would issue as either a letter or a formal report. Often the differences between a feasibility study and an evaluation report are so slight that personal preference dictates which label is used for a particular document.

Technical Proposal

There are two kinds of technical proposals: those written on a personal level between individuals, and those prepared at company level and presented from one organization to another. The former are usually informal memorandums or letters; the latter are more often formal letters or reports.

A memo requesting approval to buy new equipment or attend a conference is a proposal

Scientists, supervisors, technicians, engineers, and technologists write memorandums every day to their managers suggesting new ways of doing things. These memorandums become proposals if their writers fully develop their ideas. A memorandum that introduces a new idea without demonstrating fully how it can be applied is no more than a suggestion. A memorandum that introduces a new idea, discusses its advantages and disadvantages, demonstrates the effect it will have on present methods, calculates cost and time savings, and finally recommends what action needs to be taken, demonstrates that its originator has done a thorough research job before putting pen to paper. Only then does a suggestion become a proposal.

Semiformal Proposals

A proposal follows the basic **Summary-Background-Facts-Outcome** arrangement suggested for reports:

1 A brief **Summary** of the situation or problem, and what the writer proposes should be done to improve or resolve it.

2 An **Introduction** that defines the situation and establishes why it is undesirable or is a problem.

3 A **Discussion** of how the situation or problem can be improved or resolved, covering

- a description of the proposed solution,
- what will have to be done to put it into effect,
- what the proposed solution will achieve,
- what advantages will accrue,
- what disadvantages will be incurred, their likely effect, and how they can be resolved or ameliorated, and
- what the cost will be.

4 A **Recommendation** for action, which should be a strong, positive statement or a direct request.

Susan Jenkins used this outline to write the proposal in Figure 5-5, in which she proposes to Martin Dawes that H L Winman and Associates commission a communication consultant to determine the direction the company should take as it incorporates integrated services digital network (ISDN) technology. Her proposal contains

- a **Summary** (1),
- a lengthy **Introduction** (2), in which she establishes what ISDN is and why the company should become involved in using the technology,
- a **Discussion** (3), in which she defines what she proposes to do and what the cost will be, and
- a **Recommendation** for action (4).

When there are alternative solutions for resolving a problem or unsatisfactory condition, the **Discussion** section of the proposal becomes more comprehensive because it has to describe and evaluate all the possibilities. The writing compartment is then expanded into four subsections:

The Objective: A definition of the parameters or criteria for an ideal (or good) solution.

The margin note beside item 3 reads:

Anticipate readers' reaction by answering potential "Yes, but..." comments

H L WINMAN AND ASSOCIATES

INTER-OFFICE MEMORANDUM

To: Martin Dawes   From: Susan Jenkins
Date: February 26, 1998   Subject: Proposal for Feasibility
                                        Study: Adoption of ISDN
                                        System

① Many high-technology companies—and some that are not—are using an
Integrated Services Digital Network (ISDN) to communicate internally,
nationally, and internationally. As an ISDN system needs to be adopted as
part of an overall corporate strategy (it cannot be bought off-the-shelf and
installed quickly), I propose that we engage TransPacific Communication
Systems Ltd of Vancouver, BC, to carry out a preliminary study and recommend the approach that H L Winman and Associates should adopt.

② An ISDN system uses fibreoptic cables instead of standard wire telephone
lines, and transmits information through a digital rather than an analog
network. It costs almost twice as much to install when compared with the
standard telephone system, but offers significant advantages:

- Numerous telephone conversations, pictures, graphics, computer messages, and video signals can be sent and received simultaneously over
  the same line, and with much greater clarity and much less information loss.
- Computer messages can be transmitted instantaneously from one terminal to another, without a modem, with both users making changes
  and seeing the changes—as they are made—on their screens.
- Fax messages can be sent 10 times more rapidly and with negligible
  transmission errors.

ISDN also provides

- conference-call capability between a head office, branches, and staff
  on assignment at other ISDN-equipped locations, with the call being
  carried out by telephone, computer, or video, used either singly or in
  combination; and
- instant access to an immense range of databanks, from airline sched-
  ules and space available to library data stored on CD-ROM.

At the right moment we will have to switch to an ISDN system, so that we
will not only remain competitive but also be seen to be current in the bur-

**The bullets used here
indicate the sequence
is not important**

Figure 5-5 A technical proposal.

geoning high-technology communication field. Timing will be essential, so that we can make a rational rather than a last-minute, seat-of-the-pants decision. Our difficulty will be to know the right moment to make the changeover, how much lead time to allow, and what funds to budget.

Timing is also important because three of our branches (Kingston and Windsor, Ontario, and New Westminster, British Columbia) will be looking for new office accommodation when their current leases expire. If the branches knew for certain that we will be changing over to ISDN, they could specifically seek accommodation in buildings that have already been—or shortly will be—wired with ISDN-compatible cabling.

❸ I am proposing that we commission TransPacific Communication Systems Ltd (TCSL) to carry out a study into H L Winman and Associates' ISDN requirements. TCSL has specialized in fibreoptic data transmission for nine years, and developed the prototype ISDN network for the multi-use Washington–Alaska communication link. I have had a preliminary discussion with Anthony Symondson, TCSL's manager of development engineering, and he has agreed to assign an engineer to study our requirements. He recommends that this be done between March 15 and April 15. The objective will be to

Subparagraphs are numbered, to show there is a natural sequence

1. identify what our communication needs will be over the next decade,
2. outline the advantages of implementing ISDN, and the probable effect on H L Winman and Associates's operations if we do not,
3. develop specifications for an ISDN system that will meet our particular needs, and
4. establish a schedule and a budget for incorporating the system companywide.

I have obtained a cost proposal from TCSL quoting a fee of $20 700 to carry out the study, which will include a full investigation of H L Winman and Associates, its branches across Canada, and Macro Engineering Inc in Toronto, all travel and related expenses, and preparation of a definitive report within 28 days of the project start date.

❹ Tina Mactiere assures me that TCSL is a technically sound organization fully able to perform the study, and in an email message on February 23 stated she is willing to assign $8000 as Macro Engineering's share of the study costs. May I have your approval by March 7 to spend $12 700 as our share of the cost, so that I can authorize TCSL to start the study on March 15?

Susan

The Proposed Solution: What it is, the result it would achieve, how it would be implemented, its cost, and its advantages and disadvantages.

Alternative Solution(s): For each: what it is, the result it would achieve, how it would be implemented, its cost, its advantages and disadvantages, and why it is not as effective as the proposed solution.

Retaining the status quo (i.e. making no change) also is an alternative solution

Evaluation: A brief tradeoff analysis, with reference to the **Objective**, comparing the effects of (a) adopting the proposed solution, (b) adopting an alternative solution, and (c) taking no action.

Formal Proposals

Formal technical proposals are prepared by companies seeking to impress or convince another company, or the government, of their technical capability to perform a specific task. They are normally impressive documents prepared under extreme pressure, and often have to meet rigid format and presentation requirements spelled out in a request for proposal (generally known as an "RFP") issued by the agency asking vendors to submit bids. Preparation of such proposals is beyond the scope of this book.

ASSIGNMENTS

Project 5.1: Investigating Conditions at Sylvan Lake

You are an independent consultant working from your home, and your company is named Pro-Active Consultants Limited. Recently you bid on and won a contract from Triton Mining Corporation (TMC). The contract requires you to inspect conditions at Sylvan Lake in the northern part of your province, and to furnish TMC with an inspection/investigation report describing your findings.

You are an entrepreneur, running a one-person small business

On September 17 you attend a briefing at TMC's headquarters at 206 Hallington Street in Toronto. You meet David Yanchyn, who is contracts manager.

"That's Sylvan Lake," he announced, dropping a map on the table in front of you and pointing to a small lake (see Figure 5-6). He said it used to be a fly-in fishing resort until the lake became over-fished six years ago. Now TMC will be reopening the lodge and using it as a temporary office and living quarters while TMC does some drilling nearby, prior to setting up a full drilling operation in a year's time.

"Your primary responsibility will be to inspect the gravel road between the landing strip 4.8 kilometres east of Sylvan Lake and the fishing lodge

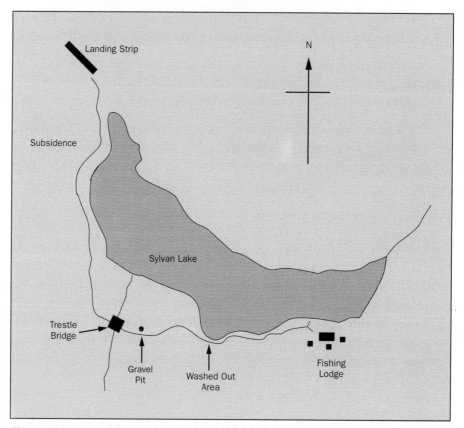

Figure 5-6 Map of Sylvan Lake (not to scale).

at the water's edge. We need to know if any repairs need to be made before the drilling crews and rigs are flown in next spring."

He instructs you also to examine the condition of the fishing lodge and identify any work that needs to be done to make it habitable. "Yours will be just a preliminary look. We'll be doing a full inspection later. Initially we'll be sending in a crew of about 30 people. I want you to tell me what has to be done before they come in."

In particular, he's interested in knowing the condition of

- sleeping accommodation,
- kitchen and bathroom facilities,
- furnishings (if there are any),
- office space (he needs two offices, each about 3 m square),
- heating equipment, for year-round operation, and
- the dock at water's edge.

As the helicopter lifts off the tarmac at Harmonsville airport on September 24, the pilot shouts across to you over the noise of the engine and blades whirring above the cabin: "I'm turning onto a heading of 072 degrees. We'll be at Sylvan Lake in 27 minutes."

You turn around to face technician Juan Martinez, whom you have hired to work with you on the project. He is seated behind you. "We'll be there at 9:34."

Juan nods and makes a note on a pad balanced on his knee.

At 9:31 the pilot (Ann) shouts, "There's the lake!"

You look down and see a crescent-shaped lake about 2.4 kilometres long completely surrounded by tall trees and thick vegetation. At the southern end of the lake there is a clearing with a cluster of buildings, one considerably larger than the others, and the remains of a wooden dock jutting out into the lake.

"That's the fishing lodge," she adds. "I'll follow the road to the landing strip." The helicopter swoops down until it is flying about 10 metres above the tree tops.

Some technical work introduces an element of adventure!

The narrow road twists and turns, skirting small, weedy ponds and lakes, and is hemmed in for most of its distance by tall trees whose branches in some places meet overhead, obscuring the road from view. At one point there is a bridge where the road crosses a stream, and at another the road seems to be partly awash as it curves around the south side of a large pond.

The landing strip is a 700-metre length of tarmac carved out of the brush and oriented northwest-southeast (to take advantage of the prevailing winds). You can see no buildings.

"I'll touch down here," Ann shouts, pointing to the northwest end of the strip. "The road starts at this end."

She helps you and Juan winch a four-wheel drive, all-terrain two-seater motorized vehicle down from the cavity in the helicopter's belly.

"I'll wait for you here," she says, reaching into the cabin for a folding stool and a book. "I want to keep an eye on the chopper."

Juan starts the engine of the vehicle and the two of you drive to the end of the landing strip and bump down onto the road.

For the first 1.8 kilometres the road is rough but usable. In several places you have to skirt fallen trees that will have to be cleared away, and at the 1.1-kilometre point you notice a subsidence on the south side of the road. You stop and measure the area: it is about 60 metres long. In parts there is only just enough space for the 1.8-metre wide vehicle to squeeze through.

After 2.9 kilometres Juan stops just short of a wooden bridge over a stream. "Better inspect it first," he says.

The bridge is 4.4 metres long and 4.8 metres wide. It is constructed of unpainted timbers, on one of which somebody has painted in barely decipherable letters: "Sylvan Creek—No Fish."

You climb down the bank of the stream and inspect the structural timbers. "They are strong enough to support our light vehicle," you call up to Juan, "but there is some rotting of the vertical supports on the north side. They'll have to be fixed before heavier vehicles drive across."

To write an effective report you have to go after specific technical data

Juan calls down to you that the planks on the bridge deck are relatively sound. "Only about 30% will have to be replaced," he adds. "That's about 25 metres of 8-inch by 2-inch lumber."

Between you, you decide that a load limit of 1.5 tonnes should be in effect until the repair work has been done.

At the 3.6-kilometre mark you encounter a 120-metre section of road that has been partly washed away where the road curves in a wide arc around the south side of the lake. ("This is the washout we saw from the air," you comment.) The road slopes toward the water and either subsidence or a high water level (or a combination of both) has caused the water to lap onto the road. In places the gravel is no more than 1.5 metres wide.

Juan stops the vehicle and hops out. He walks beside the edge of the road and with a long stick prods into the water. "The water has been high like this for three or four years," he says. "Except at the very edge the gravel has been completely washed away, probably by wave motion."

You negotiate the curve very slowly, the balloon tires of your vehicle providing the necessary traction and support on the slimy left side.

"This section of road will have to be completely rebuilt," you say, and make a note in the log.

The remainder of the road is in fairly good shape and 10 minutes later you reach the fishing camp.

"Overall, the road is not in bad condition," you say to Juan, "considering that nothing has been done to it for at least six years."

Juan agrees. "Apart from the problem spots we have identified, it'll need only some grading over its entire length, and several loads of gravel."

A useful item to include in a report

"And gravel will be easy to find," you add. "I saw a gravel pit beside the road, almost hidden by bush, at the 3.1-kilometre mark. I've made a note of it."

You approach a Y-junction, with the left fork leading toward the lake and the right fork turning toward the fishing lodge. You turn right. Ahead of you are a large two-storey building and three smaller single-storey buildings. All the windows are shuttered by sheets of plywood nailed onto the window frames. A large padlock clamps the storm door of the main building firmly in place. Or so you think.

When you insert the key, the padlock opens reluctantly, rustily. You pull harder at the lock and to your surprise the storm door opens toward you; someone has been there before and has pried the hasp loose from the door frame!

The interior, even in the dim light, is a mess. There is very little furniture (*was it stolen?* you wonder), and what furniture that does remain (a few wooden chairs and a rickety wooden table) are in bad shape.

"Shall I check the other buildings?" Juan asks.

"Sure," you say. "But make notes. I have to write a report."

You tour the building. Upstairs: four dormitory-style bedrooms, empty. Downstairs: a living room stretching across the front of the building, with a few pieces of scattered furniture, a dozen or so open, rusty cans of beans, corn, lunch meat, partly consumed, in and around the fireplace, and several empty plastic bottles of pop, lying on their sides around the room. The fireplace has ashes and the remains, half-burned, of one or two chairs. Across one wall, scrawled in large letters made with a purple marker pen: "The Rollers were here! May 6th."

You hear a motor start up, and realize that Juan has found the generator. You test the light switch and two bare bulbs, hanging at different ends of the room, light up. Then you hear a pumping sound, followed almost immediately by a hissing and swishing. In the kitchen, water is spraying across the room from a cracked pipe. Next door, in the bathroom, water is spraying from another cracked pipe. A pipe in the ceiling also must be cracked, because water is running down the wall in the main room. You run out to Juan and tell him to switch off the pump. "It's drawing water from the well and spraying it all over the place," you shout to him above the noise of the engine. "Whoever was here last didn't drain the pipes and they burst during the winter."

More information for your report!

You return to the main building and make notes:

1. Bathrooms,
   - Both toilets bowls cracked: water left in over the winter.
   - One shower stall pulled away from wall, lying on side, unusable.
   - Floor tiles loosened (from water?) many curling at edges.
2. Kitchen,
   - ceramic sink cracked (same as toilet bowls),
   - Calor gas stove removed (stolen?), pipes yanked away from wall, broken off, not disconnected properly
   - fridge still there, but insides pulled out and door twisted, loose on hinges,
   - counter okay, cupboards okay (but empty of anything useful)
3. I guess a portable kitchen (like army field kitchen) will be needed, and outhouse will have to be built until kitchen and bathroom fixed.
4. Stair treads shaky, best if whole stair replaced
5. Smell everywhere, food and musty from moisture. Carpet in living area shot.
6. Nowhere for office, not in this building.
7. Windows in some rooms cracked.

Notes to be transcribed into your report

8. No insulation. (Juan peeled off a wall panel, to check)
9. Porch outside door rotting, about 70% needs to be replaced. About 10 m × 3 m.

Juan's list is shorter:

- Small outbldg just for storage and generator and water pump. A ladder and some tools.
- Middle size outbldg has two rooms (about 3.5 × 4.5 m), probably owner's building before, when a fishing lodge; maybe could make an office. Needs paint job but okay otherwise. No means of heat.
- Larger outbldg is one big room, big hooks in ceiling, maybe stored boats in winter. Also in good shape.

You lock up the smaller buildings and drive two large nails into the door of the main building to hold the door shut. "I'll have to tell them, so they'll know how to open it," you murmur to yourself as you and Juan climb into the all-terrain vehicle.

There is little point in staying long beside the dock: only a few stumps sticking out of the reedy water show where once the dock had stood. "Must have been broken up by the ice," Juan says.

Juan turns the vehicle around and drives back to the landing strip. The inspection has taken almost three hours.

You help winch the vehicle up into the cavity of the helicopter, then you, Ann, and Juan climb into the tiny cabin.

"There's something I want to show you," Ann shouts, lifting the helicopter from the tarmac and swooping southeast along the landing strip no more than 5 metres above the ground. "While you were away I walked up to the other end of the strip. The first 300 metres can't be used."

As the helicopter swoops around the area, you see holes in the tarmac and numerous cracks through which bushes have started to grow, some 2.5 to 3 metres high.

You were asked to report on the road and fishing lodge; do you include the landing strip?

"That effectively reduces the length of the landing strip to only 400 metres, which is too short for many aircraft to use," she adds.

You nod and make a note in your logbook:

Only light aircraft; no jets.

Now you are back at your home office, writing your report (this is September 26). You draw a sketch of the road from the landing strip to Sylvan Lake, showing the distances to the areas of fallen trees (you will have to decide where they are), the subsidence at 1.1 kilometres, the bridge, and the washed-out road. Some additional facts you need are:

- You drove from your office to Harmonsville, a distance of 277 kilometres.

- The helicopter company at Harmonsville is known as Vesuvius Helicopters Inc. The pilot's name is Ann Christianson.
- The all-terrain vehicle was rented from Wheels and Skids Inc of your town. It is known as "Moonwalker III."

Write the results of your investigation as a semiformal report, and precede it with a cover letter addressed to David Yanchyn.

Project 5.2: Researching Fax Machines

You are a technician employed in H L Winman and Associates head office. Today, Susan Jenkins comes to you and says: "I've just come out of a management meeting, and I have a project for you."

Susan explains that the company has 15 fax machines spread throughout the various offices: 3 at head office in Calgary, 2 at Macro Engineering Inc in Toronto, and 10 at branch offices across Canada. The machines vary in age from six to nine years old, and in their capabilities. The management committee has decided to buy 15 identical fax machines with new technology incorporated into them, and simultaneously take advantage of a bulk purchase discount price.

"I want you to research what's new and to provide me with a report I can take back to the committee," Susan says, and then she lists eight questions articulated by members of the management committee, which she feels need to be addressed in your report:

Changes in fax technology affect a company's purchase plans

1. What are the primary differences between thermal fax machines and plain paper fax machines?
2. Will the growing impact of the Internet and electronic mail reduce the need for fax machines, or even replace them?
3. Similarly, if our office computers are equipped with fax modems, do we really need a fax machine? If so, under what circumstances?
4. What special features should the new fax machines have?
5. What about compatibility with other organizations who have older systems? Is that likely to become a problem?
6. What about faxes coming from our European clients? They print letters and reports on size A4 paper, which is about one-third of an inch longer than our 8-1/2 × 11-inch paper. That should not affect printing on roll-type thermal paper, but will it create a problem with pre-cut 11-inch plain paper?
7. Will there be a trade-in or second-hand sale value for our existing fax machines?
8. Must we have a separate, "dedicated" telephone line for each fax machine, or can the line be used for regular calls as well?

Project 5.3: Resolving a Landfill Problem

You are the assistant engineer for the Town of Quillicom in Northern Ontario. Your boss is Robert D Delorme, P.Eng, who is the town engineer.

Mr Delorme calls you into his office and announces, "I have a project for you. The Town Council has finally decided to so something about the landfill problem, and they want it done in a hurry."

You know about the landfill problem. The existing landfill site—at Lot 22, Subdivision 35—is nearly full and, recognizing that a new site will not be selected before the present site reaches its capacity, in places he has authorized dumping an additional layer of garbage on top of the compacted fill.

"Before you do anything, I want you to read this," Mr Delorme continues, placing a report in your hands. "It's a study done by H L Winman and Associates a while ago, and it affects what you will be doing. Take it away and read it, and then come back to me for further instructions."

You read the report (you will find it in Figure 5-4, on pages 137 to 142) and then go back to see Mr Delorme.

"The town councillors have decided," he says, "that before they can make a decision they need to know how much it will cost to drill a dozen boreholes north of Quillicom and analyse the results. I want you to get some quotations that I can present to them. You should also be aware that the councillors very much prefer Lot 18, Subdivision 5N, rather than either of the other locations."

"Who will plot the results of the drilling?" you ask.

"Morley Wozniak, at H L Winman and Associates in Thunder Bay. He did the previous study and wrote the report I asked you to read." Mr Delorme hands you a map (see Figure 5-7) and a list titled *Borehole Specifications for the Area North of Quillicom*, which contains the exact positions Morley has identified where the drilling must be carried out.

"And how many quotations should I get?" you ask.

"Two, as a minimum," Mr Delorme suggests. "Three would be better."

The following week you call on the only two drilling companies you know of in the area, one in Quillicom and one in Dryden, Ontario. They give you the following quotations:

Northwest Drillers, Ltd. Quillicom, Ontario	$72 760 (GST incl)
M J Peabody and Company Dryden, Ontario	$69 900 (GST extra)

You had almost given up hope that you would find a third company to give you a quotation, when Mr Delorme telephones. "Go and see Bert

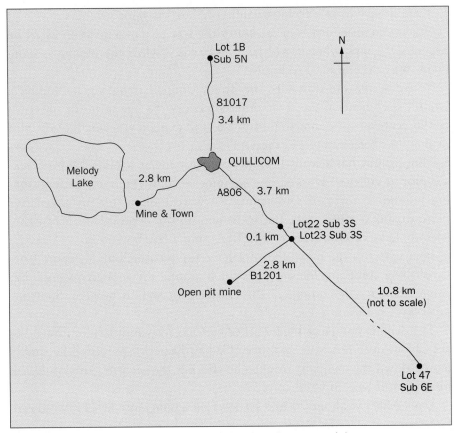

Figure 5-7 Map showing proposed landfill sites (not to scale).

Knowles," he instructs you. "He's the assistant to the superintendent at Melody Lake Mine, and he has a suggestion for an alternative landfill site."

Mr Knowles comes right to the point: "We do both open-pit and underground mining. Our open-pit mine is nearly worked out and we will finish excavating there in less than two years. The problem is that it's unsightly, and the Department of the Environment is leaning on Melody Lake Mines to do something about it. And that's where you come in."

He explains that the Town of Quillicom can use the open-pit mine for a landfill, and the mine will lease it to the town for one dollar a year. "We have only two conditions. You must spread soil over the compacted garbage, and do it progressively as you go along so there will be no obnoxious smell for the people who live near the mine to contend with. And then you must seed it and plant trees."

You agree: it's something the Town would do anyway, before closing a landfill site.

Mr Knowles drives you to the site and you stand on the lip of a shallow, roughly oval excavation varying from about 3 to 18 metres deep.

"How large is it?" you enquire.

An unusual but realistic option for a landfill site

"You'll have to talk to Inga Paullsen. She's the mine geologist."

When you visit Inga, she calculates the size of the excavation as 24.86 hectares. "That's what it will be," she adds, "when the mining is complete. Why do you need to know?"

You describe the difficulty the Town Council is having in finding a landfill site, mention the three other sites, say the one north of the town could create an environmental problem, and explain you won't know until drilling has been completed there.

A surprising piece of information introduces a new aspect

"But drilling has already been done there," Inga exclaims. "When I was a junior at college I worked one summer with an exploration crew sinking boreholes north of Quillicom. We were looking for an alternative place to sink a mine shaft, but we found no ore deposits north of either Melody Lake or Quillicom. We drilled quite a few boreholes."

Inga tells you the mine does not have the records, only a report from the drilling company. She searches for it among the geology records, but cannot find it. "It's strange," she mutters, "it should be here. Someone must have removed it."

The drilling company Inga worked for was Mayquill Explorations, but she says it does not exist anymore. "When Ernie Mays retired he simply closed down the company. Maybe he still has the records. You could ask him. He still lives in Quillicom."

Ernie Mays is about 70 and he lives in a bungalow at 211 Westerhill Crescent.

"I quit eight years ago," he tells you. "I sold some of my accounts to Northwest Drillers—those that were still active—and kept the remainder."

He remembers drilling for Melody Lake Mines. "We sank about 20 boreholes, all north of Quillicom, but we didn't find anything."

You ask if he remembers whether the bedrock slopes, but he shakes his head. "Not really, he says. "Nothing definite."

But he adds that he does remember there was evidence of a large sand esker running roughly south-southwest toward Quillicom.

"Do you still have the records?"

"No," he says. "The mine has them. Mr Caldicott came to see me himself, about three years ago, and I gave them all to him."

A second piece of information introduces still another aspect!

Suddenly, everything falls into place. Frank Caldicott is not only general manager of Melody Lake Mines, but also a very influential Quillicom town councillor. And his youngest sister, Julie, is married to the town engineer—Robert Delorme, your boss.

Because you have so much new information to include, you decide to write a semiformal report of your findings. (You will have to decide whether you will include the information you now have about the previous drilling north of Quillicom, and the location of the records.)

Here is some additional data you obtain:

1. You are concerned about ground water contamination problems if the open-pit mine is used as a landfill, so call Morley Wozniak at H L Winman and Associates in Thunder Bay. He tells you that it will not be a problem. "Both the lake and the mining community are north of the pit, and the bedrock slopes to the south."
2. You calculate that costs to develop the open-pit mine as a landfill will be only $3000, because you can use the buildings and approach roads that are already there.
3. The open-pit mine is 4.1 kilometres directly south of Quillicom, but 6.6 kilometres by road (3.8 kilometres southeast along highway A806, then 2.8 kilometres southwest along highway B1201).
4. The annual operating cost for using the open-pit mine as a landfill will be $49 500, which is $2500 more than the cost for operating the current landfill.
5. You obtain a third drilling estimate from Quattro Drilling and Exploration Company in Kenora, which quotes $77 600, GST included.

Before starting to write your report you visit Thunder Bay on other business. On a hunch you visit the Land Titles Office and look up the surveys for the area north of Quillicom. Against Lot 18, Subdivision 5N, you find the owner listed as *Julie Sarah Caldicott, 207 Northern Drive, Quillicom, Ontario*.

A question of ethics: do you mention the "family" connection?

Now write your report.

Project 5.4: Identifying a Power Plant Problem

You are an independent consultant and operate a business known as Pro-Active Consultants Limited from your home. Four days ago you received a telephone call from Paullette Machon, who is vice president, operations, of Baldur Agricultural Chemicals (BAC), a company with manufacturing plants across Canada. She said she has a task for you and invited you to visit her at the BAC office at 2120 Disraeli Crescent (of the town or city where you live).

"I want you to drive over to our plant in Rossburn," she announced, "to look into a technical problem in the power house." (Rossburn is 43 kilometres from your city, has a population of 5700, and its primary employer is the BAC plant.)

"I'm concerned that power house costs are rising at Rossburn just at the moment when world fertilizer prices are dropping," Ms Machon continues. "This is causing the company to be uncompetitive in both national and international markets."

Ms Machon explains that BAC requires a lot of hot water and steam in its manufacturing operations. However, over the past two years fuel consumption at Rossburn has risen by 18%, numerous breakdowns have occurred that have interfered with production, and there has been a sharp rise in production costs. She has visited the power house repeatedly, but has never found anything that could be attributed to poor operation. In fact, the power house has always been immaculate.

Now Ms Machon wants an independent consultant to take a look, talk to the people in the power house, and see if you can identify any production problems.

A technical problem affected by the personalities involved

She also hinted that the problem may not only be technical. "The present chief engineer at the BAC power house is Curt Hänness, and he is to retire in three months. BAC management has to decide whether to promote Harry Markham, the existing senior shift engineer, or to bring in a new chief engineer from outside the company. On paper, Markham is ideal for the job. He has worked in the power house for 15 years (he is now 36) and always under Hänness, so his knowledge of the plant and its operations cannot be challenged. Yet the rising costs indicate that all is not as it should be in the plant, and we want to be sure that the new chief engineer does not perpetuate the present conditions."

She says she will inform Hänness and Markham that she has engaged you to study the hot water and power generating system in their power house, and that they are to expect you.

You visit the BAC power plant in Rossburn today. During your talks to plant staff and tours of the plant you make the following notes:

1. Housekeeping excellent—whole place shines (but is this only surface polish for impression of visitors?)

2. Maintenance logs are inadequately kept—need to be done more often. Need more detail. Equipment files not up to date and not properly filed.

3. Boiler cleaning badly neglected. Firm instructions re boiler cleaning need to be issued by head office.

These are the "technical" details...

4. Flow meters are of doubtful accuracy. May be overreading. Not serviced for three years. Manufacturer's service department should be contacted (these are Weston meters). Manufacturer needs to be called in to do a complete check and then recalibrate meters.

5. Overreading of meters could give false flow figures—make plant seem to produce more steam than is actually produced.

6. Good housekeeping obviously achieved by neglecting maintenance. Incorrectly placed emphasis probably caused by frequent visits from company president, who likes to bring in important visitors and impress them. Hänness likes reflected glory (so does Markham).

7. Shift engineers are responsible for maintenance of pumps and vacuum equipment. Not enough time given over to this. They seem to

prefer straight replacement of whole units on failure rather than preventive maintenance. Costly method! Obviously more breakdowns: they wait for a failure before taking action. A preventive maintenance plan is needed.

8. Markham seems O.K. Genial type; obviously knows his power house. Proud of it! But seems to resist change. Definitely resents suggestions. Does he lack all-round knowledge? Is he limited only to what goes on in his plant? Is he afraid of new ideas because he doesn't understand them? Young staff hinted at this: too loyal to say it outright, but I felt they were restive, hampered by his insistence that they use old techniques that are known to work but are slow. Nothing concrete was said—I just "felt" it.

9. Hänness has done a good job training Markham. Made him a carbon copy. Hänness doesn't do much now. Markham runs the show, and has for over a year. He *expects* to get the job when Hänness retires. It'll be a real blow to him if he doesn't! BAC might even lose a good company man.

...and these are the "personal" details

10. Discussed microprocessor-controlled CORLAND 200 power panel with staff. Young engineers had read about it in "Plant Maintenance"—eager to have one installed (I described the one I'd seen at Pinewood Paper Mill). But Hänness and Markham knew nothing about it—didn't seem to be interested. Are they not keeping up-to-date with technical magazines?

When you return to your office you write an evaluation report for Ms Machon. You can either address both the technical problems and the personnel difficulties within the one report, or write two separate reports.

Proposals
www.io.com/~hcexres/tcm1603/acchtml/props.html

This document is one chapter from the online textbook used in Austin Community College's online course, Online Technical Writing (www.io.com/~hcexres/tcm1603/acchtml/acctoc.html). It describes types of proposals, their organization and format, and the common sections in a proposal. Included are several sample proposals and a revision checklist.

Proposals
darkstar.engr.wisc.edu/zwickel/397/397refs.html

The online technical writing handbook of the Department of Engineering Professional Development at the University of Wisconsin has several pages on the writing of proposals.

Recommendation and Feasibility Reports
www.io.com/~hcexres/tcm1603/acchtml/feas.html

Also from Online Technical Writing, this document describes feasibility reports in detail and includes several samples.

Chapter 6
Formal Reports

Formal reports require more careful preparation than the informal reports described in previous chapters. Because they will be distributed outside the originating company, their writers must consider the impression the reports will convey of the entire company. Harvey Winman recognized long ago that a well-written, esthetically pleasing report can do much to convince prospective clients that H L Winman and Associates should handle their business, whereas a poorly written, badly presented report can cause clients to question the company's capability. Harvey also knows that the initial impression conveyed by a report can influence a reader's readiness to plow through its technical details.

The presentation aspect must convey the originating company's "image," suit the purpose of the report, and fit the subject it describes. For instance, a report by a chemical engineer evaluating the effects of diesel fumes on the interior of bus garages would most likely be printed on standard bond paper, and its cover, if it had one, would be simple and functional. At the other end of the scale, a report by a firm of consulting engineers selecting a college site for a major city might be printed professionally and bound in an artistically designed book-type folder. But regardless of the appearance of a report, its internal arrangement will be basically the same.

Formal reports are made up of several standard parts, not all of which appear in every report. Each writer uses the parts that best suit the particular subject and the intended method of presentation. There are six major and several subsidiary parts on which to draw, as shown in Table 6-1. Opinions differ throughout industry as to which is the best arrangement of these parts. The two arrangements suggested in The Complete Formal Report later in this chapter, and illustrated in the mini-reports in Figures 6-5 and 6-10, are those most frequently encountered. Knowledge of these parts and the two basic arrangements will help you adapt quickly to any variation in format preferred by your employer.

Table 6-1 Traditional arrangement of formal report parts.

Cover or Jacket

Title Page

SUMMARY

Table of Contents

INTRODUCTION

DISCUSSION

CONCLUSIONS

RECOMMENDATIONS

References or Bibliography

APPENDICES

Notes:

1. MAJOR PARTS are in capital letters; subsidiary parts are in lower case letters.
2. A Cover Letter or Executive Summary normally accompanies a formal report.

The acronym SIDCRA is formed from the first letter of each major report part

Major Parts

The six major parts form the central structure of every formal report, and in the traditional arrangement are known by the acronym SIDCRA.

Summary

The summary is a brief synopsis that tells readers quickly what the report is all about. Normally it appears immediately after the title page, where it can be found easily. It identifies the purpose and most important features of the report, states the main conclusion, and sometimes makes a recommendation. It does this in as few words as possible, condensing the narrative of the report to a handful of succinct sentences. It also has to be written so interestingly—so enthusiastically—that it encourages readers to read further.

The criteria for a Summary are difficult to meet

The summary is considered by many to be the most important part of a report and the most difficult to write. It has to be informative, yet brief. It has to attract the reader's attention, but must be written in simple, nontechnical terms. It has to be directed to the executive reader, yet it must be readily understood by almost any reader.

Generally, the first person in an organization who sees a report is a senior executive, who may have time to read only the summary. If the executive's interest is aroused, he or she will pass the report down to the technical staff to read in detail. But if the summary is unconvincing, the executive is more likely to put the report aside to read later; if this happens, the report may never be read.

Because a summary is so important, always write it last, after the remainder of your report has been written. Only then will you be fully aware of the report's highlights, main conclusions, and recommendations. To write the summary first can prove frustrating, because you will not yet have hammered out many of the finer points. You need the knowledge of a soundly developed report fixed firmly in your mind if you are to fashion an effective summary.

For a summary to be interesting, it must be informative; if it is to be informative, it must tell a story. It should have a beginning, in which it states why the project was carried out and the report was written; a middle, in which it highlights the most important features of the whole report; and an end, in which it reaches a conclusion and possibly makes a recommendation. The example below illustrates how the interest is maintained in an informative summary:

Informative Summary

We have tested a specimen of steel to determine whether a job lot owned by Northern Railways could be used as structural members for a short-span bridge to be built at Peele Bay in the Yukon. The sample proved to be G40.12 structural steel, which is a good steel for general construction but subject to brittle failure at very low temperatures.

An informative summary answers readers' questions...

Although the steel could be used for the bridge, we consider that there is too narrow a safety margin between the –51°C temperature at which failure can occur, and the –47°C minimum temperature occasionally recorded at Peele Bay. A safer choice would be G40.8C structural steel, which has a minimum failure temperature of –62°C.

Other informative summaries preface the two formal reports presented at the end of this chapter, and both the semiformal evaluation report in Figure 5-4 and the proposal in Figure 5-5 in Chapter 5.

Some writers prefer to write a topical summary for reports that describe history or events, or that do not draw conclusions or make recommendations. As its name implies, a topical summary simply describes the topics covered in the report without attempting to draw inferences or captivate the reader's interest:

Topical Summary

Construction of the Minnowin Point Generating Station was initiated in 1994, and first power from the 1340 MW plant is scheduled for 2001. A general description of the structures and problems peculiar to the construction of this large development in an arctic climate is pre-

...whereas a topical summary only suggests what can be found within a report

sented. The river diversion program, permafrost foundation conditions, and major equipments are described. The latter include the 16 propeller Turbines, among the largest yet installed, each rated at 160 000 horsepower.

Because they are less results-oriented, topical summaries are *not* recommended for most formal reports.

In a formal report the summary should have a page to itself, be centred on the page, and be prefaced by the word "Summary" (or, sometimes, "Abstract"). If it is very short, it may be indented equally on both sides to form a roughly square block of information. For examples, see pages 196 and 212.

Introduction

The introduction begins the major narrative of the report by preparing readers for the discussion that follows. It orients them to the purpose and scope of the report and provides sufficient background information to place them mentally in the picture before they tangle with technical data. A well-written introduction contains exactly the correct amount of detail to lead readers quickly into the major narrative.

Knowing your reader influences the depth of detail required

The length of an introduction and its depth of detail depend mostly on the reader's knowledge of the topic. If you know that the ultimate reader is technically knowledgeable, but at the same time you have to cater to the executive reader who is probably only partly technical, write the introduction (and conclusions and recommendations) in semitechnical language. This permits semitechnical executives to gain a reasonably comprehensive understanding of the report without devoting time and attention to the technical details contained in the discussion.

Most introductions contain three parts: **purpose, scope,** and **background information.** Frequently the parts overlap, and occasionally one of them may be omitted simply because there is no reason for its inclusion. The introduction normally is a straightforward narrative of one or more consecutive paragraphs; only rarely is it divided into distinct sections preceded by headings. It always starts on a new page (normally identified as page 1 of the report) and is preceded by the report's full title. The title is followed by the single word "Introduction," which can be either a centre heading or a side heading, as shown in Figure 6-1.

The **purpose** explains why the project was carried out and the report is being written. It may indicate that the project has been authorized to investigate a problem and recommend a solution, or it may describe a new concept or method of work improvement that the report writer believes should be brought to the reader's attention.

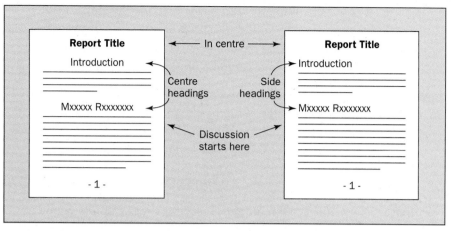

Figure 6-1 Headings used at the start of the introduction and discussion.

The **scope** defines the parameters of the report. It describes the ground covered by the report and outlines the method of investigation used in the project. If there are limiting factors, it identifies them. For example, if 18 methods for improving packaging are investigated in a project but only 4 are discussed in the report, the scope indicates what factors (such as cost, delivery time, and availability of space) limited the selection. Sometimes the scope may include a short glossary of terms that need to be defined before the reader starts to read the discussion.

Place a short glossary in the introduction, a longer glossary in an appendix

Background information comprises facts readers must know if they are to fully understand the discussion that follows. Facts may include descriptions of conditions or events that caused the project to be authorized, and details of previous investigations or reports on the same or a closely related subject. In a highly technical report, or when a significant time lapse has occurred between it and previous reports, background information may also provide a theory review and references to other documents. If the theory review is lengthy and there is a long list of documents, they are often placed in appendices, with only a brief summary of the theory and a quick reference to the list appearing in the introduction.

A good introduction "sets the scene"

The introduction shown here, plus those forming part of the two sample reports at the end of this chapter, are representative of the many ways a writer can introduce a topic.

Introduction 1

Background

Northern Railways plans to build a short-span bridge 1 kilometre north of Lake Peele in the Yukon and has a job lot of steel the company wants to use for constructing the bridge. In letter NR-70/LM dated March 20, 1998, Mr David L Harkness, Northern Area Manager, requested that

Purpose

H L Winman and Associates test a sample of this steel to determine its properties and to assess its suitability for use as structural members for a bridge in a very low temperature environment.

Scope

Two Charpy impact tests were performed, one parallel to the grain of the test specimen, and the other transverse to the grain, at 10° increments from +22°C down to –50°C.

Introduction 2

Purpose

Say why you are writing...

H L Winman and Associates was commissioned by Ms Rita M Durand, president and general manager of Auto Drive-Inns Inc of Toronto, to select a Windsor, Ontario, site for the first of a proposed chain of computerized, automatic drive-in fast-food outlets to be built in Canada. Windsor was chosen as the test site because it represents an average community in which to assess customer acceptance of such a service.

Background

...describe what has gone before, and...

Aside from exterior façade and foundation details, Auto Drive-Inns are built to a standard 8.5 by 5.3 metre pattern with an order window at one end of the longer wall, and a delivery window at the other end. Auto Drive-Inns carry only a limited selection of ready-to-serve foods, but boast a 60-second service from the time an order is placed to its delivery at the other end of the building. For this reason, Auto Drive-Inns attract people hurrying home from work, the impulse buyer, and the late-night traveller, rather than the selective meal buyer. The Inns depend more on the volume of customers than on the volume of goods sold to each individual.

Scope

...explain factors affecting both the study and the report

The chief consideration in selecting a site must therefore be a location on the homeward-bound side of a main trunk road in a large residential area. The site must have quick and easy entry onto and exit from this road, even during peak rush hour traffic. The residential area should be occupied mainly by single persons and younger families in which both parents work. And there should be little competition from walk-in fast-food outlets.

Discussion

The discussion, which normally is the longest part of a report, presents all the evidence (facts, arguments, details, data, and results of tests) that readers need to understand the subject. The writer must organize this evidence logically to avoid confusing readers, and present it imaginatively to hold their interest.

In the discussion, a writer such as Dan Skinner (see Chapter 1) describes fully what he set out to do, how he went about it, what he actually did, and what he found out as the result of his efforts. He may consider the writing a chore, and simply record his facts dutifully but unenthusiastically, or he can organize the information so interestingly that his readers feel almost as though they are reading a short story.

There is no reason why the discussion should not have a plot. It can have a beginning, in which Dan describes a problem and outlines some background events; a middle, which tells how he tackled the problem; dramatic effect, which he can use to hold readers' interest by letting them anticipate his successes, as well as share his disappointments when a path of investigation results in failure; a climax, in which he permits readers to share his satisfaction as he describes how the problem was resolved; and a denouement (a literary word for final outcome), in which he ties up loose ends of information and evaluates the results. Storytelling and dramatization can be such important factors in holding reader interest that no report writer can afford to overlook them.

An effective discussion is carefully orchestrated to hold readers' attention

There are three ways you can build the discussion section of a report:

By **chronological development**—in which you present information in the sequence that the events occurred.

By **subject development**—in which you arrange information by subjects, grouped in a predetermined order.

By **concept development**—in which you organize information by concept, presenting it as a series of ideas that reveal imaginatively and coherently how you reasoned your way to a logical conclusion.

Reports using the chronological or subject method offer less room for imaginative development than those using the concept method, mainly because they depend on a straightforward presentation of information. The concept method can be very persuasive. Identifying and describing your ideas and thought processes helps your readers organize their thoughts along the same lines.

As a report writer, you must decide early in the planning stages which method you intend to use, basing your choice on which is most suitable for the evidence you have to present. Use the following notes as a guide.

Chronological Development

A discussion that uses the chronological method of development is simple to organize and write. It requires minimal planning: you simply arrange the major topics in the order they occurred, and eliminate irrelevant topics as you go along. You can use it for very short reports, for laboratory reports showing changes in a specimen, for progress reports showing cumulative effects or describing advances made by a project group, and for reports of investigations that cover a long time and require visits to many locations to collect evidence.

Chronological development is essentially factual...

But the simplicity of the chronological method is offset by some major disadvantages. Because it reports events sequentially it tends to give emphasis to each event regardless of importance, which may cause readers to lose interest. If you read a report of five astronauts' third day in orbit, you do not want to read about every event in exact order. It may be chronologically true to report that they were wakened at 7:15; breakfasted at 7:55, sighted the second stage of their rocket at 9:23, carried out metabolism tests from 9:40 to 10:50, extinguished a cabin fire at 11:02, passed directly over Thunder Bay, Ontario, at 11:43, and so on, until they retired for the night. But it can make dull reading. Even the exciting moments of a cabin fire lose impact when they are sandwiched between routine occurrences.

When using chronological development, you must still manipulate events if you are to hold your readers' attention. You must emphasize the most interesting items by positioning them where they will be noticed, and deemphasize less important details. Note how this has been done in the following passage, which groups the previous events in descending order of interest and importance:

...but there still is room for some orchestration

> The highlight of the astronauts' third day in orbit was a cabin fire at 11:02. Rapid action on their part brought the fire, which was caused by a short circuit behind panel C, under control in 38 seconds. Their work for the day consisted mainly of metabolism tests and...

> They sighted the first stage of their rocket on three separate occasions, first at 9:23, then at...and..., when it passed directly over Thunder Bay. Their meals were similar to those of the previous day.

Over a five-year period H L Winman and Associates has been investigating the effects of salt on concrete pavement. Technical editor Anna King has suggested that the final report should have chronological development, because the investigation recorded the extent of concrete erosion at specific intervals. She wanted the engineering technologist writing the report to describe how the erosion increased annually in direct relation to the amount of salt used to melt snow each year, and for the final conclusion to demonstrate the cumulative effect that salt had on the concrete.

Subject Development

If the previous investigation had been broadened to include tests on different types of concrete pavements, or if both pure salt and various mixtures of salt and sand had been used, then the emphasis would have shifted to an analysis of erosion on different surfaces or caused by various salt/sand mixtures, rather than a direct description of the cumulative effects of pure salt. For this type of report Anna King would have suggested arranging the topics in *subject* order.

The subject order could be based on different concentrations of salt and sand. The technologist would first analyse the effects of a 100% concentration of salt, then continue with salt/sand ratios of 90/10, 80/20, 70/30, and so on, describing the results obtained with each mixture. Alternatively, the technologist could select the different types of pavement as the subjects, arranging them in a specific order and describing the effects of different salt/sand concentrations on each surface.

Barry Brewster, who is head of H L Winman and Associates's design and drafting department, has been investigating color printers for use with the company's CAD system, and plans to recommend the most suitable model for installation in his department. He writes his report using the subject method. First, he establishes selection criteria, defining what he needs in an ideal printer, such as its purchase price, speed, economy of operation, and quality of printout. Then, as his tests have already determined the most suitable printer, he discusses the best machine either first or last. If he chooses to describe it first, he can state immediately that it is the best printer, and say why by comparing it to the selection criteria. He can then discuss the remaining printers in decreasing order of suitability, also comparing each to the selection criteria to show why it is less suitable. If he prefers to describe the best printer last, he can discuss the printers in increasing order of suitability, again comparing each against the selection criteria and then stating why he has rejected each one before describing the next.

Also suitable for subject development is an analysis of insulating materials being conducted by laboratory technician Rhonda Moore. In her report, she groups materials having similar basic properties or falling within a specific price bracket. Within the groups she describes the test results for each material and draws a conclusion as to its relative suitabil-

Subject development sorts information into groups

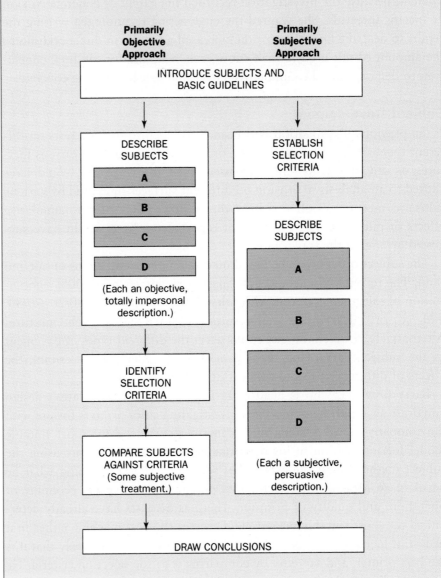

Primarily Objective Approach

Primarily Subjective Approach

INTRODUCE SUBJECTS AND BASIC GUIDELINES

DESCRIBE SUBJECTS

A

B

C

D

(Each an objective, totally impersonal description.)

ESTABLISH SELECTION CRITERIA

DESCRIBE SUBJECTS

A

B

C

D

IDENTIFY SELECTION CRITERIA

COMPARE SUBJECTS AGAINST CRITERIA
(Some subjective treatment.)

(Each a subjective, persuasive description.)

DRAW CONCLUSIONS

Figure 6-2 Alternative methods for describing a comparative analysis.

Both objective and subjective development methods provide limited opportunities to orchestrate information

ity for a specified purpose. Like Barry, she describes the materials in increasing or decreasing order of suitability.

The subject method of development permits Barry and Rhonda either to hide their personal preferences until almost the end of their reports or, if they prefer, to let their preferences show all the way through. They have the choice of using an impersonal, objective approach, or a personal, subjective approach similar to that for the concept method. These alternative approaches are illustrated in Figure 6-2.

Rhonda prefers the primarily objective approach. She discusses each insulating material (i.e. one of the subjects) without using any words that might make her readers think she believes the material to be a good or bad choice:

> Material A has an R-17 rating. It is a dense, brown, tightly packed fibrous substance enclosed in a continuous, one-metre envelope, which is supplied in 10-, 20-, and 50-metre rolls. To cover a standard 4 × 3 metre area would cost $108. The fire resistance of Material A is...

A writer has to decide whether to be objective...

Rhonda has only to introduce the insulating materials, and possibly outline how they were selected, before she describes them. She need not identify her selection criteria until almost the end of the discussion, after she has objectively described each material but before she subjectively compares their main features.

Barry prefers the more colorful subjective approach. He wants to persuade his readers to accept or reject each printer (subject) as they read about it:

> The Image-copy is the only printer that meets all our technical requirements. It has a high speed of 11 copies per minute, an extended continuous-roll feature with automatic paper cutter, and both manual and automatic stacking capabilities. Although its purchase price of $2300 is $350 above budget, its high reliability (reported by users such as Multiple Industries and Apex Architectural Services) will probably cause less equipment downtime and consequent production frustrations than other models. Specific features are...

...or to allow subjectivity (the writer's presence) to intrude

Barry must establish the criteria or parameters that most affect his choice, and must convince his readers why they are important, *before* he describes each printer; if he does not, readers will not readily accept his subjective statements.

Both Rhonda and Barry also have to remember never to compare one product directly with another. They should only compare each product against the criteria.

Whichever approach you use for a comparative analysis, your readers should find that the discussion leads comfortably and naturally into the conclusions you eventually draw. The conclusions should never contain any surprises.

Concept Development

By far the most interesting reports are those using the concept method of development. They need to be organized more carefully than reports using either of the previous methods, but they give the writer a tremendous opportunity to devise an imaginative arrangement of the topic—which is the best way to hold a reader's attention.

They can also be very persuasive. Because the report is organized in the order in which you reasoned your way through the investigation, your readers will much more readily appreciate the difficulties you encountered, and frequently will draw the correct conclusion even before they read it. This helps readers feel they are personally involved in the project.

Concept development has the greatest potential for orchestrating information

You can apply the concept approach to your reports by thinking of each project as a logical but forceful procession of ideas. If you are personally convinced that the results of your investigation are valid, and remember to explain in your report *how* you reached the results and *why* they are valid, then you will probably be using the concept method properly.

Always anticipate reader reaction. If you are presenting a concept (an idea, plan, method, or proposal) that readers are likely to accept, then use a straightforward four-step approach:

1. Describe your concept in a brief overview statement.
2. Discuss how and why your concept is valid; offer strong arguments in its favor, starting with the most important and working down to the least important.
3. Introduce negative aspects, and discuss how and why each can be overcome or is of limited importance.
4. Close with a restatement of your concept, its validity, and its usefulness.

But if you are presenting a controversial concept, or need to overcome reader bias, then modify your approach. Try to overcome objections by carefully establishing a strong case for your concept *before* you discuss it in detail.

Special projects department head Andy Rittman used this approach in a report he prepared for Mark Dobrin, owner/manager of an Edmonton company making extruded plastic and metal parts. Manufacturing costs had risen steeply over the past two years and Mark's prices had become uncompetitive. Mark thought he should replace some of his older, less efficient equipment, so he asked H L Winman and Associates to evaluate his needs.

Andy quickly realized that if Mark was to avoid going out of business he would have to replace much of his equipment with microprocessor-controlled machines, and do it soon. Because the cost would be high, he would have to lease rather than buy the new equipment.

Here, Andy had a problem. Mark was as old-fashioned as some of his extruders and shapers, and throughout his life he had steadfastly refused to purchase anything that he could not buy outright. He was unlikely to change now.

The concept method leads the reader to the right answer...

In his report, Andy used a carefully reasoned argument to prove to Mark that he needed a lot of new equipment and that the only feasible way he could acquire it would be to lease it. Throughout, Andy wrote objectively but sincerely of his findings, hoping that the logic of his argument would swing Mark around to accepting his recommendation. Very briefly, here is the step-by-step approach Andy used:

- He opened with a summary that told Mark that to avoid bankruptcy he would have to invest in a lot of expensive equipment and make extensive changes in his operating methods.
- Andy then produced financial projections to prove his opening statement, and discussed the productivity and profitability necessary for Mark to remain in business.
- He discussed why Mark's equipment and methods were inefficient, introduced the changes Mark would have to make, established why each change was necessary, and demonstrated how each would improve productivity. (Andy referred Mark to an appendix for specific equipment descriptions, justifications, and costs.)
- Andy then introduced two sets of cost figures: one for making the minimum changes necessary for Mark's business to survive, and the second for more comprehensive changes that would ensure a sound operating basis for the future. He commented that both would require capital purchases likely to be beyond Mark's financial resources.
- He outlined alternative financing methods available to Mark, the implications and limitations of each, and the financial effect each would have on Mark's business. (Although he introduced leasing, Andy made no attempt to persuade Mark that he would have to lease; he let the figures speak for themselves.)
- Andy concluded by summarizing the main points he had made: that new equipment *must* be acquired; that to buy even the minimum equipment was beyond Mark's financial resources; and that, of the financing methods available, leasing was the most feasible.
- In his recommendation, Andy suggested that Mark should make comprehensive changes and lease the new equipment. (By then, Andy had become so involved with Mark's predicament that he wrote strongly and sincerely.)

...carefully tracing a logical, persuasive, and sometimes intricate path

Even though the concept method challenges a writer to fashion interesting reports, it is not always the best reporting medium. For instance, the concept method could possibly have been used for Barry Brewster's printer report mentioned earlier, but it is doubtful whether the topic would have warranted full analysis of the author's ideas. When a topic is fairly clear-cut, there is no need to lead the reader through a lengthy "this is how I thought it out" discussion. Reserve the concept method for topics that are controversial, difficult to understand, or likely to meet reader resistance.

Whichever method you use, avoid cluttering the discussion with detailed supporting information. Unless tables, graphs, illustrations, photographs, statistics, and test results are essential for reader understanding while the report is being read, banish them to an appendix. But always refer to them in the discussion, like this:

> The test results attached as Appendix C show that aircraft on a bearing of 265°T experienced considerably weaker reception than aircraft on any other bearing. This was attributed to...

Readers *interested only in results* will consider that this statement tells them enough and continue reading the report. Readers interested in *knowing how the results were obtained*—who want to see the overall picture—will turn to Appendix C to find out how the report writer went about performing the tests. They probably will not do so immediately, but will turn to it when they have finished the report or have reached a suitable stopping place. In neither case are readers slowed down by supporting information that, though relevant, is not immediately essential to their understanding of the report.

Sometimes, if an illustration or table is essential, extract the key points from an appendix and use them to create a miniature illustration or table that can be inserted beside or embedded in the narrative without impeding reading continuity.

Unless the discussion is very short, divide it into a series of sections each preceded by an informative heading. If a section is long or naturally subdivides into subtopics, also use subordinate headings. Major topics normally demand centre or side headings, while minor topics need only side or paragraph headings, as shown in Figure 6-3 and Figure 11-2 of Chapter 11. These headings eventually form the table of contents for the report.

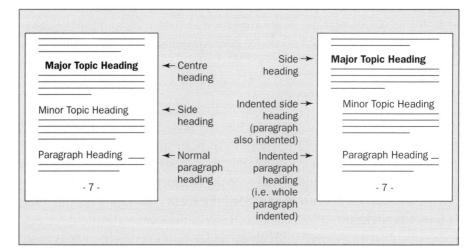

Figure 6-3 Headings show subordination of ideas.

After each heading, start the section with an overview statement to describe what the section is about and suggest what conclusion will be drawn from it. Overview statements are miniature summaries that direct a reader's attention to the point you want to make. If the section is short, the overview statement may be a single sentence; if the section is long, it will probably be a short paragraph.

At the end of each major section insert a concluding statement that summarizes the result of the discussion within that section. From these section conclusions you can later draw your main conclusions.

Conclusions

Conclusions briefly state the major inferences that can be drawn from the discussion. You must base them entirely on previously stated information and *never* surprise readers by introducing new material or evidence to support your argument. If there is more than one conclusion, state the main conclusion first and follow it with the remaining conclusions in decreasing order of importance. This is shown in the two examples below, which present the same conclusions in both narrative and tabular form.

There should be no surprises here

Narrative Conclusion

If we upgrade to version 4.1 of the *Mosaic* software, we will also have to upgrade to a 32-bit operating system. The one-time cost will be $7600, but we will increase our multimedia production capability by 65%.

Tabular Conclusion

If we upgrade to version 4.1 of the *Mosaic* software, we will
- experience a 65% increase in our multimedia production capability,
- have to upgrade to a 32-bit operating system, and
- incur a one-time cost of $7600.

Because conclusions are *opinions* (based on the evidence presented in the discussion), they must never tell the reader what to do. This must always be left to the recommendations.

Recommendations

Recommendations appear in a report when the discussion and conclusions indicate that further work needs to be done, or you have described several ways to resolve a problem or improve a situation and want to identify which is best. Write recommendations in strong, definite terms to convince readers that the course of action you advocate is valid. Use the first person and active verbs, as has been done here:

Now you can let your voice be heard

<table>
<tr><td>**Strong**</td><td>I recommend that we build a five-station prototype of the Microvar system and test it operationally.</td></tr>
</table>

Compare this with the same recommendation written in the third person, using passive verbs:

<table>
<tr><td>**Weak**</td><td>It is recommended that a five-station prototype of the Microvar system be built and tested under operational conditions.</td></tr>
</table>

Unfortunately, in many industries, and often in government circles, using the first person is severely frowned upon. Consequently, many reports fail to contain really strong recommendations because their writers hesitate to say "I recommend...," and so use the indefinite "It is recommended...." If you feel you cannot use the personal "I," try using the plural "we," to indicate that the recommendations represent the company's viewpoint. For example:

<table>
<tr><td>**Strong**</td><td>We recommend building a five-station prototype of the Microvar system. We also recommend that you
1. install the prototype in Railton High School,
2. commission a physics teacher experienced in writing programmed instruction manuals to write the first programs, and
3. test the system operationally for three months.</td></tr>
</table>

And no surprises here, either!

Because recommendations must be based solidly on the evidence presented in the discussion and conclusions, they must *never* introduce new evidence or new ideas.

Appendix

Related data not necessary to an immediate understanding of the discussion should be placed further back in the report, in the appendix. The data can vary from a complicated table of electrical test results to a simple photograph of a blown transistor. The appendix is a suitable place for manufacturers' specifications, graphs, comparative data, drawings, sketches, excerpts from other reports or books, cost analyses, and correspondence. There is no limit to what can be placed in the appendix, providing it is relevant and reference is made to it in the discussion.

The appendix is *not* a storage place for information that only might be useful

The term "appendix" applies to only one set of data. In practice there can be many sets, each of which is called an appendix and assigned a distinguishing letter, such as A, B, or C. The sets of data are collectively referred to as either "appendices" or "appendixes," the former term being preferred in Canada and Great Britain, and the latter in the United States.

The importance of an appendix has no bearing on its position in the report. Whichever set of data is mentioned first in the discussion becomes Appendix A, the next set becomes Appendix B, and so on. Each appendix is considered a separate document complete in itself and is paginated separately, with its front page labelled 1. Examples of appendices appear at the end of Report 1, later in this chapter.

Subsidiary Parts

In addition to the six major parts of a formal report, there are several additional parts that perform more routine functions. Although referred to here as "subsidiary" parts, they nevertheless contribute much to a report's effectiveness. Because they directly affect the image conveyed of both you and your company, they must be prepared no less carefully than the rest of the report.

Cover

Almost every major formal report has a cover. It may be made of glossy cardboard printed in multiple colors and bound with a dressy plastic binding, or it may be only a light cover of colored fibre material stapled on the left side. The cover not only informs readers of the report's main topic but also conveys an image of the company that originated it. This "matching" of subject matter and company image plays an important part in setting the correct tone.

The cover should reflect the company's image

The cover should contain the report title, the name of the originating company and, perhaps, the name of its author. The title should be set in bold letters well-balanced on the page and separated from any other information.

Choice of a title is particularly important. It should be short yet informative, implying that the report has a worthwhile story to tell. Compare the vague title below with the more informative version beneath it.

Original vague title

RADOME LEAKAGE

Revised informative title

POROSITY OF FIBREGLASS CAUSES RADOME LEAKS

Technical officers at remote radar sites would glance at the first title with only a muttered, "Looks like somebody else has the same problem we've got." But the second title would encourage them to stop and read the report. They would recognize that someone may have found the cause of (and, hopefully, a remedy for) a trouble spot that has bothered them for some time.

Note how the following titles each *tell* something about the reports they precede:

The report title should capture readers' interest

REDUCING AMBIENT NOISE IN AIR TRAFFIC CONTROL CENTRES
EFFECTS OF DIESEL EXHAUST ON LATEX PAINTS
SALT EROSION OF CONCRETE PAVEMENTS

These titles may not attract every potential reader who comes across them, but they will certainly gain the attention of anyone interested in the topics they describe.

Title Page

The title page normally carries four main pieces of information: the report title (the same title that appears on the cover); the name of the person, company, or organization for whom the report has been prepared; the name of the company originating the report (sometimes with the author's name); and the date the report was completed. It may also contain the authority or contract number, a report number, a security classification such as CONFIDENTIAL or SECRET, and a copy number (important reports given only limited distribution are sometimes assigned copy numbers to control and document their issue). All this information must be tastefully arranged on the page, as has been done in the full sample report later in this chapter.

Table of Contents

If the T of C seems sparse or disjointed, check that the report narrative has sufficient *descriptive* headings

All but very short reports contain a table of contents (T of C). The T of C not only lists the report's contents, but also shows how the report has been arranged. Just as a prospective book buyer will scan the contents page to discover what a book contains and whether it will be of interest, potential readers will scan a report's T of C to identify how the author has organized the work.

The pleasing arrangement for a T of C on page 197 uses the single word CONTENTS rather than "Table of Contents," which we recommend. The contents page also contains a list of appendices, with each identified by its full title. In long reports you may also insert a list of illustrations and their page numbers between the T of C and the list of appendices.

References (Endnotes), Bibliography, and Footnotes

A report writer who refers to another document, such as a textbook, journal article, report, or correspondence, or to other persons' data or even a conversation, must identify the source of this information in the report. To avoid cluttering the report narrative with extensive cross-references, the reference details are placed in a storage area at the end of the report or at the foot of the page. This storage area is known as a list of references or a bibliography. Specifically:

A **List of References** is the most convenient and popular way to list source documents. The references are typed as a sequentially numbered list at the end of the narrative sections of the report (usually immediately ahead of the appendix). Such numbered references are sometimes referred to as *Endnotes*. For a short example, see page 207.

A **Bibliography** is an alphabetical listing of the documents used to research and conduct a project. The documents are listed in alphabetical order of authors' surnames, and the list is placed at the end of the report narrative. A bibliography may list many more documents than are referred to in the report. An example appears on page 184.

(Footnotes, which are printed at the foot of the page on which the particular reference appears, are seldom used in contemporary reports.)

Preparing a List of References

References should contain specific information, arranged in this sequence:
(a) Author's name (or authors' names).
(b) Title of document (article, book, paper, report).
(c) Identification details, such as:

For a book: city and province, state or country of publication, publisher's name, and year of publication.

For a magazine article or technical paper: name of magazine or journal; volume and issue number; date of issue.

For a report: report number; name and location of issuing organization; date of issue.

For correspondence: name and location of issuing organization; name and location of receiving organization; the letter's date.

For a conversation or speech: name and location of speaker's organization; name, identification, and location of listener; the date.

(d) The page number (if applicable) on which the referenced item appears or starts.

Referencing a Book

If you are referring to information in a book authored by only one person, the entry in your list of references should contain:

(a) Author's name (in natural order: first name and/or initials, and then surname).

(b) Book title (either set in italic type or underlined*).

(c) City of publication, publisher's name, and year of publication (all within one set of brackets).

(d) Page number (the first page of the referenced pages).

If it is your first reference, and you are referring to an item on page 174 of the book, your entry would look like this:

1.   Laurinda K Wicherly, *Fibreoptic Modes of Communication* (Toronto, Ontario: The Moderate Press Inc, 1997), p 174.

If a book has two authors, both are named:

2.   David B Shaver and John D Williams, *Management Techniques for a Research Environment* (Edmonton, Alberta: Witney Publications, 1998), p 215.

But if there are three or more authors, only the first-named author is listed and the remaining names are replaced by "and others":

3.   Donald R Kavanagh and others,... *(etc)*.

If a book is a second or subsequent edition (as this book is), the edition number is entered immediately after the book title:

4.   Ron Blicq and Lisa Moretto, *Technically-Write!* Canadian 5th ed (Scarborough, Ontario: Prentice Hall Canada, 1998).

Some books contain sections written by several authors, each of whom is named within the book, with the whole book edited by another person. If your reference is to the whole book, then identify it by the editor's name:

5.   Donna R Linwood, ed, *Seven Ways to Make Better Technical Presentations* (Vancouver, BC: Bonus Books Ltd, 1998).

But if your reference is to a particular section of the book, identify it by the specific author, enclose the section title in quotation marks, set the book title in italic type, and then name the editor:

6.   Kevin G Wilson, "Preparation: The Key to a Good Talk," *Seven Ways to Make Better Technical Presentations*, ed Donna R Linwood (Vancouver, BC: Bonus Books Ltd, 1998), p 71.

(In examples 5 and 6, "ed" means "editor" or "edited by.")

---

* Set in *italics* for a word-processed report; <u>underlined</u> in typewritten reports

Referencing a Magazine or Journal Article

Similarly, if you are referring to an article in a magazine or journal, list these details:

(a) Author(s)'s name(s) (in natural order).
(b) Title of article (always in quotation marks, and *not* underlined).
(c) Title of journal or magazine (set in italics or underlined).
(d) Volume and issue numbers (shown as two numbers separated by a colon).
(e) Journal or magazine issue date.
(f) Page on which article or excerpt starts (optional entry).

If an article is your seventh reference, it would look like this:

> 7. Chelva Kanaganayakam, "Electronic Conferencing, Its Form, Function and Potential for Technical Writing Instruction," *Technostyle*, 8:3, Winter 1989, p 16.

If a magazine article does not show an author's name, then the entry starts with the title of the article:

> 8. "Selling to the EEC: Challenge of the 90's," *Technical Marketing*, 11:5, May 1997, p 113.

Referencing a Report

To refer to a report written by yourself or another person, list this information:

Reports, memos, and email messages can be listed as reference sources...

(a) Author's name (or authors' names), if the author is identified on the report.
(b) Title of report, in italics or underlined.
(c) Report number, or other identification (if any).
(d) Name and location (city, plus state or province) of organization issuing report.
(e) Report date.
(f) Page number (if a specific part of the report is being referenced).

Here is an example:

> 9. Derek A Lloyd, *Effective Communication and Its Importance in Management Consulting*. Report No. 61, Smyrna Development Corporation, Montreal, Quebec, February 18, 1998.

Referencing an Email Message, Letter, or Memorandum

For an email message, the entry should look like this:

10. Christine Lamont, Macro Engineering Inc, Toronto, Ontario. Email to Wayne Kominsky, No. 7 Design Group, Winnipeg, Manitoba, December 12, 1997.

For a letter or memo, replace the word "Email" with "Letter" or "Memorandum."

Referencing a Conversation or Speech

For a conversation or speech follow these examples:

...as can talks and telephone or face-to-face conversations

11. David R Phillips, Lakeside Power and Light Company, Thunder Bay, Ontario, in conversation with Anna King, H L Winman and Associates, Calgary, Alberta, January 7, 1998.

12. Francis R Cairns, Elwood Martens and Associates, Fredericton, New Brunswick, speaking to the 8th Potash Producers' Conference, Regina, Saskatchewan, September 16, 1997.

Additional Factors to Consider

Remember that when a magazine article, technical paper, or report is published as one of several documents bound into a volume, then it is listed within quotation marks (only the title of the volume is underlined). But if the article, technical paper, or report is published as a *separate* document, the quotation marks are omitted and the title of the article, paper, or report is set in italic type (or underlined), as in entry No. 9.

Every entry in a list of references must have a corresponding reference to it in the Discussion section of your report. At an appropriate place in the narrative you should insert a superscript (raised) number to identify the particular reference. It should look like this:

Numbering each entry simplifies cross-referencing between the text and the list of references

Earlier tests[3] showed that speeds higher than 2680 rpm were impractical.

(Alternatively, the raised 3 could go here.)

If your report is being prepared using a word-processing program and printer that cannot create superscript numbers, state the reference number in brackets:

Earlier tests (3) showed that speeds...

If you refer to the same document several times, your list of references need show full details for that document only the first time you refer to it. Subsequent references can be shown in a shortened form containing only the author's name (or authors' names) and the page number. For example, if the first reference you make is to an item on page 48 of the particular book described below, the entry in the list of references would be

1. Wayne D Barrett, *Management in a Technical Domain* (Halifax, Nova Scotia: Martin-Baisley Books, 1998), p 48.

Now suppose that your second and third sources are other documents, but for your fourth source you again refer to *Management in a Technical Domain*, this time quoting from page 159. Now you need list only the author's surname and the new page number.

4. Barrett, p 159.

You would do the same for each future reference to the same document, simply changing the page number each time. (Note that the Latin terms *ibid.* and *op. cit.* are no longer used.) You can even make repeated references to several different documents by the same author by simply inserting the year of publication for the particular document between the author's name and the page number:

9. Barrett, 1998, p 159.

Preparing a Bibliography

A bibliography lists not only the documents to which you make direct reference, but also many other documents that deal with the topic. The major differences between a list of references and a bibliography are:

- Bibliography entries are *not* numbered 1, 2, 3, etc.
- The name of the *first-named* author for each entry is reversed, so that the author's surname becomes the first word in the entry. (If there is a second-named author, his or her name is *not* reversed.)
- The *first* line of each bibliography entry is extended about 12 mm to the left of all other lines (see Figure 6-4).
- The entries are arranged in alphabetical sequence of first-named authors.
- Punctuation of individual entries is significantly different, with each entry being divided into three compartments separated by periods: (1) author identification (name, etc); (2) title of book or specific article; and (3) publishing details. (How the periods are positioned is shown in Figure 6-4.)
- Page numbers usually are omitted, since generally the bibliography refers to the whole document. (Reference to a specific page is made within the narrative of the report.)

Figure 6-4 shows how to list bibliography entries from various sources. The entries for this bibliography have been created from some of

In a bibliography, authors' surnames are used for easy cross-referencing

Most bibliography entries contain three distinct groups of information

Bibliography

1. Barrett, Wayne D. *Management in a Technical Domain*. Halifax, Nova Scotia: Martin-Baisley Books, 1998.

2. Blicq, Ron, and Lisa Moretto. *Technically-Write!*, Canadian 5th ed. Scarborough, Ontario: Prentice Hall Canada, 1998.

3. Cairns, Frances R, Elwood Martens and Associates, Fredericton, New Brunswick. Speaker at the 8th Potash Producers' Conference, Regina, Saskatchewan, September 16, 1997.

4. Kanaganayakam, Chelva. "Electronic Conferencing, Its Form, Function and Potential for Technical Writing Instruction." *Technostyle*, 8:3, Winter 1989.

5. Lamont, Christine, Macro Engineering Inc, Toronto, Ontario. Email to Wayne Kominsky, No. 7 Design Group, Winnipeg, Manitoba, December 12, 1997.

6. Lloyd, Derek A. *Effective Communication and Its Importance in Management Consulting*. Report No. 61, Smyrna Development Corporation, Montreal, Quebec, February 18, 1998.

7. Phillips, David R, Lakeside Power and Light Company, Thunder Bay, Ontario. Conversation with Anna King, H L Winman and Associates, Calgary, Alberta, January 7, 1998.

8. "Selling to the EEC: Challenge of the 90's." *Technical Marketing*, 18:5, May 1997.

9. Shaver, David B, and John D Williams. *Management Techniques for a Research Environment*. Edmonton, Alberta: Witney Publications, 1998.

10. Wilson, Kevin G. "Preparation: The Key to a Good Talk." *Seven Ways to Make Better Technical Presentations*, ed Donna R Linwood. Vancouver, BC: Bonus Books Ltd, 1998.

Extending the first line of each entry to the left helps readers find specific source references

Figure 6-4 A typical bibliogaphy.

the "reference" entries listed earlier. The number to the right of each entry is cross-referenced to the explanatory list below:

1 Book by one author.

2 Book by two authors, fifth edition.

3 Conference speech.

4 Journal (i.e. magazine) article.

5 Email.

6 Report.

7 Conversation.

8 Magazine article, with no author identification.

9 Book by two authors.

10 Section of book with section written by one author and whole book edited by another.

Because a bibliography is not numbered, you cannot cross-refer directly to it simply by inserting a superscript number in the report narrative, as can be done with a list of references. The most common method is to insert a parenthetical reference in the narrative, and in it include the author's name (or authors' names) and the specific page number:

> Although the tests conducted in the Northwest Territories (Faversham, p 261) showed only moderate decomposition...

The full descriptive listing for Faversham's book or report would be carried in the bibliography.

If several publications by the same author are listed in the bibliography, then the date of the particular publication is included as a parenthetical reference to identify which document is being mentioned:

> The most significant tests were those conducted 22 kilometres south of Old Crow, in the Yukon (Crosby, 1996, p 17), which showed that...

The text states author's name and relevant page number

Cover Letter

A cover letter is a brief letter that identifies the report and states why it is being forwarded to the addressee. The following letter accompanied the second report in this chapter (see pages 212 to 216).

Dear Mr Merrywell

We enclose our report No. 8-23, "Selecting New Elevators for the Merrywell Building," which has been prepared in response to your letter LDR/71/007 dated April 27, 1997.

If you would like us to submit a design for the enlarged elevator shaft, or to manage the installation project on your behalf, we shall be glad to be of service.

Sincerely

Barry V Kingsley

H L Winman and Associates

Some cover letters include comments that draw attention to key factors described in the report or that evolve from it, and sometimes summarize or interpret the report's main findings. This is done in the cover letter preceding the first report in this chapter (see Figure 6-7).

Executive Summary

A cover letter often is only one short paragraph...

An executive summary is an analytical summary of the purpose of the report, its main findings and conclusions, and the author's recommendations. Unlike the normal report summary prepared for all readers, the executive summary can present detailed information on aspects of particular concern to senior executives, and often may discuss financial implications. An executive summary can be presented in two ways: externally, as a letter pinned to the front of the report; or internally, as an integral part of the report.

...whereas an executive summary usually has two or more paragraphs

If the executive summary is attached to the report like a cover letter, the recipient may remove it before circulating the report to other readers. Hence, an external executive summary is written like a letter and is addressed to a specific reader, or group of readers, which permits its author to make comments that are intended only for that reader's eyes, or to deal with sensitive issues that for political reasons should not be discussed in the body of the report. The executive summary preceding Report No. 1 serves this purpose (see Figure 6-7 on page 194).

If, however, the executive summary is bound within the report so that it will be read by everyone, its purpose becomes more general and its author simply summarizes and perhaps comments on the report's major findings. Rather than being written like a letter, the words EXECUTIVE SUMMARY are centred at the top of the page and the summary is presented like a semiformal report. Such an executive summary may be positioned immediately inside the report cover, after the title page (in which case it replaces the normal short summary), or after the table of contents.

An integral executive summary rarely contains sensitive or confidential information

When an executive summary is bound inside the report, a brief cover letter may also be prepared as a transmittal document and attached to the front of the report.

An executive summary often may precede a major technical proposal, in which a company describes how it can successfully tackle a task at an economical price for the government or another company (see Chapter 5 for a more detailed description of a technical proposal). In this case the executive summary chiefly discusses financial and legal implications, such as specifications and statements that may be open to more than one interpretation, or that will have a marked effect if the proposed task is implemented.

The Complete Formal Report
The Main Parts

Two formal reports are included in this chapter, each typical of the quality of writing and presentation that the technical business industry expects of engineering, science, and computer graduates. The first report is presented in the traditional arrangement discussed so far, and the second in an alternative, pyramidal arrangement. In each case the parts of the report remain the same, but their sequence changes. Capsule descriptions of the main parts are listed in Table 6-2.

The same information, but two ways to arrange it

The text preceding each report discusses the report's sequence, identifies how the project was initiated and the report came to be written, and comments on both the report and the author's approach. The reports are typed single spaced with a clear line between paragraphs, which is the style preferred by industry. In comparison, academic institutions tend to prefer double spacing throughout.

Traditional Arrangement of Report Parts

(Conclusions and Recommendations *after* the Discussion)

In the traditional arrangement there is a logical flow of information: the **introduction** leads into the **discussion**, from which the writer draws **conclusions** and makes **recommendations** (the two latter parts sometimes are referred to jointly as the **terminal summary**), as illustrated in the minireport in Figure 6-5. This arrangement is used for most technical and business reports.

Table 6-2 The main parts of a formal report.

<table>
<tr><td>Cover:</td><td>Jacket of report, contains title of report and name of originating company; its quality and use of color reflect company image.</td></tr>
<tr><td>Title Page:</td><td>First page of report; contains title of report, name of addressee or recipient, author's name and company, date, and sometimes a report number.</td></tr>
<tr><td>Summary:</td><td>An abridged version of whole report, written in non-technical terms; *very* short and informative; normally describes salient features of report, draws a main conclusion, and makes a recommendation; always written last, after remainder of report has been written.</td></tr>
<tr><td>Table of Contents:</td><td>Shows contents and arrangement of report; includes a list of appendices and, sometimes, a list of illustrations.</td></tr>
<tr><td>Introduction:</td><td>Prepares reader for discussion to come, indicates purpose and scope of report, and provides background information so reader can read discussion intelligently.</td></tr>
<tr><td>Discussion:</td><td>A narrative that provides all the details, evidence, and data needed by the reader to understand what the author was trying to do, what he or she actually did and found out, and what he or she thinks should be done next.</td></tr>
<tr><td>Conclusions:</td><td>A summary of the major conclusions or milestones reached in the discussion; conclusions are only opinions so can never advocate action.</td></tr>
<tr><td>Recommendations:</td><td>If the discussion and conclusions suggest that specific action needs to be taken, the recommendations state categorically what must be done.</td></tr>
<tr><td>References:</td><td>A list of reference documents that were used to conduct the project and that the author considers will be useful to the reader; contains sufficient information for the reader to correctly identify and order the documents.</td></tr>
<tr><td>Appendices:</td><td>A storage area at the back of the report that contains supporting data (such as charts, tables, photographs, specifications, and test results) that rightly belong in the discussion but, if included with it, would disrupt and clutter the major narrative.</td></tr>
</table>

The traditional arrangement provides a continuous narrative

Sometimes the appendices are bulkier than the rest of the report

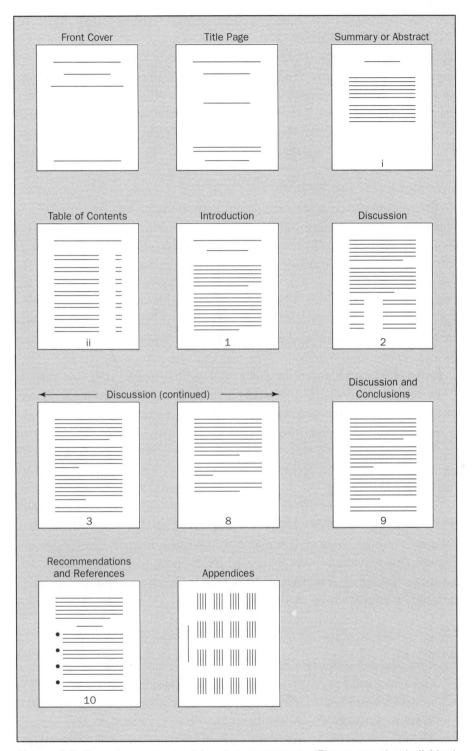

A birdseye view shows report parts

Figure 6-5 Formal report—traditional arrangement. (These are the individual pages of Formal Report 1; to conserve space some of the discussion pages have been omitted.)

Formal Report 1: Installing a Radiant Energy Heating System for Hartwell Enterprises Ltd

The author of this report (Figures 6-7 and 6-8) is Karen Woodhouse, an engineer working in the Scarborough, Ontario, branch of H L Winman and Associates. She typed and edited the report on a personal computer, using Microsoft Word 7.0 in Windows 95. She configured the word-processing program to automatically justify the right margin (make it vertically straight), insert page numbers, and create the contents page.

Comments on the Report

To understand the circumstances leading up to the report, first read Mark Hesseltine's letter of authorization in Figure 6-6. Mark is an architect with No. 5 Design Group in Burlington, Ontario. He has designed a new building for Hartwell Enterprises Ltd, a specialty manufacturer in Peterborough, Ontario, and his client has expressed interest in having radiant energy heating installed in the new building. Because No. 5 Design Group does not have radiant energy expertise in-house, Mark contracts with H L Winman and Associates to evaluate the practicability and cost of installing such a system.

Karen admitted that at first she had difficulty focusing her report: "The problem was that initially Mark seemed to be my primary reader: after all, he commissioned the report. Yet when I realized that he would want to attach my report to his proposal to Hartwell Enterprises, then I had to recognize that *they* would be my primary readers: they would be the people who will be deciding whether to install a radiant energy heating system."

Consequently, Karen tailored her approach to suit nontechnical readers (the management at Hartwell Enterprises) and included more information than she would have done if she had been writing solely for No. 5 Design Group. This is evident in her report, which can be understood by any businessperson. **Clearly identifying the reader is an essential first step when writing a comprehensive report.**

"I also wrote the report *backwards*," she explained. "Doing so really helped me organize the information." This was her approach:

> Check before you place your hands on the keyboard: the person you are writing to may not be your primary reader

1. First, she collected data on the three heating systems she would be evaluating, and researched comparative costs with experts in the radiant energy field: Darryl Berkowski in Winnipeg, Manitoba, and Vincent Harding in London, Ontario.
2. Then Karen created a table showing the relative costs of each system (this table became the report's Appendix, on page 208). From this table she also developed smaller tables identifying specific cost factors (these appear in the report's Discussion, on page 8).

> Writing in reverse order may seem unnatural, yet it results in a better report

No. 5 Design Group

240 Victoria Drive – Suite 300
Burlington, Ontario L7R 1R5
Tel: 905 234 1786; Fax: 234 1807
email: 5design@aol.com

November 27, 1997

Vern Rogers, Manager
H L Winman and Associates
970 Birchmount Road
Scarborough ON M1P 3J6

Dear Mr Rogers

As we discussed by telephone earlier today, we are commissioning H L Winman and Associates to prepare a report on the efficacy of installing a radiant heating system in the new office and assembly plant we are designing for Hartwell Enterprises Ltd of Peterborough, Ontario. The plant is to be built at the northwest corner of the intersection of Seymour Drive and Graveley Street, with construction starting on April 1, 1998 (see attached preliminary design).

Our client has indicated interest in radiant heating but needs substantive information before deciding on installing such a system. Consequently, in your report will you please describe

- how radiant heat works and how it differs from traditional heating methods,
- the advantages of installing and using radiant heat,
- the cost to install radiant heat, compared to traditional heating systems, and
- the savings to be accrued by Hartwell Enterprises over, say, a five-year period.

The four bulleted items became Karen's project criteria

Your contact at Hartwell Enterprises will be operations manager Vincent Correlli. He is aware that you are preparing a study for us and will be ready to answer questions about their operation.

I would appreciate receiving your report by January 8, 1998, because we will be submitting our design to Hartwell Enterprises on January 15. Please call me if you have any questions.

Sincerely

Mark Hesseltine

Mark Hesseltine
Design Associate
No. 5 Design Group

Figure 6-6 Letter authorizing the report in Figure 6-8.

3. Next, she created two sets of illustrations depicting installation and operating costs at one, three, and five years: first a series of bar charts and then a series of graphs. From these she chose the graph in Figure 3 (on report page 9) as the most descriptive and simplest to read.

4. Karen's fourth step was to write the Introduction, to "set the scene." Here, she drew on information in No. 5 Design Group's letter of authorization to establish the background to the report, and its purpose and scope. She knew her primary readers would not have seen the letter.

5. Next, Karen wrote a preliminary outline, which really was an early version of the Table of Contents on page 197. She used this as a loose guide for structuring the report, making changes to it as she wrote. "Organizing the outline was surprisingly straightforward," she said, "once I had done my initial research, developed the cost analysis tables, and made the charts. All I had to do was identify what preliminary information my readers would need before they got into the system comparisons and cost analyses."

6. That preliminary information became the first four pages of the Discussion, in which she defined what a radiant energy heating system is like and how it is installed (see report pages 1 to 4).

7. Next she wrote the comparisons and analyses. "These were easy to write," she said, "because my charts and tables gave me a clear direction to take."

8. Now Karen wrote the Conclusions. "They fell naturally into place," she explained. "I went back to the Introduction and identified the three factors we were particularly asked to describe, and then provided brief answers for them."

Inserting a recommendation sometimes is optional

9. She deliberated whether to write a Recommendation. The authorizing letter did not specifically ask for a recommendation yet she felt that, because she had researched the information and as such was the local expert, she should identify what route she felt Hartwell Enterprises should take.

10. And *last*, Karen wrote the Summary, drawing principally on the Conclusions to compose it. "If I had tried to write it first, before writing the report," she said, "I would have had much more difficulty writing it. Probably I would have had to go back and rewrite most of it, after the remainder of the report was finished."

Remember Karen's "backwards" approach when you have to write your next long report. By documenting all the details *first*, you will find you can organize your ideas more easily and write more fluidly.

Comments on additional aspects of Karen's report follow, with specific references to the individual pages of the report in Figure 6-8.

- Karen's cover letter in Figure 6-7 is equivalent to an executive summary because she comments on the report's contents. Because she knows the cover letter will be seen only by Mark Hesseltine, she uses a friendly tone and the first person "I."
- A quick glance at the Contents (page ii) tells Karen's readers that she has organized her information into a logical, coherent flow, and that there are three main components: background information on radiant energy systems; a plan for installing a radiant energy heating system in the client's building; and a cost projection and analysis.
- In the Introduction (report page 1), the Background is in paragraph 1, the Purpose is in paragraph 2, and the scope is in the three bulleted points.
- For the Discussion, Karen has adopted an overall "concept" arrangement of information, but internally it breaks into a subject arrangement when describing the proposed installation and some of the costs.
- She has written the entire report in the first person plural. "When I am presenting the results of a study I have done personally," she said, "and am writing directly to my client, then normally I would use the first person singular: 'I.' But when I am writing on behalf of the company, and simultaneously am writing for my client's client, and don't know the client personally, then I use the first person plural: 'We'."

Writing in the first person means making a decision: "I" or "we"?

- The Conclusions, on report pages 9 and 10, are longer than normal, but necessarily so to cover all the points the client requested. They show the advantages and disadvantages of radiant energy heating but do *not* advocate what action should be taken. Karen has taken care not to introduce new information into her Conclusions.
- The Recommendation advocates action and does so in strong definite terms, using "We recommend..." rather than "It is recommended...."
- The Appendix is a "landscape-view" page, and as such has been turned correctly so that it is read from the right (see page 208).

Pyramidal Arrangement of Report Parts

(Conclusions and Recommendations *before* the Discussion)

In recent years, more and more report writers have altered the organization of their reports so they more effectively meet their readers' needs. The pyramidal arrangement brings the conclusions and recommendations forward, positioning them immediately after the introduction so that executive readers do not have to leaf through the report to find the terminal summary (the report's outcome). The advantages of the pyramidal approach are immediately evident: Busy readers have only to read the initial pages to learn the main points contained in the report, and the writer

A way to address three different levels of reader within a single document

H L WINMAN AND ASSOCIATES

475 Lethbridge Trail, Calgary AB T3M 5G1

email: kwoodhouse@winman.on.ca

January 7, 1998

Mark Hesseltine
Design Associate
No. 5 Design Group
240 Victoria Drive, Suite 300
Burlington ON L7R 1R5

Dear Mark

I am enclosing our report *Installing a Radiant Energy Heating System for Hartwell Enterprises Ltd*, as requested in your letter of November 22, 1997. The report shows that in the long term radiant energy will be the most economical heating system for Hartwell Enterprises' new building.

Inserting this second paragraph converted Karen's cover letter into an executive summary

To some extent I am concerned that Hartwell Enterprises may hesitate when they see the high front-end cost, particularly in comparison to electric heat. Hence, I have taken care to include a graph which shows clearly that radiant energy heating will be particularly efficient from a cost viewpoint after the fourth year. If the graph on page 9 were to be extended for another five years, the considerably lower operating cost of radiant energy heating would be even more noticeable. You may want to address this factor in your proposal.

Please call me if you need further information on any of the points addressed by the report. I'll be glad to supply it.

Sincerely

Karen Woodhouse

Karen Woodhouse, P.Eng
enc

Figure 6-7 Cover letter accompanying the formal report in Figure 6-8. This cover letter also is an executive summary.

H L WINMAN AND ASSOCIATES

The appearance of the title page subtly comments on the quality of the information you are presenting

INSTALLING A RADIANT ENERGY HEATING SYSTEM FOR HARTWELL ENTERPRISES LTD

Prepared for

No. 5 Design Group
Burlington, Ontario

Prepared by

Karen Woodhouse, P.Eng
H L Winman and Associates
Scarborough, Ontario

January 7, 1998

Figure 6-8 Formal Report 1: traditional arrangement (13 pages)

The Summary: the full story in a capsule — difficult to write!

SUMMARY

We have evaluated three methods for heating the proposed Hartwell Enterprises Ltd plant designed by No. 5 Design Group for construction in Peterborough, Ontario. Electric heat is the least expensive to install but the most expensive to operate. Gas-fired forced hot air is moderately expensive to install and moderately expensive to operate. Radiant energy heating is the most expensive to install and the least expensive to operate. Long term, however, radiant energy offers the most efficient and cost-effective method.

Installing a radiant energy heating system in a new building means the system can be incorporated into the overall design so that it becomes unobtrusive as well as efficient and cost-effective. It also provides more gentle warming than the other two methods, and the temperature in each area of the building can be controlled separately. The operating cost will be 39% less than for electric heating, and 30% less than for gas-fired hot air heating.

i

CONTENTS

The preliminary pages bear roman page numbers; all other pages bear arabic numbers

APPENDIX

If more than one appendix, the title changes to "Appendices" and each appendix is identified by a letter: "A," "B," "C," etc

INSTALLING A RADIANT ENERGY HEATING SYSTEM
FOR HARTWELL ENTERPRISES LTD

INTRODUCTION

The Introduction establishes why the project was undertaken and the report has been written

Hartwell Enterprises Ltd assembles and packages modules and specialty products for public and private organizations such as the Department of Defence, NavCan, Multiple Industries Limited, and Northern Paging and Cellular Systems. The company does no original manufacturing itself, confining its role to assembling components supplied by carefully chosen manufacturers. Hartwell Enterprises Ltd has built a solid reputation as a fast, high-quality producer of specialty systems, and the company's business has increased steadily since its inception in 1982. Today, it needs a larger building, but research for suitable accommodation among Peterborough area properties has failed to find a building that can be adapted to the company's special needs.

In August 1997, Hartwell Enterprises Ltd commissioned No. 5 Design Group of Burlington, Ontario, to design a building that will meet the company's particular requirements, and to find a site on which to place it. No. 5 Design Group has, in turn, asked H L Winman and Associates to evaluate the efficacy and cost to install a radiant heating system in the new building, rather than a more traditional heating method. Specifically, they asked us to describe

These requirements were copied almost verbatim from the client's letter of authorization

- how radiant heat works and how it differs from traditional heating methods,
- the advantages of and the cost to install a radiant heating system, and
- the savings to be accrued by Hartwell Enterprises Ltd over the first five years.

RADIANT HEATING VS TRADITIONAL HEATING

A traditional heating system warms air directly, which we feel on our skin as immediate heat, but its impact is transitory. Turn off the source of the heat, and the space being warmed immediately starts to cool. On a winter's day, for example, a residential furnace pumps hot air into the rooms until a preset temperature is reached, then the thermostat switches off the furnace. The

1

warming effect stops immediately and the air temperature, influenced by cooler windows and walls, begins to drop.

A radiant heating system, rather than warming air directly, radiates heat outwards in all directions until the rays contact another surface. If the surface is cooler than the radiant panel, the surface begins to warm up and the air near to it also warms, but gently, and so our skin feels the warmth as a gentle, comfortable heating. On a cool winter's day, a furnace pumps heat into the radiant panels as hot water or they are heated electrically until a preset room temperature is reached, when the source of the heat is switched off. The warming effect, however, does not stop immediately because the radiant panel continues to radiate residual heat for a considerable time. Consequently, the air in the room cools much more slowly than with hot air heating.

Compare the difference in heat produced by a gas ring and an electric hotplate. The gas ring provides immediate heat to the surrounding air, but the heat stops immediately when the gas is switched off. The electric hotplate builds up its heat more slowly, but *continues to radiate heat* for 10 to 15 minutes after the electricity is switched off.

Comparing a complex technical concept to a familiar everyday event helps reader understanding

In locations that are unlikely to be affected by external sources introducing sudden changes in temperature, a 100% radiant heating system is ideal. Even in areas that are subjected to marked changes in temperature, such as an aircraft hangar, radiant heat also is effective. Lawrence Drake, executive director of the Radiant Panel Association, explains that

> Thermal mass in a heated shop or hangar floor responds rapidly to the change of air temperature when a big overhead door is opened. All the heat that has been "trickled" into the slab over time is released quickly to combat the cold air rolling in over the floor. This happens because of the sudden, dramatic increase in temperature difference between the slab and the new air. Once the door is closed the building returns to its normal comfort setting almost immediately.[1]

He adds, however, that under such conditions a combination of radiant energy and a back-up hot air heating system can be even more effective, because occupants of the space immediately feel the heated air.

2

A statement made in an email message or during a conversation may be documented as a source reference

Darryl Berkowski, who installs radiant heating systems in Manitoba and Northwestern Ontario, points out that electric or hot water baseboard heaters may *appear* to produce radiant heat, but in effect they release only a small amount of radiant energy. Primarily, they heat the air.[2]

A particular advantage of radiant heating is that *it can be controlled locally*. In a traditional heating system, hot air is supplied to vents, or hot water to radiators, from a central furnace, the operation of which is controlled by a single thermostat mounted on a wall of only one of the rooms. Thus, the temperature in the other rooms cannot be controlled separately. If, for example, one of the rooms tends to be cooler than the others because it has large windows on a north wall, the heating system is unable to compensate for the variation. Conversely, in a radiant heating system, the heat supplied to the radiant panels can be controlled separately in each room. This is true of radiant panels heated either electrically or by hot water.

RADIANT PANEL INSTALLATIONS

Radiant heating panels may be installed in the wall, ceiling or floor. Because they heat all objects within their line of sight, floor panels heat the ceiling and walls, wall panels heat the opposite wall, and ceiling panels heat the floor and walls. Similarly, all three heat furniture they can "see" and, of course, the people in the room. The radiant energy is felt as a very gentle, subtle warming, never as a searing blast.

Generally, wall panels tend to be smaller than ceiling panels, and ceiling panels tend to be smaller than floor panels, which often use the whole floor as their radiating surface. Also generally, the smaller the panel the greater the temperature at which it must operate to gain a measurable effect. Consequently, wall panels may be as hot as 65°C, ceiling panels as warm as 40°C, but floor panels rarely exceed 25°C. Lawrence Drake writes that

Introducing excerpts from an existing document adds credibility and reduces writing effort

> A heated floor normally "feels" neutral. Its surface temperature is usually less than our body temperature, although the overall sensation is one of comfort. Only on very cold days when the floor is called on for maximum output will it actually "feel" warm.
>
> Heat coming from a wall radiator can be felt the closer you get to it because its surface is much warmer than your body. Radiant ceiling

3

panels are also generally warmer than your body so you will feel some warmth on your head and shoulders.[3]

Wall and ceiling panels usually come preassembled and are fixed onto the surface of the wall or ceiling. Floor panels normally are embedded in the floor, as part of the floor construction, and then are covered with a layer of concrete or similar floor material. Consequently the whole floor becomes a single, large radiant panel.

When installed as an integral part of a new installation, either electric elements or water pipes are laid in continuous parallel rows in the floor (see Figure 1). The pipes are made of a strong, durable, light, cross-linked polyethylene (PEX). Connectors are made of noncorroding copper, brass, or plastic.

Although it's possible to install a radiant heating system as a retrofit in an existing building, the ideal arrangement is to design a radiant heating system specifically for a new building. It can then be installed as an integral part of the new structure.

A cutaway illustration like this helps readers visualize the technology

Figure 1. Various types of radiant energy panels
(Illustration courtesy of Radiant Panel Association)

4

Hartwell Enterprises' operation is unique, in that it works on a just-in-time method of delivery for the components that are assembled into products. The company's contracts with its suppliers stipulate that components are to be delivered in small lots only a few hours before they are to be assembled. This has two major effects:

1. Because components are moved rapidly from the delivery semitrailer to the assembly line, and then to the shipping area for packing and storing in a second semitrailer ready for shipping, Hartwell Enterprises requires only a small warehouse area.

2. Because the loading bay doors have to be opened frequently, the loading bay demands special attention from a heating viewpoint.

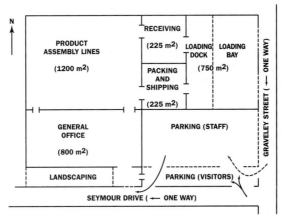

Figure 2. Design of proposed new building for Hartwell Enterprises Ltd

Proposed Heating System

The Proposed Hartwell Enterprises building particularly lends itself to heating by radiant energy. The diverse activities in the various parts of the building call for different types and methods of heating. As Figure 2 shows, there will be five main areas of activity, each requiring a different level and form of heating, and each with its own thermostat. The floor in each area will be concrete, but there will be some variations:

5

Research into a client's business practices can help you focus information accurately

The scene has been set: now for the plan!

- The general office will have a 0.20 metre thick slab with its upper surface 0.35 metres above grade. It will be covered by Orlando carpet and underlay.

- The product assembly, receiving, and shipping areas will be set on a 0.20 metre thick slab with the floor level at 1.3 metres above grade. There will be basement under these three areas, with a 2.2 metre high ceiling.

- The unloading and loading bays will be at grade level, built on a 0.35 metre reinforced concrete slab.

As the primary source of heat for these areas, we propose installing hot-water pipes embedded in the concrete with a separate circulation system and thermostat for each area. The mechanical room for the hot water boiler and the controls will be set up in the basement. (We are not recommending electrically heated panels, because in southern Ontario it is on average 20% more expensive to heat the radiant panels by electricity than by natural-gas-fired hot water.[4])

The product assembly, receiving, and shipping areas will require no supplementary heating. However, the general office and the loading bays will. For these areas we propose the following additional heating arrangements.

- The ceiling in the general office will slope upward, toward the north (the back of the office), which will tend to draw warm air aloft, into the higher part of the ceiling and away from the south side of the area. To maintain an even warmth on the south side, we propose installing two 1.5×0.8 metre horizontal radiant panels under the windows along the south wall. These will be heated by hot water and will be controlled by a separate thermostat.

- The loading bays will cool rapidly in midwinter when the doors are opened to admit and exit semitrailers. Here we propose installing two overhead unit heaters, one above each door. They also will be hot-water heaters, and will be triggered to start up when a door opens and to shut down when a predetermined ambient temperature is reached. (In effect, the radiant energy from the floor panels will restore the temperature quite quickly. The unit heaters will provide a supplementary, readily noticeable, immediate source of heat during extreme cold conditions.)

6

The main paragraphs contain the plan...

...while the bulleted subparagraphs introduce variations and exceptions

Proposed Cooling System

Although radiant energy panels can provide moderate cooling in the summer, particularly in dry climates, in moist or semimoist conditions they tend to be less efficient.[5] Consequently we propose that all cooling be carried out through a separate system, comprising

- a Hyperion Model 2000 air conditioning unit for cooling the general office, product assembly, and receiving and shipping areas, and
- a Hyperion Model 2720 air conditioning unit for cooling the loading bay.

The two air conditioners will be located in the basement, with the model 2000 at the west end and the model 2720 at the east end. The model 2000 will feed cooled air through ductwork concealed in the ceiling of the office, assembly, and receiving/shipping areas. The model 2720, which will be a fast-response unit designed to recover quickly from rapid changes in temperature, will feed cooled air through ductwork under the roof of the loading bay.

COSTS: RADIANT VS TRADITIONAL HEATING

Two cost factors have to be considered when comparing a radiant energy heating system with a traditional heating system: the cost of installation and the cost of operation. The cost of installing and operating air conditioning from May to September also has to be taken into account.

For this study, we have examined the cost of installing and operating three systems:

1. Radiant energy heating, plus a separate air-conditioning system.
2. All-electric heating, plus a separate air-conditioning system.
3. Forced hot-air heating fueled by natural gas, with integral air-conditioning.

Installation Costs

The costs to install these three systems in the proposed Hartwell Enterprises building are listed in Table 1, which at first glance shows that electric heating is the least expensive and radiant energy is the most expensive. However, when air conditioning is included, gas-fired hot air becomes the least expensive.

A section opening paragraph should act like a minisummary, identifying aspects to be discussed

7

Table 1. Installation Costs

System	Heating System	Air-Conditioning	Total
Radiant energy	$47 200	$21 600	$68 800
All-electric plus air-conditioning	$26 600	$21 600	$48 200
Gas-fired hot air with integral air-conditioning	$38 700	(incl)	$38 700

The table lists key cost figures, provides a ready comparison

Annual Operating Costs

To calculate potential operating costs, we referred to a 1997 study carried out by Vincent Harding of V Harding Associates, in which he averaged the annual heating costs of 30 industrial buildings in Ontario, Manitoba, and Saskatchewan for the years 1992 to 1996.[6] From these we culled the heating costs for eight light-industry buildings in Southern Ontario, each of a similar size to the proposed 3200 m^2 Hartwell Enterprises Building. Two are heated by electricity, three by gas-fired forced hot air, and three by radiant energy. The results are summarized in Table 2, which shows that radiant energy heating has the lowest annual operating cost and all-electric heating has the highest annual operating cost. A detailed breakdown is shown in the Appendix.

Table 2. Average Annual Operating Costs

System	Heating Only	With Air-Conditioning
Radiant energy	$17 300	$20 800
All-electric	$31 700	$35 200
Gas-fired forced hot air	$27 600	$31 100

A simple table with key cost figures can be embedded conveniently into the narrative

8

Projected Costs Over Five Years

In Figure 3, the combined installation and operating costs for both heating and cooling are shown for year one, and then the projected operating costs for heating and cooling are shown for years two through five. The graph shows that in the first year the installation and operating cost for electric heating is the least expensive, and radiant energy is the most expensive. However, after 1 year and 5 months the positions are reversed, with the cumulative costs of installing and operating radiant energy heating becoming less expensive than electric heating.

Installing and operating a gas-fired forced-air heating and air-conditioning system remains less expensive than radiant energy for the first 3 years and 2 months, after which the cumulative cost of radiant energy heating plus a companion air-conditioning system becomes more economical.

Both this graph and the two previous tables are supported by the detailed cost comparison in the appendix

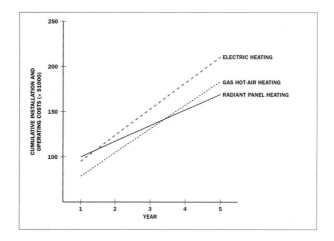

Figure 3. Comparison of installation and operating costs over five years, for radiant energy, hot air, and electric heating systems

CONCLUSIONS

Conclusions sum up key outcomes, never specifically advocate action

For the proposed Hartwell Enterprises Ltd office and product assembly plant planned for 1650 Seymour Drive in Peterborough, Ontario, a radiant energy heating system offers several advantages:

9

- Its operating cost will be 39% less expensive than for electric heating, and 30% less expensive than for gas-fired forced hot-air heating.

- It will provide a softer, less obtrusive, and more stable source of warmth than either electric or hot-air heating.

- It can be controlled separately for each area of the building.

- Because it will be a new building, it can be designed and installed as an integral part of the structure and thus be less obtrusive.

Its chief disadvantage is that its installation cost will be 77% higher than for an electric heating system, and 22% higher than for a gas-fired hot-air system.

However, in the long term, the combined installation and operating cost of radiant energy heating becomes less than that of electric heating in 17 months, and less than that of forced hot-air heating in 38 months.

They may, however, *imply* the course to be taken

RECOMMENDATION

Viewing Hartwell Enterprises' move into a new building as a long-term venture, we recommend installing a radiant energy heating system to take advantage of the long-term low operating expenses it will incur.

REFERENCES

1. Lawrence V Drake, *Radiant Panel Heating and Cooling*. Report: Radiant Panel Association, Hyrum, Utah, 1995, p 3.
2. Darryl Berkowski, D & M Innovators, Winnipeg, Manitoba. Email to Karen Woodhouse, H L Winman and Associates, December 10, 1997.
3. Drake, p 2.
4. Berkowski, p 2.
5. Drake, p 4.
6. Vincent Harding, *Comparison of Costs: Electric, Gas, and Radiant Energy Heating in 30 Industrial Buildings, 1992-1996*. Report: V Harding Associates, London, Ontario, February 23, 1997.

The reference section lists all written and spoken information sources

10

Placing a detailed table like this within the report narrative would distract a reader's attention

APPENDIX

COMPARISON OF COSTS: THREE HEATING/COOLING SYSTEMS FOR HARTWELL ENTERPRISES LTD

Heating/Cooling System	Installation Cost	Annual Operating Cost	First Year Costs	Cumulative Three-Year Costs	Cumulative Five-Year Costs
Radiant Energy:					
• Heating	$47 200	$17 300			
• Air-Conditioning	$21 600	$ 3 500			
Combined Systems:	$68 800	$20 800	$89 600	$131 200	$172 800
Electric Heat:					
• Heating	$26 600	$31 700			
• Air-Conditioning	$21 600	$ 3 500			
Combined Systems:	$48 200	$35 200	$83 400	$153 800	$224 200
Gas-fired Forced Hot Air:					
• Heating	$38 700	$27 600			
• Air-Conditioning		$ 3 500			
Combined Systems:	$38 700	$31 100	$68 700	$130 900	$193 100

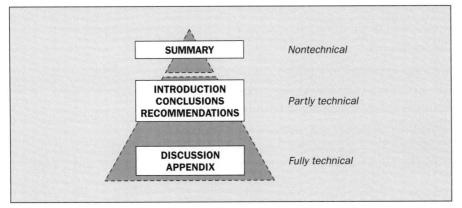

Figure 6-9 The formal report arranged pyramid style.

Three reports in one, each a complete story in itself

can help them along by gradually increasing the technical content of the report, catering to semitechnical executive readers up to the end of the recommendations, and to fully technical readers in the discussion and appendix. Although the natural flow of information that occurs in the traditional arrangement is disrupted, Figure 6-9 shows there is now a reader-oriented flow, with the three compartments each containing progressively more technical details.

This gradually increasing development of the topic in three separate stages is similar to the newspaper technique described earlier, in which the first one or two paragraphs contain a capsule description of the whole story, the next three or four paragraphs contain a slightly more detailed description, and the final eight or nine paragraphs repeat the same story, but this time with more names, more peripheral information, and more details. Newspapers cater to both the busy reader who may not have time to read more than the opening synopsis, and the leisurely reader who wants to read all the available information.

A minireport showing the pyramidal arrangement of report parts is illustrated in Figure 6-10. Segments of a sample report written using the pyramidal approach are shown in Figure 6-11.

Excerpts from Formal Report 2: Selecting New Elevators for the Merrywell Building

Before reading these excerpts, read the client's letter authorizing H L Winman and Associates to initiate an engineering investigation:

Dear Mr Bailey

The elevators in the Merrywell Building are showing their age. Recently we have experienced frequent breakdowns and, even when the elevators are operating properly, it has become increasingly evident that they do not provide adequate service at the start of work,

The report writer used some of these words in the Introduction on page 213 at noon, and at the end of the working day. I have therefore decided to install a complete range of new elevators, with work starting in mid-August.

Before I proceed any further, I would like you to conduct an engineering investigation for me. Specifically, I want you to evaluate the structural condition of my building, assess the elevator requirements of the building's occupants, investigate the types of elevators available, and recommend the best type or combination of elevators that can be purchased and installed within a proposed budget of $800 000.

Please use this letter as your authority to proceed with the investigation. I would appreciate receiving your report by the end of June.

Regards

David P Merrywell, President

Merrywell Enterprises Inc

By comparing this letter with the conclusions and recommendations, you can assess how thoroughly Barry Kingsley (the report's author) has answered the client's requests.

Comments on the Report

The summary is short and direct because it is written primarily for one reader: the president of Merrywell Enterprises Inc. It encourages him to read the report immediately, and to accept its recommendations, by offering the opportunity to save $50 000.

A report may be directed to one reader, but also must consider other readers who may see it

Although the background information contained in the first two paragraphs of the introduction seems to repeat details the client already knows, Barry recognizes he must satisfy the needs of other readers who may not be fully aware of the situation in the Merrywell Building. He then defines the purpose and scope of the investigation by stating the client's terms of reference in paragraph 3 of the introduction. (Note that he has copied them almost verbatim from Mr Merrywell's letter.)

The conclusions present Barry's answers to Mr Merrywell's four requests. Their order is different from that in paragraph 3 of the introduction because he has chosen to present the main conclusion first (in this case, the best combination of elevators that can be purchased within the stipulated budget), and to follow it with subsidiary conclusions in descending order of importance. Barry is aware that when using the pyramidal report format he must write conclusions that evolve naturally and logically from the introduction, *because his readers have not yet read the discussion.*

Barry uses the first person plural to open his recommendations because, although he alone is the report's author, he is representing H L Winman and Associates' views to the client.

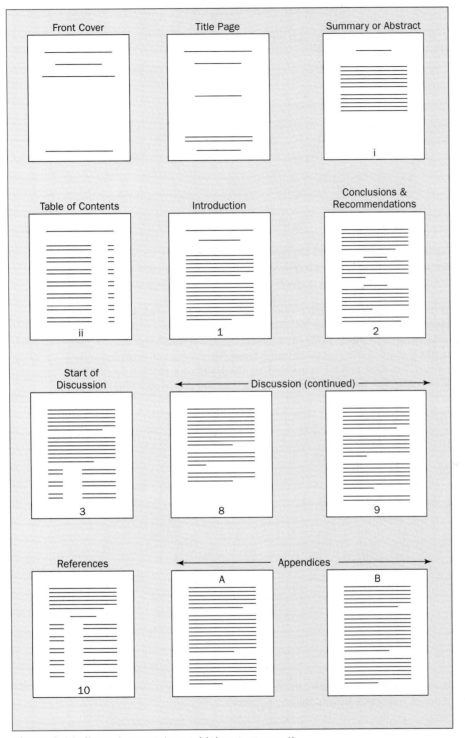

Compare this arrangement with the bird's-eye view in Figure 6-5

Figure 6-10 Formal report (pyramidal arrangement).

H L WINMAN AND ASSOCIATES

SUMMARY

The elevators in the 71-year-old Merrywell Building are to be replaced. The new elevators must not only improve the present unsatisfactory elevator service, but must do so within a purchase and installation budget of $800 000.

Of the many types and combinations of elevators considered, the most satisfactory proved to be four 2.45 × 2.15 metre deluxe passenger elevators manufactured by the YoYo Elevator Company, one of which will double as a freight elevator during off-peak traffic times. This combination will provide the fast, efficient service requested by the building's tenants for a total price of $750 000, which will be 6.25% less than the projected budget.

A simple, straightforward summary that answers the reader's most immediate question

Figure 6-11 Formal Report 2: pyramidal arrangement.

Selecting New Elevators for the Merrywell Building

Introduction

When in 1970 Merrywell Enterprises Ltd purchased the Wescon property in Montrose, Alberta, they renamed it "The Merrywell Building" and renovated the entire exterior and part of the interior. The building's two manually operated passenger elevators and a freight elevator were left intact, although it was recognized that eventually they would have to be replaced.

Recently the elevators have been showing their age. There have been frequent breakdowns and passengers have become increasingly dissatisfied with the inadequate service provided at peak traffic hours.

In a letter dated April 27, 1997, to H L Winman and Associates, the president of Merrywell Enterprises Ltd stated his company's intention to purchase new elevators. He authorized us to evaluate the structural condition of the building, to assess the elevator requirements of the building's occupants, to investigate the types of elevators that are available, and to recommend the best type or combination of elevators that can be purchased and installed within the proposed budget of $800 000.

Conclusions

The best combination of elevators that can be installed in the Merrywell Building will be four deluxe 2.45 × 2.15 metre passenger models, one of which will serve as a dual-purpose passenger/freight elevator. This selection will provide the fast, efficient service desired by the building's tenants, and will be able to contend with any foreseeable increase in traffic. Its price at $750 000 will be 6.25% less than the proposed budget.

Installation of special elevators requested by some tenants, such as a full-size freight elevator and a small but speedy executive elevator, would be feasible but costly. A freight elevator would restrict passenger-carrying capability, while an executive elevator would elevate the total price to at least 20% above the proposed budget.

The quality and basic prices of elevators built by the major manufacturers are similar. The YoYo Elevator Company has the most attractive quantity price structure and provides the best maintenance service.

1

The Introduction, Conclusions, and Recommendations stand alone as a composite section

The primary conclusion comes first, followed by subsidiary conclusions

The building is structurally sound, although it will require some minor modifications before the new elevators can be installed.

Recommendations

Recommendations
should be written in the
first person, singular or
plural

We recommend that four Model C deluxe 2.45 × 2.15 metre passenger elevators manufactured by the YoYo Elevator Company be installed in the Merrywell Building. We further recommend that one of these elevators be programmed to provide express passenger service to the top four floors during peak traffic hours, and to serve as a freight elevator at other times.

2

Evaluating Building Condition

We have evaluated the condition of the Merrywell Building and find it to be structurally sound. The underpinning done in 1964 by the previous owner was completely successful and there still are no cracks or signs of further settling. Some additional shoring will be required at the head of the elevator drive shaft immediately above the 9th floor, but this will be routine work that the elevator manufacturer would expect to do in an old building.

The existing elevator shaft is only 7.5 metres wide by 2.5 metres deep, which is unlikely to be large enough for the new elevators. We have therefore investigated relocating the elevators to a different part of the building, or enlarging the existing shaft. Relocation, though possible, would entail major structural alterations and would be very expensive. Enlarging the elevator shaft could be done economically by removing a staircase that runs up the centre of the building immediately east of the shaft. This staircase is used very little and its removal would not conflict with fire regulations. Removal of the staircase will widen the elevator shaft by 3.35 metres, which will provide sufficient space for the new elevators.

Establishing Tenants' Needs

To establish the elevator requirements of the building's tenants, we asked a senior executive of each company to answer the questionnaire attached as Appendix A. When we had correlated the answers to all the questionnaires, we identified five significant factors that would have to be considered before selecting the new elevators. (There were also several minor exceptions that we did not include in our analysis, either because they were impractical or because they would have been too costly to incorporate.) The five major factors were:

- Every tenant stated that the new elevators must eliminate the lengthy waits that now occur. We carried out a survey at peak travel times and established that passengers waited for elevators for as much as 70 seconds. Since passengers start becoming impatient after 32 seconds,[1] we estimated that at least three, and probably four, faster passenger elevators would have to be installed to contend with peak-hour traffic.

- Although all tenants occasionally carry light freight up to their offices, only Rad-Art Graphics and Design Consultants Limited considered that a freight elevator would be essential. However, both agreed that a separate

3

The Discussion tells the reader the building is in good shape...

...and then goes on to assess what needs to be done

freight elevator would not be necessary if one of the new passenger elevators is large enough to carry their displays. They initially quoted 2.75 metres as the minimum width they would require, but later conceded that with other modifications they could reduce the length of their displays to 2.3 metres. All tenants agreed that if a passenger elevator is to double as a freight elevator, they would restrict freight movements to non-peak travel times.

- The three companies occupying the top four floors of the building requested that one elevator be classified as an express elevator serving only the ground floor and floors 6, 7, 8, and 9. Because these companies represent more than 50% of the building's tenants, we considered their request should be entertained.

- Three companies expressed a preference for deluxe elevators. Rothesay Mutual Insurance Company, Design Consultants Limited, and Rad-Art Graphics all stated that they had to create an impression of business solidarity in the eyes of their clients, and feel that deluxe elevators would help convey this image.

- The managements of Rothesay Mutual Insurance Company and Vulcan Oil and Fuel Corporation requested that a small key-operated executive elevator be included in our selection for the sole use of top executives of the building's major tenants. We asked other companies to express their views but received only marginal interest. The consensus seemed to be that an executive elevator would have only limited use and the privilege would too easily be abused. However, we retained the idea for further evaluation, even though we recognized that an executive elevator would prove costly in relation to passenger usage.[2]

We decided that the first two of these factors are requirements that must be implemented, while the latter three are preferences that should be incorporated if at all feasible. The controlling influence would be the budget allocation of $800 000 stipulated by the landlord, Merrywell Enterprises Ltd. In decreasing order of importance, the requirements are:

Identifying users' needs
helps establish clear
criteria

1. Passenger waiting time must be no longer than 32 seconds.
2. At least one elevator must be able to accept freight up to 2.3 metres long.
3. An express elevator should serve the top four floors.
4. The elevators should be deluxe models.
5. A small private elevator should be provided for company executives.

4

We have included the first two pages of the discussion to show that, early in his report, Barry establishes criteria that will subsequently influence how he selects a combination of elevators that will best meet his client's needs. By carefully identifying the five criteria and describing why each is valid, he shows his readers the direction his report will take. (In later sections of his report—not included in the sample pages—he identifies various combinations of elevators that could be installed, and demonstrates which do or do not meet the criteria, until he finally reaches an optimum configuration.)

ASSIGNMENTS

Project 6.1: Researching a New Site

Vern Rogers telephones you from the local H L Winman and Associates's branch, where he is branch manager, and asks you to drop in. (You own a small business called Pro-Active Consultants Limited, which you operate from an office in your home.)

You visit his office the next day, on the third floor of the Connor building at 444 Main Street. The building was built around 1880 and has stood up well over the years, however it is aging and clearly needs some renovations.

A company in search of a better location

"We need better and new accommodation," Vern says, "something more in line with the tenor of our engineering consulting business, and something a little bigger." He says no one on his staff has the time to research new accommodation, so he's asking you to do it. "I need your report by the end of next month, that's two months before our lease runs out: time for me to choose one of the sites you recommend, and for me to give a month's notice to the owners of this building."

Vern explains that over the years the centre of commerce has drifted away from the Connor building, toward the tree-lined Broadway area of the city. "Broadway doesn't have to be the chosen location, but it will give you some idea of the kind of area I think would be most suitable."

You ask some questions, and Vern offers detailed answers (see Table 6-3).

220 Broadway. You first look at Vern's preferred location, where there seems to be only one vacancy. You contact Camilla Loew, a sales representative with Modern Management Inc.

"The occupancy rate is very high on Broadway," Camilla says. "If you want this one, you'll have to move fast to close the deal."

As she drives you along the tree-lined avenue, she briefs you about the vacancy. "The property's in the Chancellor Building, fourth floor, 220

Table 6.3 HLW's local office requirements.

Setting criteria in a
table makes the
requirements more
accessible

Your questions:	Vern Roger's answers:
1. How many people do you have on staff? And how much increase do you see over the next five years?	Currently, 42. Over the next five years I expect the number to increase about 10% per year, to a total of, say, 60 in five years.
2. How much space will you need?	We have 410 square metres right now, and frankly it's a bit tight. I'd say we need 480 square metres immediately, and 580 in five years.
3. What rent do you anticipate paying?	Good point. We're lucky here: it's only $12.50 per month per square metre; I guess I could go up to about $15.00.
4. How many parking stalls? And what will the staff be willing to pay for them?	Between 25 and 30. Currently they pay $20 per month in summer, and $25 in winter.
5. Should there be a cafeteria in the building?	Ideally, yes. Or a restaurant or two within easy walking distance.
6. Do any of your staff travel by bus?	About 20%. A bus route is important.
7. Anything else you can think of?	False ceilings! We're getting so dependent on computers, we need false ceilings so we can bring the cables in unseen. And make changes as the technology changes.

Broadway. Actually, it's occupied right now and will be for another two-and-a-half months. That would suit your client just about right, wouldn't it?"

You agree that it would.

"The present owners are ManSask Insurance Corporation. They're moving to larger premises in the old Ashton Warehouse."

An ideal location, but it
comes at a price!

The Chancellor building lives up to its name. There is a refined, almost old-world atmosphere about it: dark walnut panelling with gold trim, high ceilings, rich gold carpet, and spacious but rather slow, sturdy, panelled elevators. The offices themselves have a panelled foyer, but inside are businesslike with off-white painted walls bearing framed prints of well-known traditional Canadian artists. There is no false ceiling. ("But that could be installed without any problem," Camilla suggests. "The high ceilings will permit it.") The 12 rooms that make up the 500 square metre space are arranged along two walls that face north and east, and so overlook the old warehouse district and the railway yards.

"Five hundred square metres is barely enough," you murmur, more to yourself than to Camilla.

"Ah, but in five years one of the adjoining offices is bound to be freed up," she says optimistically. "One is 120 square metres; the other 185 square metres." She says the rent is $16.75 per square metre per month, and the rate is firm for the first two years. After that it has to be renegotiated every year.

"Inevitably, there's always an increase in the rental rate," you say more as a statement of fact than a question. Camilla agrees that past history shows there is a modest increase "every year or so."

"It depends who defines 'modest'," you reply.

"There are 22 parking spaces in an underground car park you can inherit from the present occupant," she adds hastily. "If you need more, we have a waiting list. They rent at $40 per month."

"Cafeteria?" you ask.

"At the back of the building, on the second floor."

You don't have to ask about bus connections, because Broadway is a major thoroughfare served by several bus routes.

1650 Manor Road. Dana Wintersen calls from Provo Realty to say she has an excellent property for you, recently developed in the suburbs.

"In the suburbs?" you reply. "I don't think my client would appreciate that."

"It's worth a look," she says. "Don't 'nay say' it until you've seen it."

Dana is right: the accommodation would suit HLW perfectly—if it were only downtown. There are 750 square metres of space, available immediately; the rent is only $13.00 per square metre per month, and the lease is for one year, but with a clause guaranteeing it can be renewed for one more year at the same rate.

A not-so-ideal location, but it has everything the client wants

"You don't need all that space?" Dana asks. "You can sublet what you don't need—I'll even help you find a tenant—and then you can call it back in when you eventually need it. That way, you maintain control."

The development is part of a new mall built where there once had been a supermarket. It forms a "U," with ample parking space both within the U and behind the centre block. The three sides are numbered 1640, 1650 (the one with a vacancy), and 1660. The lower level of each side of the U is occupied by small, mostly retail, businesses: stationers, a pet store, a medical clinic, a small restaurant and bake shop, and so on. The upper level is occupied by small-to-medium-size businesses: a specialty importer, a software developer, a social services consultant, an accounting firm, a legal firm, and Provo Realty. The ceilings are false, hung from cables. ("Your client will be able to design the place to suit his specific needs," Dana says. "They'll be the first tenant.")

She says there will be no charge for parking. "We're setting aside special rows for the buildings' occupants, with other rows for visitors. We're also putting in a limited number of power plugs for the winter: $12.50 a month."

The interior has been painted only with an undercoat ("You can choose your decor," Dana explains), has large triple-pane windows ("Excellent natural light"), and central air heat and air-conditioning. The walk-up to the second floor, via a central flight of stairs, is spacious and pleasingly decorated in a light green. There are wheelchair ramps, both from the street to the sidewalk and from the sidewalk into the building, and a small elevator beside the stairwell. The building does not have a basement.

Manor Road skirts the southern edge of a residential area known as Silver Heights, which 80 years ago was the premium residential area of the city and still retains some of its early-century charm. The trees are fully grown and the boulevards are well developed with shrubs and flower beds. Manor Road has become a fast route from the city centre to a new, prestigious residential district known as Shaunessy Heights, at the edge of the city.

A heritage building, just down the street!

100 Sheridan Street. The third location you visit is the old Ashton building, which by coincidence is the same building that ManSask Insurance Company is moving into, from 220 Broadway. (They are moving into suite 702 on the seventh floor. You are shown suites 404 and 405 on the fourth floor.) You are accompanied by Laurence DeWitt of Corisand Development Corporation.

"Our development of the Ashton building is part of a long-range plan to rejuvenate the old city centre, to reverse the drift out to the suburbs. You'll be blessed by a strong, warm building—warm both physically and aesthetically—that has large rooms and high ceilings."

When you enter the building's front door, you realize the late 1800's decor has been carefully retained: there are marble floors and steps, gilt-edged "picture frame" windows, a sweeping staircase leading up to a mezzanine with a lounge and coffee shop overlooking the entrance hall.

"Elegant," you say.

"Right!" Laurence replies. "And the office decor is equally interesting."

You take an old-fashioned but completely refurbished open-cage elevator with sliding concertina-style black metal doors. You step out into a six-sided foyer in the middle of the building, with a double door set into each side. Laurence opens a solid oak door with a brass nameplate with "405" engraved on it, and you step into a "warehouse" past. The ceilings are a good 4 metres high, the walls are scrubbed *real* red brick, the pillars and beams are 30 cm square natural oak, and floors are polished hardboard planks.

"Aren't those beams a fire hazard?" you ask.

"No way! They'll only smoulder, and char slowly rather than burn. Far safer than metal beams that tend to buckle in extreme heat." Laurence reminds you that the whole building has been designated a no smoking area. (You wonder how Vern Rogers would like that: you have noticed he smokes.)

"Rent?" you ask.

"$15.50 per square metre per month. You'll have a three-year lease, so the rent is guaranteed for three years."

The price and the location are right

"How large is it?"

"This room's 410 square metres. Room 404, next door, is another 220 square metres. We'd cut a door to suit you. And if you need more space, room 505 directly above is available. We'd put in a circular staircase."

"Can we put in false ceilings?"

"No, but you can run suspended cable channels. No problem there."

"Parking?"

"A problem. There's nothing at the moment. We're converting the Carter building, diagonally across the back lane, into a car park: $30 per month summer and winter. It'll be ready in six months; until then, it's find what you can." (Later, you investigate: there are two car parks three blocks away: $40 per month.)

It's at this moment you realize the Ashton building is only two blocks from HLW's present location! Bus routes, you know, are frequent on Main Street, a three minute walk. The building has adequate wheelchair access.

Now you have a quandary: the two most promising sites are *not* in the area preferred by Vern Rogers.

You are to prepare him a formal report describing your findings (you are aware he will send a copy to HLW's head office in Calgary, for senior management approval), and to preface it with a cover letter/executive summary.

You may also, if you wish to add depth and realism to your report, research a real location in your city and include it as a fourth option. If you do, the information you present *must* be factual and provable.

Add your own dimension: a building you know about

(We recommend you start this project by creating a comparative analysis table similar to that prepared by Morley Wozniak as his attachment to the report in Figure 5-4, page 135. It will help you reach a decision and focus your writing.)

Project 6.2: Testing Highway Marking Paints

For the past six years the Department of Highways in your province has used "Centrex CL" for marking highway pavement centrelines and lanes. Recent advances in paint technology, however, have brought several new products onto the market, which their manufacturers claim are better than Centrex CL. To meet this challenge, Centrex Inc has developed a new paint ("TL") and has recommended that the Department of Highways use it in place of CL.

In a letter dated March 18 of this year, senior provincial highways engineer Morris Hordern commissioned you to carry out independent tests of the new paints. (You own a home-based consulting company known as Pro-Active Consultants Limited.)

You start your project by obtaining samples of white and yellow highway paint from six manufacturers, transferring the samples into unmarked cans and then coding the cans like this:

New technology creates new products for evaluation

	Manufacturer	Paint Coding White	Yellow
1.	Centrex Inc, Truro, Nova Scotia Paint type: CL (the "old" paint)	WA	YL
2.	Novell Paint Ltd, Guelph, Ontario Paint type: 707	WB	YM
3.	Hi-Liner Products, Winnipeg, Manitoba Paint type: HILITE	WC	YN
4.	Multiple Industries Corporation, Calgary, Alberta, Paint type: MICA	WD	YO
5.	Wishart Incorporated, Thunder Bay, Ontario, Paint type: ROADMARK 8	WE	YP
6.	Provincial Paint Company, Saskatoon, Saskatchewan, Paint Type: 81-234	WF	YQ
7.	Centrex Inc, Truro, Nova Scotia Paint type: TL (their "new" paint)	WG	YR

You then place the coding list into a sealed envelope, and lock it away in a safety deposit box at a local bank.

You decide to paint sample stripes on two regularly travelled stretches of highway and to assess the samples in four ways:

1. Spraying characteristics.
2. Drying time.
3. Visibility after three months.
4. Visibility after six months.

You assess spraying characteristics as excellent, very good, good, fair, and poor. The ratings are:

Some factors demand
personal judgment, oth-
ers are measurable

Very good: WA, WB, WC, WD, WF, YL, YR
Good: WE, WG, YM, YN, YO, YQ
Fair: YP

You assess drying time in minutes:

WA:16 WC:18 WE:14 WG:19 YM:26 YO:14 YQ:12
WB:33 WD:11 WF:13 YL:13 YN:18 YP:10 YR:15

After three months you assess visibility by day and by night. You use five drivers (one is yourself) to rate the stripes independently and to place the stripes' visibility on a scale of 1 to 10. You then average the five assessments (night readings are taken with headlights at high beam).

Paint Code	Concrete Pavement		Asphalt Pavement	
	Day	Night	Day	Night
WA	8	8	8	9
WB	7	8	7	7
WC	8	9	10	9
WD	9	9	7	8
WE	7	6	7	8
WF	8	9	9	10
WG	7	7	7	8
YL	8	9	9	9
YM	7	7	7	9
YN	7	9	7	8
YO	8	8	7	8
YP	7	8	6	8
YQ	5	6	4	6
YR	9	10	9	9

After another three months the same five drivers again assess stripe visibility, with these results:

Paint Code	Concrete Pavement		Asphalt Pavement	
	Day	Night	Day	Night
WA	6	6	5	7
WB	4	5	5	5
WC	8	8	9	9
WD	6	7	6	8
WE	6	5	6	7
WF	5	7	5	6
WG	6	7	6	7
YL	6	7	7	8
YM	6	7	6	8
YN	6	7	6	7
YO	6	7	6	8
YP	3	4	4	5
YQ	2	3	3	4
YR	8	9	8	9

You will need to create two comparison tables before writing your report

You consolidate all your results into two tables, one for white paint, one for yellow paint, and then:

- Reject any unacceptable paints (see guidelines below).
- Rank acceptable paints in order of suitability.
- Identify the best paint(s) to use for highway marking.
- Retrieve the paint coding list from the bank deposit box.
- Write your report.

Some factors you use to conduct your study and to write your report are:

1. Provincial senior highways engineer Morris Hordern's office address is 416 Inkster Building, 2035 Perimeter Road of your city.
2. The paint stripes were painted on two stretches of highway:

    2.1 Highway 101 (concrete surface), 2.4 km north of the intersection with Highway 216.

    2.2 Highway 216 (asphalt surface), 0.8 km west of the intersection with Highway 101.

3. You are unable to borrow the regular highway paint stripe applicator from the Department of Highways. Instead, you mount an applicator on a small garden tractor. The paint stripes are applied at night, between midnight and 6:00 a.m.
4. Paint Manufacturers' Association specification PMA-28H states that spraying characteristics for fast-drying highway paint should be at least "Good," and preferably "Very Good." To achieve

"Very Good" the paint must flow smoothly and evenly without forming globules or dripping from the nozzle.

5. You refer to specification ASTM D-711 to establish the maximum acceptable paint drying time, which is 20 minutes.

These factors help establish acceptability criteria

6. Guidelines you give to the drivers assessing paint visibility are:

	Distance visible	
Rating	Day	Night
10	500 m	200 m
8	400 m	160 m
6	300 m	120 m
4	200 m	80 m
2	100 m	40 m

You then average the five assessments.

7. You establish minimum acceptable visibility levels for the paints to be:

After three months' traffic wear: 7

After six months' traffic wear: 6

Note: Calculate real dates for each stage of the study and quote them in your report.

Your report should not only present the results of your tests, but also analyse them, draw conclusions, and make a recommendation.

Project 6.3: Researching Computers

The Association of Small Business Operators (ASBO) is a major society with headquarters in Chicago, Illinois. Across Canada and the US there are individual branches of the Association, each of which operates independently. The executive director of the ASBO branch in your province is Milton Lajzieks, and the branch's office is in Suite 460 of the Dorfmann Building at 245 King Street of the province's principal city.

Two days ago you had a telephone call from Milton Lajzieks. "Do you have time to carry out a study for us?" he asked.

You said you could find the time.

As an independent consultant, you need an occasional telephone call like this

"Good!" he continued. "Our members need someone who is *not* a computer salesperson to tell them what kind of computer they should buy. Not by brand name, but to focus on the features they should be looking for." He explained that the ASBO members in your province are confused: salespeople tout their products to the exclusion of others, and only the occasional salesperson seems willing or able to discuss which is the best type of computer for the individual's needs. "I want you to write a report that I can copy and send to all members, so they can go to a supplier and ask educated questions."

Milton asked what your fee would be (your hourly rate) and he accepted your proposal. Then he said he would write you a letter authorizing you to carry out the project. The letter arrives today:

Dear *(your name)*

I am confirming that the Association of Small Business Operators (ASBO) is engaging your company to investigate personal computers and identify what type of computer will best suit ASBO members' needs. Our members own a range of businesses, ranging from firms that provide "trade" services (i.e. electricians, pipefitters, carpenters, stonemasons) to people who offer personal services (i.e. accountants, hairdressers, day care operators, even engineering consultants). Some operate independently as single-operator business owners, while others employ up to 6 or 8 people.

Here are 11 questions that ASBO members asked most frequently, in their response to a questionnaire I sent out earlier this year:

You also need to ask questions, to make sure you have all the information

1. Is it better to have a desktop computer or a portable computer?

2. Why is a portable computer more expensive than a desktop computer, even though both may have the same memory, speed, and peripherals?

3. How much memory will I need in my hard drive? What about RAM?

4. Do I need a 3.5-inch disk drive or a CD-ROM drive? Or both?

5. What provision should there be for accommodating future upgrades?

6. Which kind of modem is better: external or internal? What speed do I need in a modem?

7. I'm a consultant who travels fairly frequently. Do I need both a desktop *and* a portable computer? Or is it safe to have my whole office in a portable computer and carry it with me? What are the risks? Do airport security systems pose a danger, maybe erase information?

8. How long can I expect my computer to last? (By which I mean, how soon will it become too slow, or impossible to upgrade?)

9. What new technologies are on the horizon, that I don't know about yet but should be aware of?

10. What software should I look for, either bundled (built in) or as an add-on?

11. Where can I find training locally that will help me use my computer more efficiently?

Please provide me with a formal report which I can mail out to members.

Sincerely

Milton Lajzieks

You are going to have to do some personal research to complete this project. Your company name is Pro-Active Consultants Limited, and your office is in your home. (If you prefer, you can create your own company name.) In fact, you realize now that you are a small business operator, so after completing the study you plan to join ASBO.

Project 6.4: Correcting a Noise Problem

This morning you receive a letter from Trudy Parsenon, the area manager of Mirabel Realty. (You are the owner/manager of Pro-Active Consultants Limited, which you operate from your home.)

Dear (you)

As I mentioned when I telephoned last week, my staff have been complaining for the past three months that the noise level in our office has been too high. They claim it is affecting their work and causing fatigue. I have noticed, too, that staff turnover has been higher lately.

Will you please look into the problem for me to determine whether their complaints are justified. If they are, will you suggest what can be done to remedy the problem, recommend the most suitable method, and include a cost estimate.

Sincerely

Trudy G Parsenon

Area Manager, Mirabel Realty

A letter confirming a telephone request

Part 1

At 4:00 p.m. the same day you visit Mirabel Realty (the office is in room 210, on the second floor of the Fermore Building at 381 Conway Avenue of your city). You notice a background hum, which you consider to be caused by motors in the electric typewriters, computers, and printers. You are still there when the office staff quits at 4:30. After they go, you notice you can still hear the hum, but at a lower level.

You walk around the office with Trudy, who plagues you with questions. "What do you think?" she asks. "Seems like the same noise level you get in any business office, don't you think?"

You suspect she is hoping for a good report from you, which she can use to prove to her staff that their complaints are imaginary.

"I can't tell you without taking readings," you hedge. "Noise is a pretty tricky thing. What some people think is too noisy, others hardly notice."

But you do notice that the hum gets significantly louder near the north wall of the office. Then suddenly it stops; or rather, dies away. The time is 4:45.

An initial visit shows there may well be a problem

"What's on the other side of this wall?" you ask.

"Oh, that's Superior Giftware," Trudy replies. "They distribute cheap imports—that sort of thing."

"Have you talked to them about the hum?"

"Yes. I asked Saul Ferguson about it—he's the manager next door. Pretty hostile, he was."

"And what time do they quit work?" you ask.

"Right now," Trudy replies. "You can always tell, because they switch their machines off."

You arrange with Trudy that you will take sound-level measurements one week from today. You want to find out how much of the noise in the realty office is generated by normal office activity and how much by the company next door.

You consider that a visit to Superior Giftware is essential, since you suspect that the machines Trudy mentioned may be the problem. You want to know the sound levels on both sides of the wall between the two companies, to assess the extent of soundproofing you may want to recommend.

Write to Saul Ferguson, manager of Superior Giftware, to ask permission to carry out sound-level measurements in his offices one week from today. His business is in room 208, 381 Conway Avenue.

Part 2

It is now one week later. You take a Nabuchi Model 1300 sound-level meter with you and return to 381 Conway Avenue. You plan to measure sound levels at various locations in the Mirabel Realty office under four conditions:

- When both businesses are empty.
- When only Superior Giftware is working (4:30–4:45).
- When only Mirabel Realty is working (8:00–8:15).
- When both business are working.

You also plan to take readings in Superior Giftware's office.

Since Mr Ferguson has not replied to your letter, yesterday afternoon you telephoned him to ask if you could come in today to take the mea-

Table 6-4 Average sound levels second floor, 381 Conway Avenue, Montrose, Alberta.

Location	Both Offices Working (dB)	Only Superior Working (dB)	Only Mirabel Working (dB)	No One* Working (dB)
Mirabel Realty				
A	74	73	48	27
B	71	69	51	27
C	66	65	50	27
D	64	61	52	26
E	63	59	51	28
F	59	53	49	26
G	54	49	44	28
Superior Giftware				
H	86	—	—	25
I	83	—	—	26

*Mostly air-conditioner noise.
Note: Measurements made with Nabuchi Model 1300 Sound-Level Meter set to "A" scale.

surements. He said he was "terribly busy" and that it was "damned inconvenient," but he somewhat reluctantly agreed. (It was worthwhile being persistent. When you visit Superior Giftware, almost right away you notice a packaging and sealing machine only 2.5 metres from the wall separating the two business offices.)

You record the measurements you take (see Table 6-4) and compare them to the general ratings for office noise, which you obtain from City of Montrose standard SL2020, dated January 20, 1995. The recommended sound levels for an urban office are:

Quiet office: 30–40 dB
Average office: 40–55 dB
Noisy office: 55–75 dB

You note that the sound level in Mirabel Realty's office increases as you move toward the dividing wall between the two offices (see Figure 6-12).

You also notice there seem to be two components of noise in Mirabel Realty's office, some being transmitted through the air and some being transmitted through the structure (from Superior Giftware's machines,

You suspect the next door neighbor is being defensive

through the floor). A hand placed on the walls or floor feels the vibration. Floors in both offices are tiled.

Before leaving you tell Trudy Parsenon there seems to be a noise problem, but it can be corrected. You warn her, however, that it may prove expensive. She says it will be difficult to justify the costs to her head office.

On returning to your office you summarize your findings in a brief progress report, which you mail to Trudy Parsenon.

You prepare your client for news she does not want to hear

Part 3.
You consider possible ways to reduce the sound levels in Mirabel Realty's office:

1. You could erect a false wall, insulated internally with Corrugon, from floor to ceiling on Mirabel Realty's side of the wall between the two companies.
2. Black cork panels, 18 mm thick, could be glued on the Mirabel Realty side of the wall.
3. You could install carpeting throughout Mirabel Realty's office.
4. Superior Giftware's machine could be mounted on Vib-o-Rug (insulating rubber that eliminates transmission of vibration from machine to building structure).

You recognize that remedies 1 and 2 are alternatives (they both deal with sound transmitted through the air). Remedies 3 and 4 also are alternatives (they both dampen vibrations and sound carried through the structure). Remedy 3 also quite effectively dampens internal office noise.

You consider the approximate costs:

Remedy 1 – $8700
Remedy 2 – $1600
Remedy 3 – $14 900
Remedy 4 – $950

You consider possible problems each remedy may present:

You have several options, some of which can be used in tandem

1. Corrugon is in short supply; delivery time would be a minimum of 3 months.
2. To some people, cork has an offensive smell; this can be partly corrected by treating the cork with polymethynol.
3. The carpet must be dense and have a good quality rubber underlay (included in the approximate cost above).
4. Depends on cooperation of Superior Giftware's manager.

You calculate probable noise reductions for each method:

Remedy 1 – 6 to 10 dB
Remedy 2 – 4 to 7 dB

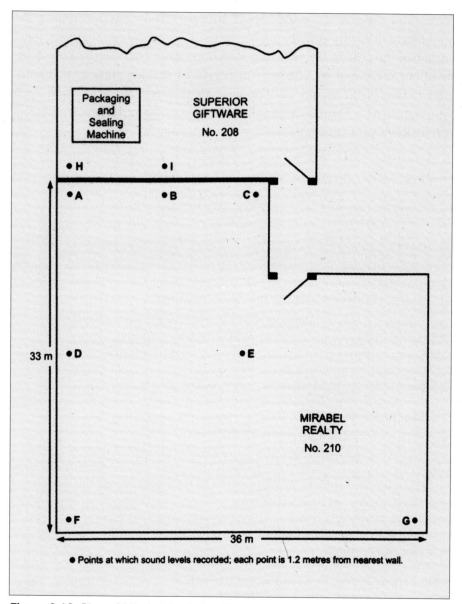

The packaging machine is the problem, but difficult to remedy at source

Figure 6-12 Plan of Mirabel Realty's office.

Remedy 3 – 8 to 12 dB
Remedy 4 – 3 to 5 dB

Unfortunately, predicted sound reductions cannot be added together

(These anticipated reductions apply only to the Mirabel Realty office, when both businesses are working.)

You consider which alternatives to recommend to Mirabel Realty, and then write an investigation report describing your findings and suggested corrective measures. You also write a brief cover letter to Trudy Parsenon summarizing your findings.

Note: You prepare a *formal* report because Trudy has mentioned she might have difficulty convincing her head office executives that they must authorize the cost of the remedial measures. And, because she knows little about noise and its effects, you decide to include some explanatory information. (You would also be wise to research and document such information at a library, to establish positive evidence for the statements you make in your report.)

WEBLINKS

Technical Reports
www.io.com/~hcexres/tcm1603/acchtml/techreps.html

This document is one chapter from the online textbook used in Austin Community College's online course, Online Technical Writing (www.io.com/~hcexres/tcm1603/acchtml/acctoc.html). It describes types of technical reports and their general characteristics and audience, and provides a checklist that can be used when writing technical reports.

Technical Report Writing
www.lerc.nasa.gov/WWW/STI/editing/vidoli.htm

Scientists at NASA's Lewis Research Center must write reports that are both technically correct and easy to read. This NASA guide was written to make writing reports easier. Separate chapters deal with the stages of report preparation, report style, the introduction, experiment and analysis descriptions, results and discussions, concluding and supporting sections, reviewing reports, and references. An author's checklist and reporting aids provide quick guidelines for technical report writers.

References and Notes for Tech Writing Students
darkstar.engr.wisc.edu/zwickel/397/397refs.html

The numerous documents from this University of Wisconsin technical writing handbook cover many aspects of technical writing, including abstracts; executive summaries; footnotes and fractions; heads, subheads, and tables of content; jargon; proposal writing; presentation graphics; job search material; editing and revising; and many other related topics.

Bibliography Styles Handbook
www.english.uiuc.edu/cws/wworkshop/bibliostyles.htm

The Bibliography Styles Handbook, one section of the University of Illinois's Writers' Workshop, provides information about the bibliographic styles of the American Psychological Association (APA) and the Modern Languages Association (MLA), and about the old MLA style.

Chapter 7
Other Technical Documents

This chapter describes how to write a user's manual, provides detailed guidelines for writing a technical instruction, offers suggestions for writing a scientific paper, and describes how to convert your knowledge of a process, equipment, or new technique into an interesting magazine article or technical paper.

User's Manual

In our technological era, with its increasingly complex range of hardware and software, there is a growing need for manufacturers to write clear technical manuals to accompany what they sell. Most manufacturers issue a user's manual with each of their products, which contains (1) a brief description of the product, (2) instructions on how to use it, and (3) suggestions for remedying problems that may occur. For qualified repair specialists they may also produce a set of maintenance instructions containing detailed service and repair procedures. Both publications perform the same task for different readers: user manuals assume the reader has only slight technical knowledge, while maintenance instructions assume the reader is a technical expert.

The suggestions that follow apply to any basic user's manual. Whether you are writing a manual to accompany heavy construction equipment, a delicate instrument, or a new version of a software program, you must organize your description so that it follows a coherent pattern.

Identify the Audience

The problem with many user's guides is that they are written from an engineer's or a technical person's point of view and often are too complex or incomplete from the user's point of view. To avoid this you need to identify your audience before you start writing, and understand what level of knowledge and experience they have with the product. The language you use must be appropriate for the intended audience: be careful not to use jargon they are not familiar with.

The first step in *any* writing situation

We often suggest to engineers that they write a brief description of who the user is, so that as they write they can refer back to it and keep the focus of their writing on that audience. This audience analysis often becomes the first section in the user's guide, where it's called "Who Should Read This Document."

Writing Plan

The writing plan for most user manuals has four compartments, as shown in Figure 7-1. The two top compartments describe the product, while the two lower compartments tell the reader how to use it.

The following sections demonstrate how these four compartments are used. The product in this instance is electronic mail software.

Describing the Product

The **Summary Statement** briefly describes the product and its main purpose:

> This electronic mail software allows you to communicate with other email users by sending and receiving messages. It allows you to connect to a remote computer, called a server, and access messages other people have sent to you. The server computer will also send any messages you have written to other people.

Online documentation is preferred for most software programs

The **Product Description** identifies each part and describes its components. For example, a user manual accompanying a word-processing program would describe the program's contents and application. For

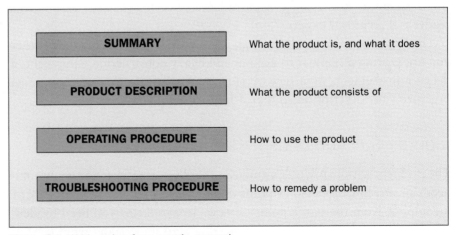

SUMMARY	What the product is, and what it does
PRODUCT DESCRIPTION	What the product consists of
OPERATING PROCEDURE	How to use the product
TROUBLESHOOTING PROCEDURE	How to remedy a problem

Figure 7-1 Writing plan for a user's manual.

equipment that has several discrete components or parts, the Product Description may be subdivided into two sections: Main Parts and Detailed Description.

- The **Main Parts** section simply lists the main components:

  The electronic mail program contains four components:
  - The In Tray
  - The Out Tray
  - The File Cabinet
  - The Address Book

  List the components in the same sequence you will describe them

  If there are three or less points, they can be presented as one paragraph.

- The **Detailed Description** provides a lot more information about each part, and in particular draws attention to items the user will operate or use. The parts must be described *in the same sequence* that they have been presented in the Main parts section:

  The In Tray is identified by a square icon with an arrow pointing down. This indicates that messages are directed to you. Any messages that are sent to you are automatically placed in this area. You can view your messages in the In Tray by pointing your mouse on the icon and clicking.

  The Out Tray is identified by a square icon with an arrow pointing up. This indicates that messages are from you to someone else. This is the area where messages you have written are stored until you are ready to send them.

  The File Cabinet is identified by an icon that looks like a traditional two or three drawer filing cabinet you might find in an office. This is the area where you can store or file messages that are important or that you want to keep. You can create different folders for different situations and thus create a filing system for your messages.

  Clear, simple language helps *all* readers understand

  The Address Book is identified by an icon that looks like a small book. This is the area where you store the electronic addresses of people you frequently send messages. When you write a message you can select the address of the person or persons you are sending it to directly from the Address Book.

There must be a logical pattern to the Detailed Description, which may be presented in either a spatial ("arranged in space") or a sequential arrangement. In a spatial arrangement, the sequence depends on the shape of the equipment or product:

	Vertical Subjects	Tall, narrow subjects lend themselves to a vertical description, each item being described in order from top to bottom, or bottom to top. Vertically arranged control panels and electronic equipment racks fall into this category.
	Horizontal Subjects	Long, relatively flat subjects demand a horizontal description, with each part being described in order from left to right, or right to left.
The physical shape or arrangement of controls dictates the writing pattern	**Circular Subjects**	Round subjects, such as a clock, a revolution counter, a pilot's altitude indicator, or a circular slide rule, suit a circular arrangement, with the description of markings starting at a specific point (e.g. "12" on a clock) and continuing clockwise until the entire circle has been described. In effect, this is the same as the horizontal description, if you consider that the circle can be broken at one point and unwound into a straight line. Alternatively, if the subject has a series of items that are arranged more or less concentrically (such as the scales of a circular slide rule), they can be described in sequence starting at the centre and working outwards, or from the outermost item inwards.
	Conceptual Subjects	Conceptual or intangible subjects like a software interface can be described in any of the methods explained so far. You can start at the top of the screen and move down to the bottom, or you can move from left to right. You could also describe an interface in a circular fashion. The important thing is to pick a pattern that seems logical and make sure the reader can follow you.

A sequential arrangement introduces the parts or items in the sequence in which the reader will use them. In this arrangement the description may end up sounding like one long procedure. Here's a Detailed Description written in a sequential arrangement for an LCD computer projection panel:

A technical description tells what is done, not what to do

The overhead projector is set up the recommended distance from the screen so the audience can see the projected image. The LCD panel is placed on the glass plate on top of the overhead projector. The portable computer is connected to the LCD panel by plugging the cable into the serial port on the back of the computer. The other end of the cable is connected to the LCD panel in the port labelled

COMPUTER. When the computer is turned on, the images displayed on the computer screen are projected through the LCD panel onto the large screen, so the entire audience can view them.

Using the Product

The **Operating Procedures** provide step-by-step instructions for each task that is possible with the product. One of the major problems with many user manuals is that they are not written from the user's perspective: instead of describing *how* to do something with the product, the manual describes *what* can be done with it. If you follow these three steps, your User's Manual will be user-oriented:

1. Perform a Task Analysis.
2. Group and label the tasks.
3. Write the steps for each task.

Step 1—Perform a Task Analysis

When you have decided who your audience is, you need to list everything *they might want to do* with the product. Make sure you don't list everything *the product can do*. Instead, focus on the tasks the user will perform. For example, a task analysis for a simple electronic mailing package might look like this:

Tasks:

Sending messages
Receiving messages
Forwarding messages
Addressing messages
Printing messages
Adding names to an address book
Storing messages in folders
Creating folders
Downloading messages
Assigning a password
Deleting messages
Creating messages
Connecting to the server
Checking spelling
Attaching documents
Installing
Customizing
Calling manufacturer for support
Using the Help system

When writing a task analysis, use verbs that end with "...ing" (*sending, receiving*, etc)

This list is developed while brainstorming and is not meant to be in any particular order. Later, you can add items; for now, just list the tasks.

Step 2—Group and Label Tasks

Looking at the above task list, you can see that certain items are related. For example, Receiving messages and Forwarding messages are related, just as Creating messages, Addressing messages and Checking spelling form a second group of related tasks or topics. We suggest grouping the related topics and giving them a letter to identify each group. For example, in the following revised list, each task has been assigned a letter (A, B, C, etc) to show which group it belongs to.

Tasks:

Sending messages	F
Receiving messages	A
Forwarding messages	A
Addressing messages	B
Printing messages	C
Adding names to an address book	B
Storing messages in folders	A
Creating folders	A
Downloading messages	A
Assigning a password	D
Deleting messages	A
Creating messages	B
Connecting to the server	F/A
Checking spelling	B
Attaching documents	F
Installing	D
Customizing	D
Calling manufacturer for support	E
Using the Help system	E

So far, no organization has been done with the list of topics. The first organizational step is to assign a label to each group. Choose a label that *describes what the user is doing* or is trying to accomplish with the tasks. Use "ing" words whenever you can in your labels, because they *indicate an action performed by the user.* Here are labels for the above groups:

A	–	Handling Incoming Messages
B	–	Writing Messages
C	–	Printing Messages
D	–	Getting Started
E	–	Getting Additional Help
F	–	Sending Messages

Identify which tasks seem related

This becomes the first step toward organizing the information

238   Chapter 7

Now is the time to organize the groups of topics into a logical sequence for the intended audience. User manuals are usually structured in a sequential arrangement listing what needs to be done first, or by introducing easy tasks first. In this example the structure might look like this:

An outline emerges naturally, almost painlessly...

Getting Started *(D)*
  Installing Your Software
  Customizing Your Software
  Assigning a Password to the System

Writing Messages *(B)*
  Creating Messages
  Addressing Messages
  Adding Names to an Address Book

Sending Messages *(F)*
  Sending Messages You Have Written
  Connecting to the Server
  Attaching Documents

Handling Incoming Messages *(A)*
  Connecting to the Server
  Receiving Messages
  Forwarding Messages
  Creating Folders
  Storing Messages in the Folders
  Downloading Messages
  Deleting Messages

Printing Messages *(C)*

Getting Additional Help *(E)*
  Using the Help System
  Calling the Manufacturer for Support

We now have a user-focused, task-oriented structure that describes how to use the product. This process is similar to the stages described in Chapter 1. Figure 1-3 shows how to plan the writing task.

Step 3—Write the Steps

Each of the identified tasks now becomes a procedure, and you can write the steps it takes to accomplish each task. Here are two examples from the electronic mail tasks:

...from which a logically flowing procedure can be written

Installing Your Software
1. Unpack the contents of the box and make sure you have
    • this manual

- six 3.5-inch disks
- the software license.

2. Turn your computer on and start Windows.
3. Put the disk labelled Disk 1 into your A drive.
4. Click on the Windows Start menu .
5. Select Run...
6. Type **A:INSTALL**
7. Follow the directions on the screen.

Connecting to the Server
1. Click on the File menu.
2. Choose the *Connect to Server* command.
3. When a dialog box appears, enter your ID and password.
4. Click OK.
5. Wait while the system dials the telephone number configured in the Server Settings dialog box.
6. When the message **You are Now Connected** appears on screen, you have successfully connected to the remote server.

You can now send and receive your electronic mail messages.

Note that each step is short, has a number, and uses verbs in the imperative mood. (For more information on writing instructional steps, see "Give Your Reader Confidence" on page 243 of this chapter.)

The **Troubleshooting Procedure** tells the reader what to do if having followed the Operating Procedure correctly, the equipment does not work. It also has short, numbered steps and verbs in the imperative mood.

When all else fails, call for help

If the message **Server Connection Failed** appears, follow this procedure:
1. Click on the Settings Menu.
2. Choose Server Settings... command
3. Make sure your ID, password, and modem are correctly identified.
4. Click OK.
5. Try connecting to the server again.
6. If the problem continues, call the manufacturer for technical support.

Technical Instruction

When H L Winman and Associates' special project engineer, Andy Rittman, wants a job done, he issues instructions in clear, concise terms: "Take your crew over to the east end of the bridge and lay down control points 3, 4, and 7," he may say to the survey crew chief. If he fails to make himself clear, the crew chief has only to walk back across the bridge to ask questions. But Fred Stokes, chief engineer at Macro Engineering Inc, seldom gives spoken instructions to his electrical crews. Most of the time they work at remote sites and follow printed instructions, with no opportunity to walk across a project site to clarify an ambiguous order.

A technical instruction tells somebody to do something. It may be a simple one-sentence statement that defines what has to be done but leaves the time and the method to the reader. Or it may be a step-by-step procedure that describes exactly what has to be done and tells them when and how. It is the latter type of technical instruction that will be described here.

Before attempting to write an instruction, you must first define your readers, or establish their level of technical knowledge and familiarity with your subject. Only then can you decide the depth of detail you must provide. If they are familiar with a piece of equipment, you may assume that the simple statement "Open the cover plate" will not pose a problem. But if the equipment is new to them, you may have to broaden the statement to help them first identify and open the cover plate:

An instruction *must* be written from the reader's point of view

> Find the hinged cover plate at the bottom rear of the cabinet. Open it by inserting a Robertson No. 2 screwdriver into the narrow slot just above the hinge and then rotating the screwdriver half a turn counterclockwise.

Start with a Plan

A clearly written instruction contains four main compartments, as shown in Figure 7-2. These compartments contain the following information:

- A **Summary Statement** outlines briefly what has to be done.

  > The 28 Vancourt Model AL-8 overhead projectors in rooms A4 and A32 are to be bolted to their projection tables...

- The **Purpose** explains why the work is necessary:

  > ...to reduce the current high damage rate caused by projectors being accidentally knocked onto the floor.

  (A technician who understands *why* a job is necessary will much more readily follow an instruction.)

- A short paragraph or list describes the **Tools and Materials** that technicians will need to perform the task (they can use this as a checklist to ensure they have accumulated everything they need before they start work).

To carry out the modification you will require:

- Modification kit OHP4, comprising
    1 template, OHP4-1
    4 bolts, flat head, 50 mm long, 3 mm dia
    4 washers, 25 mm dia, with 4 mm dia central hole
- A 6 mm drill with a 3.5 mm drill bit
- A Phillips No. 2 screwdriver
- A slot-head No. 3 screwdriver
- A sharp pencil

- The Steps that readers must follow take them through the whole process.

    Proceed as follows:

    1. Disconnect the projector's power cord from the wall socket, then take the projector to a table and turn it on its side.
    2. Using a slot-head No. 3 screwdriver, unscrew the four bolts that hold the feet onto the base of the projector. Remove but retain the bolts and feet for future use.
    3. Place template OHP4-1 onto the projection table and position it where the projector is to stand. Using a sharp pencil, mark the table through each of the four holes in the template.
    4. Drill four 3.5 mm dia holes through the table top, at the places marked in step 3.

    5. Place the overhead projector, right side up and with the lens assembly facing the screen, so that the four screwholes identified in step 2 coincide with the four holes in the table top.
    6. From beneath the table, place a washer under each hole and insert a 50 mm flat head bolt up through the washer and hole until it engages the corresponding hole in the projector base.

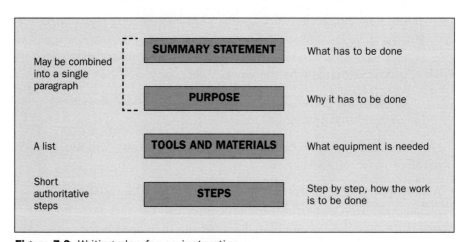

Figure 7-2 Writing plan for an instruction.

Tighten the four bolts in place, using a Phillips No. 2 screwdriver.

The writing plan embodying these four compartments, when combined with the suggestions below, will consistently ensure that any instructions you write will be clear, direct and convincing.

Give Your Reader Confidence

A well-written technical instruction automatically instills confidence in its readers. They feel they have the ability to do the work even though it may be new to them and highly complex. Consider these examples:

Vague	Before the trap is set, it is a good idea to place a small piece of cheese on the bait pan. If it is too small it may fall off and if it is too big it might not fit under the serrated edge, so make sure you get the right size.
Clear and Concise	Cut a 0.62 m length of 10-gauge wire and strip 20 mm of insulation from each end. Solder one end of the wire to terminal 7 and the other end to pin 49.

An instruction is not the place for weak, wishy-washy words

The first excerpt is much too ambiguous. It only suggests what should be done, it hints where it should instruct (almost inviting readers to nip their fingers), and in 31 explanatory words it fails to define the size of a "small" piece of cheese. The second excerpt is assertive and keeps strictly to the point. The verbs *cut*, *strip*, and *solder* make readers feel they have no alternative but to follow the instructions. Such clear and authoritative writing immediately convinces them of the accuracy and validity of the steps they have to perform.

The best way to be authoritative is to write in the imperative mood. This means beginning each step with a strong verb, so that your instructions are commands:

Ignite the mixture...	*Connect* the green wire...
Mount the transit on its tripod...	*Excavate* 1.4 metres down...
Apply the voltage to...	*Measure* the current at...
Cut a 40 mm wide strip of...	*Insert* the PCMCIA card...

The imperative mood in the clear, concise excerpt quoted above keeps the instruction taut and definite. The vague excerpt would have been equally effective (and much shorter) if it also had been written in the imperative mood:

Before setting the trap, wedge an 8 mm cube of cheese firmly under the serrated edge of the bait pan.

(Note how the vague word "place" has been replaced by the image-conveying verb-adverb combination "wedge…firmly.")

The following two statements clearly show the difference between an instruction written in the imperative mood and one that is not:

Make each step a command, not a broad statement of intent

A. Disengage the gear, then start the engine.
   (*Definite: uses strong verbs*)

B. The gear should be disengaged before starting the engine.
   (*Indefinite: uses weaker verbs*)

Statement A is strong because it *tells readers to do something*. Sentence B is weak because it neither instructs nor insists that anything need be done ("should" implies it is only *preferable* that the gear be disengaged before the engine is started).

In the imperative mood, the first word in a sentence almost always is a strong verb:

Position the pointer on File, then select Print.

Sometimes, however, the verb may be preceded by an introductory or conditional clause:

Before connecting the meter to the power source, *set* all the switches to "zero."

The imperative mood is maintained here because the main verb starts the statement's primary clause (the clause that describes the action to be taken).

If you want to check whether a sentence you have written is in the imperative mood, ask yourself whether it *tells* the reader to *do* something. If it does, then you have written an *instruction*.

Avoid Ambiguity

There is no room for ambiguity in technical instructions. You have to assume that the person following your instructions cannot ask questions, and so you must never write anything that could be interpreted more than one way. This statement is open to misinterpretation:

Align the trace so that it is inclined approximately 30° to the horizontal.

Write specific details; never generalize

Each technician will align the trace with a different degree of accuracy, depending on his or her interpretation of "approximately." How accurate does "approximately" require the technician to be? Within 5°? Within 2°? Within 1/2°? Maybe even 10° either side of 30° is acceptable, but the reader does not know this and is left feeling doubtful. Worse still, the reader's confidence in the technical validity of the whole instruction is undermined. Replace such vague references with clearly stated tolerances:

Align the trace so that it is inclined 30° (±5°) to the horizontal.

More subtle, but equally open to misinterpretation, is this statement:

Adjust the capstan handle until the rotating head is close to the base.

Here the offending word is "close," which needs to be replaced by a specific distance:

Adjust the capstan handle until the distance between the rotating head and the base is 2.5 mm.

Similarly, replace vague references such as "*relatively* high," "*near* the top," and "*an adequate* supply" with clearly stated measurements, tolerances, and quantities.

Never leave a statement open to misinterpretation

Avoid weak words such as "should," "could," "would," "might," and "may," because they weaken the authority of an instruction and reduce the reader's confidence in the writer. For example:

Set the meter to the +300 V range. The needle should indicate 120 V (±2 V).

Here "should" implies that it would be nice if the needle indicated within 2 V of 120 V, but it is not essential! No doubt the writer meant it *must* read the specified voltage, but has failed to say so. Neither has the reader been told to note the reading. The writer has forgotten the cardinal rule of instruction writing: *Tell* the reader to *do* something. To be authoritative, the instruction needs very few changes:

Set the meter to the +300 V range, then check that its needle indicates 120 V (±2 V).

Notice how the steps in the sample instructions in Figure 7-3 are clear, concise, and definite. You need not be a specialist in the subject to recognize that they would be easy to follow.

Write Bite-Size Steps

Technicians working on complex equipment in cramped conditions need easy-to-follow instructions. You can help them by writing short paragraphs, each containing only one main step. If a step is complicated and its paragraph grows unwieldy, divide it into a major step and a series of substeps, numbering the paragraphs and subparagraphs:

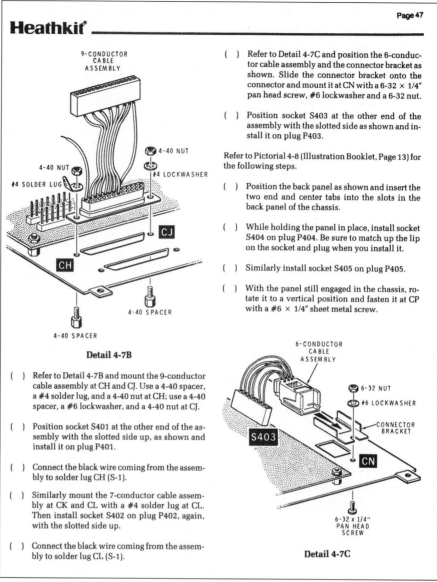

9-CONDUCTOR
CABLE
ASSEMBLY

4-40 NUT

4-40 NUT

#4 LOCKWASHER

#4 SOLDER LUG

CJ

CH

4-40 SPACER

4-40 SPACER

Detail 4-7B

( ) Refer to Detail 4-7B and mount the 9-conductor
cable assembly at CH and CJ. Use a 4-40 spacer,
a #4 solder lug, and a 4-40 nut at CH; use a 4-40
spacer, a #6 lockwasher, and a 4-40 nut at CJ.

( ) Position socket S401 at the other end of the assembly with the slotted side up, as shown and
install it on plug P401.

( ) Connect the black wire coming from the assembly to solder lug CH (S-1).

( ) Similarly mount the 7-conductor cable assembly at CK and CL with a #4 solder lug at CL.
Then install socket S402 on plug P402, again,
with the slotted side up.

( ) Connect the black wire coming from the assembly to solder lug CL (S-1).

( ) Refer to Detail 4-7C and position the 6-conduc-
tor cable assembly and the connector bracket as
shown. Slide the connector bracket onto the
connector and mount it at CN with a 6-32 × 1/4″
pan head screw, #6 lockwasher and a 6-32 nut.

( ) Position socket S403 at the other end of the
assembly with the slotted side as shown and install it on plug P403.

Refer to Pictorial 4-8 (Illustration Booklet, Page 13) for
the following steps.

( ) Position the back panel as shown and insert the
two end and center tabs into the slots in the
back panel of the chassis.

( ) While holding the panel in place, install socket
S404 on plug P404. Be sure to match up the lip
on the socket and plug when you install it.

( ) Similarly install socket S405 on plug P405.

( ) With the panel still engaged in the chassis, rotate it to a vertical position and fasten it at CP
with a #6 × 1/4″ sheet metal screw.

6-CONDUCTOR
CABLE
ASSEMBLY

6-32 NUT

#6 LOCKWASHER

CONNECTOR
BRACKET

S403

CN

6-32 × 1/4″
PAN HEAD
SCREW

Detail 4-7C

Figure 7-3 Excerpts from an instruction manual. (Courtesy the Heath Company,
Mississauga, Ont.) As each step is completed, the user inserts a check mark
beside the appropriate paragraph.

3. List the documentary evidence in block J of Form 658. Check that
    blocks A to G have been completed correctly, then sign the form
    and distribute copies as follows:

Offer spoon-size helpings of information

3.1 Attach the documentary evidence to Copies 1 and 2 and mail them to the Chief Recording Clerk, Room 217, Civic Centre, Thompson, Manitoba.

3.2 Mail Copy 3 to the Computer Data Centre, using one of the special preaddressed envelopes.

3.3 File Copy 4 in the "Hold—Pending Receipt" file.

3.4 When Copy 2 is returned by the Chief Recording Clerk, attach it to Copy 4 and file them both in the "Action Complete" file.

Insert Fail-Safe Precautions

Insert precautionary comments into instructions whenever you need to warn readers of dangerous conditions, or of damage that may occur if they do not exercise care. There are two precautionary notices you can use:

Warning: To alert readers of an element of personal danger (such as unprotected high voltage terminals).

Caution: To tell readers when care is needed to prevent equipment damage.

Draw attention to a precautionary comment by placing it in a box in the middle of the text, indenting the box from both margins. Precede the cautionary note with the single word WARNING or CAUTION.

> ### WARNING
>
> ### Disconnect the power source before removing the cover plate.

Use word-processing technology to design a simple but noticeable warning

Ensure that every precautionary comment *precedes* the step to which it refers. This will prevent an absorbed reader who concentrates on only one step at a time from acting before reading the warning. Never assume that mechanical devices, such as indentation and the box drawn around the precautionary note, are enough to catch the reader's attention.

Use warnings and cautions sparingly. A single warning will catch a reader's attention. Too many will cause a reader to treat them all as comments rather than as important protective devices.

Insist on an Operational Check

Usability testing is an essential part of instruction writing...

The final test for any technical instruction is the reader's ease in following it. Since you cannot always peer over a reader's shoulder to correct mistakes, you should find out whether users are likely to run into difficulty *before* you send an instruction out. To obtain an objective check, give the instruction to someone of roughly equal competence to the people who eventually will be using it, and observe how well that person performs the task.

...but, sadly, it's often overlooked in the rush to get a product out

Note every time the user hesitates or has difficulty, and then, when he or she has completed the task, ask if any parts need clarification. Rewrite ambiguous steps and then recheck your instruction with another person. Repeat these steps until you are confident your readers will be able to follow your instruction easily.

Scientific Paper

Earlier chapters described how technical reports should be planned, organized, written, and presented by engineers and computer specialists working for industry, business, and government. Within this context one additional report remains to be described: the research report prepared by scientists and technologists working in industrial and university laboratories. Research reports are most often prepared and published as *scientific papers* that, although their parts are similar to those of an investigation report, differ in style, organization, and emphasis.

A scientific paper either identifies and attempts to resolve a scientific problem, or it tests (validates) a scientific theory. It does so by describing the four main stages of the research:

1. Identifying the problem or theory.
2. Setting up and performing the tests.
3. Tabling the test results (the findings).
4. Analysing and interpreting the findings.

A scientific paper is like a highly professional lab report

These four stages represent the major divisions of a scientific paper, with each stage preceded by a descriptive heading: **Introduction, Materials and Methods, Results,** and **Discussion.** These stages are similar to those used for the laboratory report in Chapter 4 and the investigation report in Chapter 5. There are, however, differences in a scientific paper's appearance and writing style.

Appearance

A scientific paper straddles the borderline between semiformal and formal presentation. Normally the title is centred about 50 to 80 mm from the top of the first page (see Figure 7-4). The author's name and the name of the company or organization the author works for can also be centred at the top of the page, about 25 mm below the title. Alternatively, the author's name and affiliation can be placed at the bottom left of the first page, or at the end of the paper.

The abstract (summary) appears next, and starts about 30 mm below the title or the author's name. It should be indented about 25 mm from both side margins.

The body of a scientific paper starts about 20 mm beneath the abstract and looks much like the body of a semiformal report. Normally, a scientific paper is double-spaced throughout, including the abstract. Often, the first line of every paragraph is indented 1.0 to 1.5 cm from the left margin.

Writing Style

The rules for brevity, clarity, and directness suggested in Chapters 1, 3, and 11 for technical letters and reports apply equally to scientific papers. But there is one exception: Where we have recommended that you write in the active voice, in some technical disciplines it's more common to write scientific papers in the passive voice. For example, in a technical report we have advised you to write

A slight bending of the rules!

I placed the sample in the chamber...

or

We placed the sample in the chamber...

or

The technician placed the sample in the chamber...

However, in a scientific paper you will more likely be expected to write

The sample was placed in the chamber...

In other words, the identity of the "doer" often is concealed in a scientific paper, and use of the first person ("I" or "We") is carefully avoided.

Unfortunately, we cannot give you definitive advice, because different organizations, academic institutions, and technical disciplines adhere to different guidelines. In papers on medical research, for example, scientists tend to write predominantly in the passive voice (although this tradition is gradually changing), whereas in papers on computer technology, authors tend to write in the active voice. If you are writing a paper that may be published in a peer-reviewed scientific journal, you will need to study previous issues—or write to the journal editor—to identify whether the active voice is acceptable.

Talk to the journal editor: determine the journal's guidelines

Acid Rain Testing in the City of Feldspar, Ontario, 1997

Corrine L Danzig and Mark M Weaver
University of Feldspar Research Laboratory

ABSTRACT

Acid rain is a growing concern in Southern Ontario, with its effects becoming increasingly noticeable south of Lakes Erie and Ontario. To determine what increases have occurred in the City of Feldspar over the past nine years, acidity levels were measured and compared to measurements recorded in 1986, when the average pH level was YYY. The current tests showed that the average pH now is ZZZ, but that the acidity levels are not uniform. The greatest toxicity was found to be at the University of Feldspar, in the southeastern area of the city, with the lowest toxicity in the northwestern area of the city. The change was attributed primarily to the increase in fossil-fuel-burning heavy industries in the Poplar Heights industrial park, which is 3.8 kilometres northwest (generally upwind) of the principally affected areas.

This is the only time we suggest you write a summary in the passive voice!

INTRODUCTION

Over the past 20 years, observers in the counties adjacent to Lake Ontario have reported an increased incidence of crop spoilage and tree defoliation, which has been attributed to acid rain created primarily by fossil-fuel-burning industries and automobile emissions. Tests were taken initially in 1986 to assess the pH levels of the precipitation falling in and around the City of Feldspar, Ontario.[1] These were compared with the measurement recorded on

1

Figure 7-4 The first page of a scientific paper.

Organization

A scientific paper has six main parts, which are described in detail below.

1. **Abstract** (Summarizes the paper, emphasizing the results)
2. **Introduction** (Provides background details and outlines the problem or theory tested)
3. **Materials and Methods** (Describes how the tests were performed)
4. **Results** (States the findings)
5. **Discussion** (Analyses and interprets the findings)
6. **References** (Lists the documents consulted)

Sometimes you may receive significant assistance or guidance from other people during your research and may want to acknowledge their help. Such acknowledgements normally are placed after the discussion but before the references.

If you are accustomed to writing laboratory reports or investigation reports, you have probably noticed that the "Conclusions" heading has been omitted from this list. Where the conclusions in a laboratory or investigation report serve as a *separate* terminal summary (a summing up of the results and their analysis), in a scientific paper they are more often embedded as a closing statement at the end of the Discussion.

Abstract

The rules for writing an abstract are almost identical to those for writing the summary of an investigation report. In an abstract you (1) outline the problem and the purpose of your investigation, (2) mention very briefly how you conducted the investigation or tests, (3) describe your main findings, and (4) summarize the conclusions you have drawn. All this must be done in as few words as possible; ideally, your abstract will be about 125 words long and never more than 250 words. A typical abstract is shown in Figure 7-4.

Write an informative rather than a topical abstract

From the abstract, readers must be able to decide whether the information you provide in the remainder of the paper is of particular interest to them and whether they should read further. Because a scientific paper is addressed to readers who generally are familiar with your technical or scientific discipline, you may use technical terminology in the abstract. (This is the major difference between a report summary and a scientific abstract.) The abstract should be written last, when the remainder of the paper has been written, so that you can *abstract* the brief details you need from what you have already written.

Introduction

In the introduction you prepare readers so that they will readily understand the technical details in the remainder of the paper. The introduction contains four main pieces of information:

1. A definition of the problem and the specific purpose of the investigation or tests you conducted. (This should be a much more detailed definition than appears in the abstract.)

2. Presentation of background information that will enable the reader to fully understand and evaluate the results. Often it will include—or sometimes may consist entirely of—a review of previous scientific papers, journal articles, books, and reports on the same subject. (This is known as a literature survey.) Rather than simply list the pertinent documents, you are expected to summarize the main findings of each and their relevance to your investigation. These documents should be cross-referenced to your list of references at the back of the report. Here is an example:

> Previous measurements of acid rain in Montrose were recorded in 1994 by Gershwain (3), who reported an average acidity level of xxxx, and...

3. A short description of your approach and why you chose that particular method for your investigation.

4. A concluding statement that outlines your main findings.

Notice that the introduction of a scientific paper *includes a brief summary of the results*, which is much less common in a semiformal investigation report and rare in a laboratory report. Yet here there is a parallel with the alternative method of presenting a formal report, in which the writer presents the results three times (see Figure 6-9). In a scientific paper the results also are presented three times: very briefly in the abstract and introduction, and fully in the results section.

Materials and Methods

This section has to be thoroughly prepared and presented because, from what you write here, readers must be able to replicate (perform) an identical investigation or series of tests. The materials list must include *all* equipment used and specimens or samples tested, which may range from a whole-body nuclear radiation counter to a tiny microorganism. For ease of reference they should be listed in representative groups, such as:

Equipment	Plants
Instruments	Animals
Chemicals	Birds or fishes
Specimens	Humans

Precede the specific methods with the subheading "Method." Describe the tests chronologically in paragraph and subparagraph form, using a main

Conduct a detailed literature review

Be detailed and specific: help the reader understand exactly what you did

paragraph to introduce a test or part of a test and short, numbered sub-paragraphs to describe the specific steps you took. Avoid writing in the imperative mood, so that you do not inadvertently start writing an instruction. For example:

Write this:	The test unit was connected to the X-Y terminals of the recorder.
Or this:	We connected the test unit to the X-Y terminals of the recorder.
But not this:	Connect the test unit to the X-Y terminals of the recorder.

Results

The results section often is the shortest part of your paper. If your methods section has described clearly how the investigation or tests were conducted, the results section has only to state the result:

> Acid rain is above average in the southeastern part of the city, below average in the northwestern part, and average in the southwestern and northeastern parts. Tables 1 through 8 show the measurements recorded at the eight metering stations.

Present the facts; avoid offering opinions

(Tables and charts depicting your findings often will be a major part of the results section.)

Never comment on the results, because analysis and interpretation belong *only* in the discussion.

Discussion

Your readers now expect you to analyse and interpret your findings. You will be expected to discuss

- the results you obtained, compared to the results obtained by previous researchers,
- any significant correlation, or lack of correlation, between parts of your own findings,
- factors that may have caused the differences,
- any trends that seem to be evident or to be developing (ideally, by referring to graphs and charts you have presented in the results), and
- the conclusions you draw from your analysis and interpretation.

The conclusions should show how you have responded to the problem stated at the start of your introduction, and identify clearly what your investigation, tests, and analyses show. As such, they should form a fitting close to the narrative portion of your research paper.

Embed the conclusions in the Discussion, or introduce them with a heading

References

The final section of your paper is a list of the documents you have referred to earlier or from which you have extracted information. Chapter 6 provides general instructions for preparing a list of references, which is the preferred method for most technical reports, and a bibliography. You can use Chapter 6 as a guideline, but you should check first with the editor of the journal in which your scientific paper is likely to be published to determine (1) whether a list of references or a bibliography is preferred, and (2) the exact format the journal uses for listing authors' names, book and journal titles, publisher details, and so on.

Technical Papers and Articles

Why Write for Publication?

The likelihood that one day you may be asked to write a technical paper for publication, or even want to do so, may seem so remote to you now that you might be justified in skipping this section. Yet this is something you should think about, for getting one's name into print is one of the fastest ways to obtain recognition. Suddenly you become an expert in your field and are of more value to your employer, who is happy because the company's name appears in print beneath yours. You become of more value to prospective employers, who rate authors of technical papers more highly than equally qualified persons who have not published. And you have positive proof of your competence, and sometimes a few extra dollars from the publisher.

Mickey Wendell has an interesting topic to write about: As a senior technologist with H L Winman and Associates' materials testing laboratory, he has been testing concretes with various additives to find a grout that can be installed in frozen soil during the winter. One mixture he analysed but discarded contained a new product known as Aluminum KL. As a byproduct of his tests he has discovered that mixing Aluminum KL with cement in the right proportions results in a concrete with very high salt resistance. He reasons that such concrete could prove invaluable to builders of concrete pavements in snow-affected areas of Canada and the United States, where salt mixtures are applied in winter to melt the snow.

Mickey has been thinking about publishing this particular aspect of his findings and has jotted down a few headings as a preliminary outline. Here are the four steps he must take before his ideas appear in print.

Step 1: Solicit Company Approval

Most companies encourage their employees to write for publication, and some even offer incentives such as cash awards to those who do get into

Writing and presenting a paper enhances your professional reputation

print. However, they expect prospective authors to ask for permission before they submit their manuscripts.

To obtain permission, Mickey must write a brief memorandum outlining his ideas to John Wood, his department head. He should ask for approval to submit a paper, explain what he wants to write about and why he thinks the information should be published, and outline where he intends to send it. His proposal is shown in Figure 7-5. John Wood will discuss the matter at management level, then signify the company's approval or denial in writing. Mickey knows he must have *written* consent before he can publish his findings.

Have a clear, well-thought-out idea before requesting approval

Step 2: Consider the Market

Mickey must decide very early where he will try to place his paper. If he prefers to present his findings as a technical paper before a society meeting, as suggested by John Wood (see his comment in Figure 7-5), he will be writing for a limited audience with specialized interests. If he decides to publish in the journal of a technical society, he will be writing for a larger audience, but still within a limited field. If he plans to publish in a technical magazine, he will be appealing to a wide readership with a broad range of technical knowledge. His approach must therefore differ, depending on the type of publication and level of reader.

A guiding factor may be Mickey's writing capability. A paper to be published by a technical society requires high-quality writing. The editor of a society journal normally does not do much prepublication editing, other than making minor changes to suit the format and style of the society's publications. A technical magazine article, however, will be edited—sometimes quite fiercely—by a professional editor who knows the exact style that readers expect. Such an editor prefers authors to approximate that style and expects them to organize their work well and to write coherently; but he or she is always ready to prune or graft, and sometimes even completely rewrite portions of a manuscript. Hence, the pressure on authors of magazine articles is not so great.

Articles written for a professional journal are usually sent out for peer review

Perhaps the most important factor is for Mickey to identify a potential audience for his information. Readers of society journals and technical magazines may be the same people, but they expect different information coverage in a technical paper than they do in a magazine article.

Technical Paper

Readers of society journals are looking for facts. They neither expect nor want explanation of basic theories, and they can accept a strongly technical vocabulary. A technical paper can be very specific. It can describe a minute aspect of a large project without seeming incomplete, or it can outline in bold terms the findings of a major experiment. No topic is too large

475 Lethbridge Trail, Calgary AB T3M 5G1

INTEROFFICE MEMORANDUM

To: John Wood   From: Mickey Wendell

Date: June 24, 1997   Subject: Approval for Proposed Technical
                                        Article

May I have company approval to write an article on concrete additives for publication in a technical journal? Specifically, I want to describe our experiments with Aluminum KL, and the salt-corrosion resistance it imparted to the concrete samples we tested for the northern Quebec transmission tower project. I believe that our findings will be of interest to many municipal engineers in Canada and the northern United States, who for years have been trying to combat pavement erosion caused by the application of salt during snow removal.

I was thinking of submitting the article to the editor of "Municipal Engineering," but I'm open to suggestions if you can think of a more suitable magazine.

Your proposal could also include an outline of the proposed paper

MW

Approval granted. Let me see an outline and the first draft before you submit them.

I suggest you also consider writing a technical paper for presentation at the combined conference on Concrete to be held in Toronto next March.

John Wood
July 6, 1997.

Figure 7-5 Soliciting company approval to publish an article or paper.

or too small, too specialized or too complete, to be published as a technical paper.

Technical Article

Most readers of magazine articles are looking for information that will keep them up-to-date on new developments. Some will have definite interest in a specific topic and would welcome a lot of technical details. Others will be looking mainly for general information, with no more than just the highlights of a new idea. Magazine authors must therefore appeal to a maximum number of readers. Their articles should be of general interest; their style can be brief and informal; their vocabulary must be understandable; and they should include some background details for readers whose technical knowledge is only marginal.

Step 3: Write an Abstract and Outline

Many editors prefer to read either a summary of a proposed paper or an abstract and outline before the author submits the complete manuscript. They may want to suggest a change in emphasis to suit editorial policy, or even decline to print an interesting paper because someone else is working on a similar topic.

If you are describing a scientific breakthrough, wait until you are ready to send in the whole paper

This type of summary is much longer than the summary at the head of a technical report; the abstract, however, usually is quite short. The summary contains a condensed version of the full paper in about 500 to 1000 words. An abstract contains only very brief highlights and the main conclusion (rather like the summary of a report), since it is supported by a comprehensive topic outline.

Some authors write the complete first draft of the paper before attempting to write a summary or abstract, then leave the revising and final polishing until after the paper has been accepted by an editor. Others prepare a fairly comprehensive outline, often using the freewheeling approach suggested in Chapter 1, and leave the writing until after acceptance. Both methods leave room for the author to incorporate changes before the final manuscript is written.

Since the summary or abstract and outline have to "sell" an editor on the newsworthiness of his topic, Mickey Wendell must make sure that the material he submits is complete and informative. In addition he must indicate clearly

- why the topic will be of interest to readers,
- how deeply the topic will be covered,
- how the article or paper will be organized,
- how long it will be (in words), and
- his capability to write it.

Details like these demonstrate you have developed your idea in depth

Mickey can cover the first four items in a single paragraph. The fifth he will have to prove in two ways. He can prove his technical capability by mentioning his involvement in the topic and experience in similar projects. And he can demonstrate his writing ability by submitting a clear, well-written summary or abstract.

If Mickey later decides to prepare his paper for presentation before the Combined Conference on Concrete, as John Wood has suggested (see Figure 7-5), he will have to prepare a summary in response to a "call for papers" sent out by the society, and submit it to the chairperson of the papers selection committee. If his paper is to be accepted, his summary has to convince the committee that the subject is original, topical, and interesting, and that he has the capability to write and present the paper.

Step 4: Write the Article or Paper

A good technical paper is written in an interesting narrative style that combines storytelling with factual reporting. Articles published in general interest magazines tend to be written like feature newspaper stories, whereas technical papers more nearly resemble formal reports. If the article deals with a factual or established topic, the writing is likely to be crisp, definite, and authoritative. If it deals with the development of a new idea or concept, the narrative will generally be more persuasive, since the writer is trying to convince the reader of the logic of his or her argument.

The parts of an article or paper submitted to a magazine are similar to those of a report. Mickey's article, for example, contains four main parts:

Summary	A synopsis that tells very briefly what the article is about. It should summarize the three major sections that follow. Like the summary of a report, it should catch and hold the reader's attention.
Introduction	Circumstances that led up to the event, discovery, or concept that Mickey will describe. It should contain all the facts readers need to understand the discussion that follows.
Discussion	How Mickey went about the project, what he found out, and what inferences he drew from his findings. The topic can be described chronologically (for a series of events that led up to a result), by subject (for descriptive analyses of experiments, processes, equipments, or methods), or by concept (for the development of an idea from concept to fruition). The methods are very similar to those used for writing the discussion of a formal report (see Chapter 6).

The pyramid approach is equally valid for a technical article

| Conclusion | A summing-up, in which Mickey draws conclusions from and discusses the implications of his major findings. Although he will not normally make recommendations, he may suggest what he feels needs to be done in the future, or outline work that he or others have already started if there is a subsequent stage to the project. |

If, however, Mickey had decided to submit his paper to a peer-reviewed scientific journal, then its parts and his writing style would have had to comply with the requirements described in the section on scientific papers (pages 248 to 254).

Illustrations are a useful way to convey ideas quickly, to draw attention to an article, and to break up heavy blocks of type. They should be instantly clear and should usefully *supplement* the narrative. They should never be inserted simply to save writing time; neither should they convey exactly the same message as the written words. For examples of effective illustrating, turn to any major publication in your technical field and study how its authors have used charts, graphs, sketches, and photographs as part of the story. For further suggestions on how to prepare illustrative material, see Chapter 8.

Mickey should not be surprised when the editor who handles his manuscript makes some changes. An editor who feels the material is too long, too detailed, or wrongly emphasized for the journal's readers, will revise it to bring it up to the expected standards. Mickey may feel that the alterations have ruined his carefully chosen phrases, but readers will not even be aware that changes have been made. They will simply recognize a well-written paper, for which Mickey rather than the editor will reap the compliments.

Don't feel rebuffed if an editor cuts some of your words!

ASSIGNMENTS

Project 7.1: Performing a Task Analysis

Think of a product you use often. It might be a microwave oven, a fax machine, a vacuum cleaner, or anything else you are familiar with. Perform a task analysis for the product and then group and label the topics. Remember to use "ing" words to indicate the user's perspective. List the tasks and then group and label them.

Project 7.2: Programming a VCR and a Microwave Oven

Two days ago a retired couple who live near to you bought a new VCR. They asked you to show them how to use it to record a program. You show them the instruction manual and "walk" them through the instructions.

Part 1

Today they telephone you and say they couldn't make it work. "All those pictures!" Mr Smithson exclaims. "I'd just like a straightforward instruction—just words."

You borrow the manual and write them a step-by-step instruction, using good design and simple explanations.

(Note: Use a VCR you are familiar with to write this instruction.)

Finish with a usability test: watch a neighbor use your instructions...

Part 2

Mr and Mrs Smithson were so pleased with your instruction for programming their new VCR, they ask you to write an instruction for their microwave oven.

"Something I can glue to the side of the oven," Mrs Smithson explains. "We have no trouble operating the oven, but in the spring each year, when we change to daylight saving time, I have to reset the clock. And it's the same again in the fall. The instruction manual that came with the oven just confuses me."

...then revise them!

Write an instruction for resetting the clock on a microwave oven that you are familiar with.

Project 7.3: User's Manual for a Meeting Timer

Mechanical engineer Darwin Haraptiniuk places a drawing on your desk (see Figure 7-6) and says, "You're a good writer. Write a user's manual to go along with this."

You ask what "this" is.

"It's a meeting timer I designed several years ago. Now I'm having 10 prototypes built to test in various companies locally. If their response is positive, I'm going to redesign it using microprocessor technology and go into mass production. I want the user's manual to ensure that whoever tests the timer uses it properly."

"How does it work?" you ask.

"Like a gasoline pump. You work out the average hourly rate for the kind of people attending a particular meeting and set it in one window, and the number of people sitting around the table goes in the other window."

"How?"

"By rotating the knobs on the left. Then you just depress the buttons, one at the start of the meeting, and one at the end."

"But what's it for?" you ask.

A recipe for a short instruction

"To help speed up business meetings," Darwin says. "You put it on the table so everyone can see it. When they can *see* how long a meeting has lasted and what it has cost so far, I figure they'll concentrate on getting things done quickly."

"All right," you say. "I'll get right onto it."

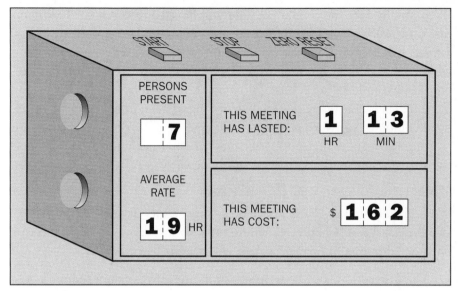

Figure 7-6 Darwin Haraptiniuk's meeting timer.

"One more thing," Darwin adds. "You'd better warn users not to change the number of people attending the meeting or the average rate while the machine is running. That tends to damage the mechanism."

"What do you do if you have a problem, then?"

"Switch the machine off, make the change, then switch it on again."

"And if that doesn't work?"

"Call me: 774-1685."

"Okay," you say. And then you ask, "What am I to call your machine?"

"Ah! Good point. What about the 'Darwin Meeting Timer Model No. 1'?"

"Fine!"

Now write the user's manual.

Project 7.4: Researching a New Manufacturing Material or Process

You are to research information on a topic allied to your technology and then prepare it for both written and oral presentation. The topic may be a new manufacturing material, method, or process. The written and spoken presentations must

1. introduce the topic,
2. state why it is worth evaluating,
3. describe the material, method, or process,
4. discuss its uniqueness and usefulness, and
5. show how it can be applied in your particular field.

A recipe for thorough research and a detailed description

To obtain data for your topic, you will have to research current literature and probably talk to industrial users, manufacturers, and suppliers. Typical examples of topics are a new oil that can be used at very low temperatures; a method for supporting the deck of a bridge during concrete-pouring by building up a base on compacted fill; a new paint for use on concrete surfaces; a new materials-handling system; and a recently developed computer software program.

You may assume that both your readers and your audience are technicians to whom the topic will be entirely new.

Project 7.5: Testing Cable Connectors

Write an instruction to all installation supervisors at sites 1 through 17 (see Project 4.8 of Chapter 4) telling them to inspect all cable connectors on site. Additional information you will need is listed below.

Consider sending this instruction by email

1. The instruction is to be written as an interoffice memorandum.
2. Don Gibbon, electrical engineering coordinator at H L Winman and Associates, will sign the memorandum.
3. Tell the site installation supervisors that they are to report the number of GLA connectors they find to Don Gibbon.
4. Tell them to replace all connectors marked GLA with connectors marked MVK.
5. All connectors on site are to be checked. (It would be best to check those in stock first, then use checked connectors to replace those in use that are found to be faulty.)
6. Inform them that faulty GLA connectors are to be sent to the contractor with a note that they are to be held for analysis under project 92A7.
7. The site supervisors are to complete their tests within seven days.

Project 7.6: Installing a Mini-Minder

You are employed in the construction department of Midprovince Telephone System, and you have been asked to write an installation instruction sheet to be shipped with the "Mini-Minder" (an electronic intrusion detection device developed by Midprovince's engineering department). The Mini-Minder is a small detection unit that is mounted above windows, doors, or any other entries, where it automatically detects movement and transmits an alarm signal to a central unit. The central unit is concealed within the building and, when activated, both sounds an audible alarm and sends a message to police headquarters.

The Mini-Minder is shown in Figure 7-7 on page 263. It is fixed to the wall above the door or window by removing the backplate and screwing the plate to the wall. The unit is then snapped onto the backplate (removing the unit from the backplate also sounds the alarm). The unit is battery operated.

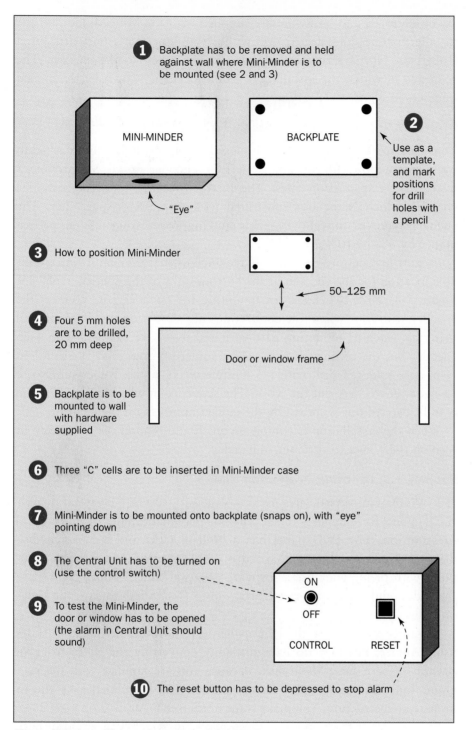

1. Backplate has to be removed and held against wall where Mini-Minder is to be mounted (see 2 and 3)

MINI-MINDER

"Eye"

BACKPLATE

2. Use as a template, and mark positions for drill holes with a pencil

3. How to position Mini-Minder

50–125 mm

4. Four 5 mm holes are to be drilled, 20 mm deep

Door or window frame

5. Backplate is to be mounted to wall with hardware supplied

6. Three "C" cells are to be inserted in Mini-Minder case

7. Mini-Minder is to be mounted onto backplate (snaps on), with "eye" pointing down

8. The Central Unit has to be turned on (use the control switch)

9. To test the Mini-Minder, the door or window has to be opened (the alarm in Central Unit should sound)

ON
OFF
CONTROL   RESET

10. The reset button has to be depressed to stop alarm

Figure 7-7 Installing and testing a Mini-Minder.

A reminder: write in the imperative mood

All materials and hardware are supplied with the Mini-Minder but the installer will need an electric drill with a 5-millimetre masonry drill bit, a Robertson No. 2 screwdriver, and a sharp pencil to do the job. The sequence in which the installation should occur is shown by the circled numbers in Figure 7-7.

Project 7.7: Installing Encoder EC7

You work for Midprovince Telephone System and you have been asked to write instructions for installing an EC7 encoder at all Midprovince's microwave transmission sites. The instructions are to accompany the encoder, which is the box illustrated in Figure 7-8 on page 265. The encoder removes unwanted signals and improves transmission performance by 8% to 10%.

In your instructions, tell the site technicians that they must find a suitable location for the encoder at the bottom right of the control panel (see Figure 7-8); drill two holes for mounting the encoder; and connect the encoder with four wires (the connections are shown on the diagram). All materials (such as mounting hardware and wire) are supplied with each encoder but the technician will need a soldering iron, some Ersin 60/40 resin core solder, a drill with a 5-millimetre bit, and a Robertson No. 2 screwdriver to carry out the work. The sequence in which the installation is to be carried out is shown by the circled numbers on the diagram.

When the installation is complete you should remind the technician to turn on the power to the control panel.

Project 7.8: Creating Projection Slides

H L Winman and Associates owns a Vancourt 300 liquid crystal display (LCD) panel for engineers to use when displaying graphics during an oral presentation. The LCD panel has a built-in CPU and a 3.5-inch disk drive: they prepare their slides on disk rather than carry a portable computer with them. They use a software program called *Amaze* to prepare the slides, and then convert the program to *Disk Show* to save the slides on a disk.

Most people who use the LCD panel have no difficulty creating slides using *Amaze*, but they do have difficulty converting the slides from an *Amaze* file to a *Disk Show* disk. Because you are familiar with the program, Tanys Young (head of Administration and Personnel) asks you to prepare an instruction. You make some notes (see Figure 7-9 on page 266).

Prepare an instruction from these notes, in which you embody both good design and clear writing principles.

You will have to extract pertinent points from the illustration

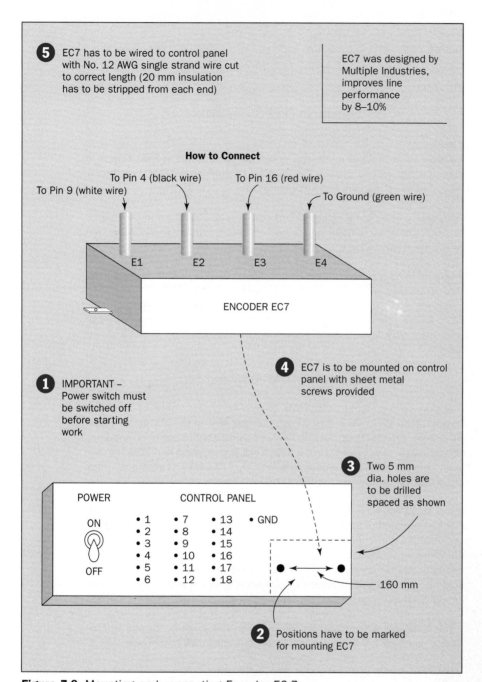

5 EC7 has to be wired to control panel with No. 12 AWG single strand wire cut to correct length (20 mm insulation has to be stripped from each end)

EC7 was designed by Multiple Industries, improves line performance by 8–10%

How to Connect

To Pin 9 (white wire)

To Pin 4 (black wire)

To Pin 16 (red wire)

To Ground (green wire)

E1   E2   E3   E4

ENCODER EC7

Consider writing in the active voice

1 IMPORTANT – Power switch must be switched off before starting work

4 EC7 is to be mounted on control panel with sheet metal screws provided

3 Two 5 mm dia. holes are to be drilled spaced as shown

POWER

ON

OFF

CONTROL PANEL

• 1   • 7   • 13   • GND
• 2   • 8   • 14
• 3   • 9   • 15
• 4   • 10   • 16
• 5   • 11   • 17
• 6   • 12   • 18

160 mm

2 Positions have to be marked for mounting EC7

Figure 7-8 Mounting and connecting Encoder EC-7.

Making a *Disk Show* Disk

Create a step-by-step instruction...

In Amaze...
The program is brought up on screen and set to slide 1. The next step is to select the appropriate printer. This is done by clicking on *File* then *Printer Setup*, and selecting *Disk Show Printer Driver*.

It's most important at this stage to carry out a spell-check. Avoid clicking on the *ABC* symbol on the tool bar; instead, select *Text* then *Spell-Check*, and then *Entire Presentation*.

Make sure there is a fresh disk (a clean, unused disk) in drive A, and print the slide program on it. This is done by clicking on *File* then *Print*. **

NOTE: *Before* printing the program, set the margins by selecting *Slide Page Setup* after clicking *File\Print*. Check that the margins are set to 0.50" all round.

In *Disk Show for Windows...*

After you have brought *Disk Show* on screen (you do this by selecting *Program*, then *Tools*, then *Disk Show*), the first step is to click on *View* and select *Show Effects*. Then click on *File* and select *Fast Unpack*.** (The slides will appear on your screen, 12 at a time. Each is numbered and labelled randomly with a transition effect such as "Up," "Down," "Left/Right," "Box In/Out" above the slide, in random order.)

...using good design to make the information readily accessible

At this stage you have to examine the transitions to check that each is suitable for bringing the next slide on screen. To do this, you select each slide in turn and place the pointer on the appropriate transition in the *Transitions* box. (Note: when selecting an effect, avoid using *Random*, *Fade*, or *Dazzle* – they are not effective transitions for HL Winman presentations.) When complete, click on *File* and select *Pack*.** Your disk is now ready for presentation through your LCD panel.

** The computer will prompt you to insert a disk in drive A, and in *Fast Unpack* and *Pack* it will remind you to use the same disk both times.

Figure 7-9 Notes: making a *Disk Show* disk.

User Friendly Manuals' Website
pip.dknet.dk/~pip323/index.html

This site is sponsored by PRC of Denmark, an international company specializing in user friendly manuals. Included are links to related sites.

Writing Report Abstracts
owl.english.purdue.edu/Files/88.html

The Online Writing Lab at Purdue University has information about writing report abstracts. Included are sections on types of abstracts, qualities of a good abstract, and steps for writing effective abstracts.

Abstracts
darkstar.engr.wisc.edu/zwickel/397/abstract.html

This page on writing abstracts comes from the online technical writing handbook of the Department of Engineering Professional Development at the University of Wisconsin.

How to Have Your Abstract Rejected
parcftp.xerox.com/pub/popl96/vanLeunenLipton

This is a tongue-in-cheek guide describing how to guarantee the rejection of an abstract submitted for a conference. The comments can be generalized to technical papers of any length.

Chapter 8
Illustrating Technical Documents

If you open any well-known technical magazine you will immediately notice that illustrations are an integral part of most articles. Some are photographs that display a new product, a process, or the result of some action; others are line drawings that illustrate a new concept; some demonstrate how a test or an experiment was tackled; still others are charts and graphs that show the progress of a project or depict technical data in an easy-to-visualize form.

Illustrations serve an equally useful purpose in technical reports, where their primary role is to help readers understand the topic. Interesting illustrations attract readers' eyes and encourage them to read a report. They can also break up dull-looking pages of narrative that lack eye appeal.

This chapter discusses the illustrations seen most often in technical reports, indicates the overlaps that exist between illustration types, suggests occasions when they can be used most beneficially, and provides guidelines for preparing them.

Primary Guidelines

The criterion for any illustration is that it should help to explain the narrative—the narrative should *never* have to explain an illustration. Hence, an illustration must be simple enough for readers to understand quickly and easily. When readers turn over the pages of *Scientific American* and stop at a particular page, usually they look first at the drawings or photographs. If an illustration immediately tells a story and captures the readers' interest, they are encouraged to read the article.

To help readers readily understand the illustrations you insert into your reports, follow these seven guidelines:

1. Before selecting or designing an illustration, first consider the specific audience for whom you are writing, and what you want your readers to learn from each illustration.

Good illustrations
attract readers

2. Keep every illustration simple and uncluttered.
3. Let each illustration depict only one main point.
4. Position each illustration as near as possible to the narrative it supports (see "Positioning the Illustrations," at the end of this chapter).
5. Label each illustration clearly with a figure or table number and a title (with the figure number and title centred *beneath* a figure or chart, and the table number and title centred *above* a table).
6. Add a caption (that is, comments or remarks) beneath a figure title, to draw attention to significant aspects of the illustration.
7. Refer to every illustration at least once in the report narrative.

Computer-Designed Graphs and Charts

Computer software has
taken much of the
drudgery out of preparing graphs and charts

There are a number of simple-to-use yet powerful computer software programs that will help you create illustrations and graphics. Software programs now allow a report writer to enter the quantities of each function and then choose the type of graph or chart. It is possible to view the same information in several formats; for example, as a line graph, a bar chart, or a histogram. The following sections describe the different types of graphs and charts available. Understanding the benefits of each type will help you decide which is best for your information and audience.

Computer-designed graphics still need to follow the above guidelines. In addition you need to be aware of how the computer truncates and positions the labels, because the results may make a chart incomprehensible. Many software programs create very attractive two- and three-dimensional illustrations, often in multiple colors on a computer screen, which may look fine in a glossy magazine or daily newspaper but look out of place in a business report. Knowing the guidelines will help you plan effective illustrations to insert into your reports.

Graphs

Graphs are a simple means for showing a change in one function in relation to a change in another. A function used frequently in such comparisons is time. The other function may be temperature, erosion, wear, speed, strength, or any of many factors that vary as time passes.

Engineering technologist John Greene wants to determine how long it takes a newly painted manometer case to cool down after it comes out of the drying oven. He goes to the paint shop armed with a stopwatch and a special thermometer. When the next manometer case comes out of the oven he starts taking readings at half-minute intervals, and records the results as in Table 8-1.

Table 8-1 Cooling rate, manometer case MM-7.

Time Elapsed (min:sec)	Temperature (°C)	Time Elapsed (min:sec)	Temperature (°C)
:30	152.9	5:30	43.9
1:00	123.4	6:00	41.1
1:30	106.7	6:30	38.9
2:00	91.2	7:00	37.2
2:30	77.8	7:30	35.6
3:00	69.5	8:00	34.5
3:30	61.7	8:30	33.4
4:00	55.6	9:00	32.8
4:30	51.5	9:30	31.7
5:00	47.3	10:00	31.1
Ambient temperature 22.8°C		Oven temperature 180°C	

A table provides specific details

These readings are part of a study he is undertaking into the cooling rate of different components manufactured by Macro Engineering Inc. The information will also be used by the production department to establish how long manometer cases must cool before assemblers can start working on them with bare hands (the maximum bare-hand temperature has been established by management/union negotiation to be 38°C).

A quick inspection of this table shows that the temperature case drops continuously, is within 8.3° of the ambient temperature after 10 minutes (*ambient* means "surrounding environment"), and is down to the bare-hand temperature after 7 minutes. A closer examination identifies that the temperature drops rapidly at first, then progressively more slowly as time passes.

Single Curve

John Greene can make the data he has recorded in Table 8-1 much more readily understood if he converts it into the graph in Figure 8-1. Now it is immediately evident that the temperature drops rapidly at first, then slows down until the rate of change is almost negligible. (There is no point in John measuring or showing any further drop in temperature unless he wants to demonstrate how long it takes the component to cool right down to the ambient temperature, which probably would be 30 minutes or more.)

A graph translates the details into an easily understood form

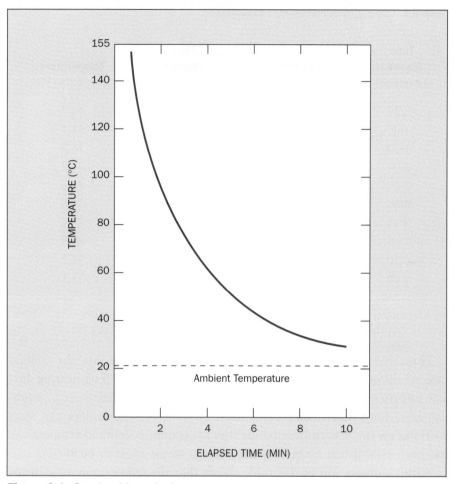

Figure 8-1 Graph with a single curve.

Multiple Curves

In Table 8-2 on page 273 John compares the temperature readings he has recorded for the cooling manometer case with measurements he has taken under similar conditions for a cover plate and a panel board. This time, however, he simplifies the table slightly by showing the temperatures at one-minute intervals.

What can we assess from this table? The most obvious conclusion is that in 10 minutes the cover plate has cooled down less than the manometer case, and even less than the panel board. We can also see that the initial rate at which the components cooled varied considerably: the panel board, very quickly; the manometer case, fairly quickly; the cover plate, seemingly quite slowly. But it is difficult to assess whether there were any *changes in the rate of cooling* as time progressed.

This data again can be shown more effectively in a graph, which John Greene has plotted in Figure 8-2. The rapid initial drop in temperature is

Table 8-2 Cooling rates for three components.

Time Elapsed (minutes)	Temperature (°C)		
	Cover Plate	Panel Board	Manometer Case
0:30	154.5	145.1	152.9
1	136.7	97.3	123.4
2	112.3	67.8	91.2
3	95.1	51.7	69.5
4	82.3	42.3	55.6
5	71.7	36.1	47.3
6	63.9	32.2	41.1
7	55.6	29.5	37.2
8	49.5	27.2	34.5
9	43.4	26.1	32.8
10	38.9	25.0	31.1

A multi-factor table can be even more difficult to interpret

evident from the initial steepness of the three curves, with each curve flattening out to a slower rate of cooling after 2 to 4 minutes. The difference in cooling rates for the three components is much more obvious from the curves than in the table. (The lines in Figures 8-1 and 8-2 are commonly referred to as curves, even though in some cases they may be straight lines or a series of short straight lines joining points plotted on the graph.)

Eventually John will have to present this data in a report. If he plans to present only a general description of temperature trends, his narrative can be accompanied by graphs like these. But if he also wants to discuss exact temperatures at specific times for each material, then the narrative and graphs will have to be supported by figures similar to those in Table 8-2.

Constructing a graph usually offers no problems to technical people because they recognize a graph as a logical means for conveying statistical data. But if they are to construct a graph that both tells a story *and* emphasizes the right information, they must know the tools they will be working with.

Scales

The two functions to be compared in the graph are entered on two scales: a horizontal scale along the bottom and a vertical scale along the left side. (On large graphs the vertical scale is sometimes repeated on the right side to simplify interpretation.) The scales meet at the bottom left corner, which normally—but not always—is designated as the zero point for both.

Both functions may be variable, but only one depends on the other

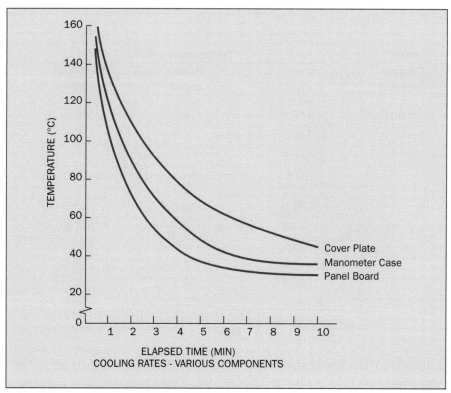

Figure 8-2 Graph with multiple curves.

The two functions are commonly known as the dependent and independent variables, so named because a change in the dependent variable *depends* on a change in the independent variable. For example, if we want to show how the fuel consumption of a car increases with speed, we will enter speed as the independent variable along the bottom scale, and fuel consumption as the dependent variable along the left side, as in Figure 8-3. (Fuel consumption *depends* on speed; speed does not depend on fuel

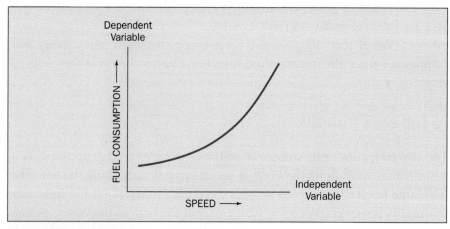

Figure 8-3 The dependent variable depends on the independent variable.

Table 8-3 Identifying dependent and independent variables.

Graph Illustrates	Dependent Variable (vertical scale)	Independent Variable (horizontal scale)
1. The effect that frequency has on the gain of a transducer	Gain	Frequency
2. How attendance at a ball game varies with temperature	Attendance	Temperature
3. How much a motor's speed affects the noise it produces	Noise	Speed
4. The changes in temperature brought about by changes in pressure	Temperature	Pressure
5. How much an increase in payload reduces an aircraft's range by limiting the amount of fuel it can carry	Aircraft range (or fuel load)	Payload
6. How much increasing the fuel load of an aircraft to achieve greater range reduces its effective payload	Payload	Fuel load (or aircraft range)

Note: A function can be either dependent or independent, depending on its role in the comparison (see temperature in examples 2 and 4, and both functions in examples 5 and 6).

consumption.) The same applies to John Greene's temperature measurement graphs: temperature is the dependent variable because it depends on *the time that has elapsed* since the components came out of the oven (the independent variable).

When you construct a graph, the first step is to identify which function should form the horizontal scale and which the vertical scale. Table 8-3 lists some typical situations that show that the same function (e.g. temperature) can be an independent variable in one situation and a dependent variable in another. Selection of the independent variable depends on which function can be more readily identified as influencing the other function in the comparison.

Select scale intervals
that will give a *bal-
anced* appearance

The second factor to consider is scale interval. Poorly selected scale intervals, particularly scale intervals that are not balanced between the two variables, can defeat the purpose of a graph by distorting the story it conveys. Suppose John Greene had made the vertical scale interval of his time vs temperature graph in Figure 8-1 much more compact, but had retained the same spacing for the horizontal scale. The result is shown in Figure 8-4(a). Now the rapid initial decrease in temperature is no longer evident; indeed, the impression conveyed by the curve is that temperature dropped only moderately at first, remained almost constant for the last three minutes, and will never drop to the ambient temperature. The reverse occurs in Figure 8-4(b), which shows the effect of compressing the horizontal scale: now the curve seems to say that temperature plummets downward and it will be only a minute or two until the ambient temperature is reached. Neither curve creates the correct impression, although technically the graphs are accurate.

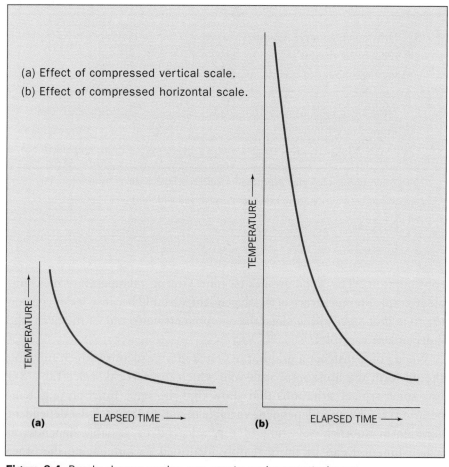

Figure 8-4 Poorly chosen scales can create an inaccurate image.

Normally both scales start at zero, which would be the case when the curve is comfortably balanced in the graph area. If it is crowded against the top or right-hand side, then a zero starting point is unrealistic. In Figure 8-1 the curve occupies the top 75% of the graph area. Since no points will ever be plotted below the ambient temperature (which will hover around 23°C), the bottom portion of the vertical scale is unnecessary. This can be corrected by starting the vertical scale at a higher value (say 20°, as in Figure 8-5), or by breaking the scale to indicate that some scale values have been omitted (Figures 8-2 and 8-6).

Adjust a scale's starting point to centre the curve(s)

Multiple-curve graphs should have no more than four curves otherwise they will be difficult to interpret, particularly if the curves cross one another. You can help a reader identify the most important curve by making it heavier than the others (Figure 8-6), and can differentiate among curves that cross by creating different weight lines (Figure 8-7):

Most important curve	–	bold line
Next most important curve	–	light line
Third curve	–	dashes
Least important curve	–	dots

Emphasize the most significant curve(s)

Avoid using colored lines because the average copier reproduces all the lines in one color (usually black).

Simplicity

Simplicity is important in graph construction. If a graph illustrates only trends or comparisons, and the reader is not expected to extract specific

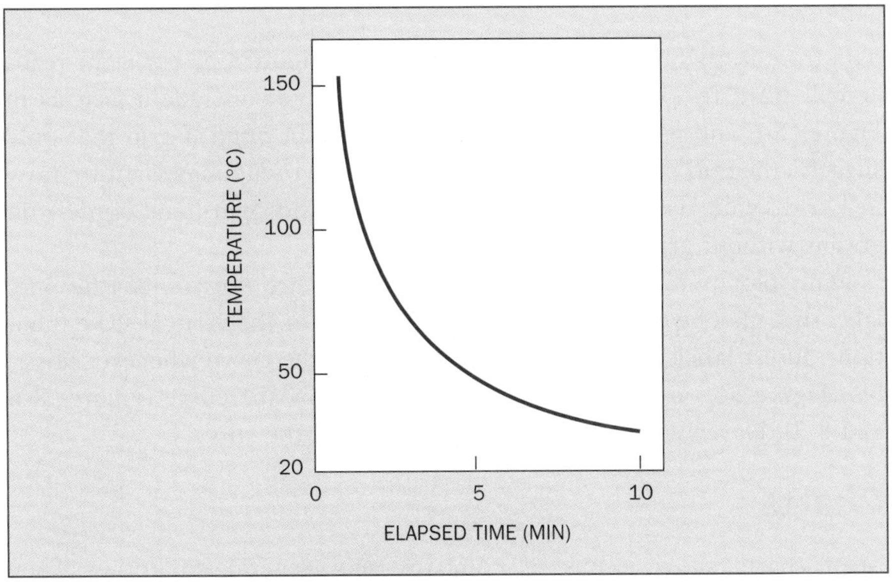

Figure 8-5 A correctly centred curve.

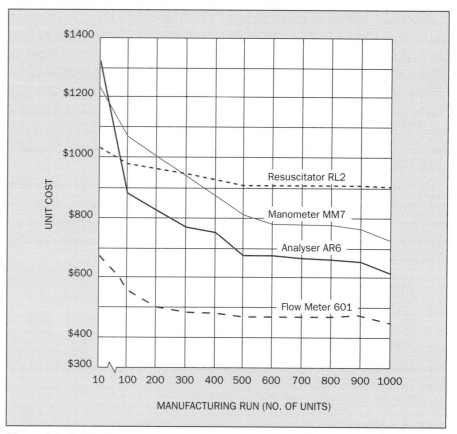

Avoid confusion: limit the number of curves

Figure 8-6 Bolder lines draw attention to the most important curve (those showing maximum benefit from quantity manufacturing). The grid permits readers to draw reasonably accurate data from the graph. (Courtesy Macro Engineering Inc, Toronto, Ontario.)

data from it, then you may omit the grid as in Figure 8-5. But if the reader will want to extrapolate quantities, you should include a grid as in Figures 8-6 and 8-7. Note that Figure 8-1 has an *implied* grid that only suggests the grid pattern for the occasional reader who may want to draw in a grid. Note also that you may omit the top and right-hand borders on graphs without grids, as in Figure 8-3.

Omit plot points and ensure that all labels are *horizontal* (the only label that may be entered vertically is the label for the vertical scale function). Insert labels for the curves at the end of the curve whenever possible (Figure 8-2) or, alternatively, above or below the curve (Figures 8-6 and 8-7). Never write a label along the slope of the curve.

Charts

Most charts show trends or compare only general quantities. They include bar charts, histograms, surface charts, and pie charts.

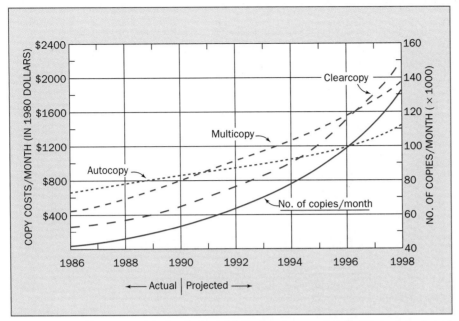

Avoid writing along the slope of the curve

Figure 8-7 Different symbols distinguish between curves showing current and projected copying costs for three copiers. Note the two vertical scales, which permit three functions to be shown on one graph. (Courtesy H L Winman and Associates, Calgary, Alberta.)

Bar Charts

You can use bar charts to compare functions that do not necessarily vary continuously. In the graph in Figure 8-1, John Greene plotted a curve to show how temperature decreased continuously with time. He could do this because both functions were varying continuously (time was passing and temperature was decreasing.) For the production department, however, he has to prepare a report on how long it takes various components coming from the oven to cool to a safe temperature for bare-hand work. He prepares a bar chart to depict this because he knows the report will be read by both management and union representatives, and some of the readers may need easy-to-interpret data. He also has only one continuous variable to plot: elapsed time. The other variable is noncontinuous because it represents the various components he has tested. In this case elapsed time is the dependent variable, and the components are the independent variable. The bar chart John constructs is shown in Figure 8-8.

Scales for a bar chart can be made up of such diverse functions as time, age groups, heat resistance, employment categories, percentages of population, types of soil, and quantities (of products manufactured, components sold, software programs used, and so on). Charts can be arranged

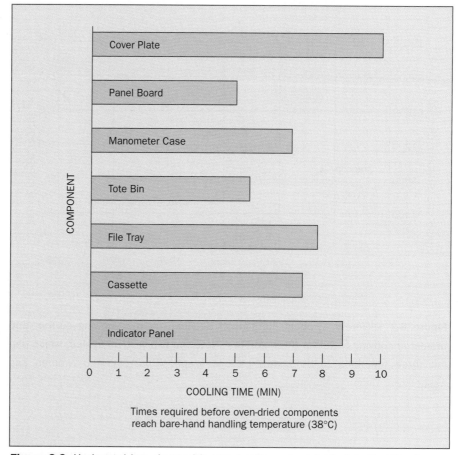

Figure 8-8 Horizontal bar chart with one continuous variable (cooling time).

Use horizontal bar charts to depict elapsed time

with either vertical or horizontal bars depending on the type of information they portray; when time is one of the variables, usually it is plotted along the horizontal axis, as in Figure 8-8. The bars normally are separated by spaces the same width as each of the bars.

In a complex bar chart, the bars may be shaded to indicate comparisons within each factor being considered. The vertical bar chart in Figure 8-9 uses two shades to describe two factors on the one chart, individual bars can also be shaded to show proportional content, as has been done in Figure 8-10, in which case a legend must be inserted beside or below the graph to show readers what each shading represents. Alternatively, each segment may be labelled as in Figure 8-11. A computer-generated 3-D vertical bar graph showing the same information as presented in Figure 8-11 is shown in Figure 8-12.

Some computer software provides only a narrow space between the bars

Horizontal bar charts can be used in an unconventional way by arranging the bars on either side of a zero line. This can be done, for example, to compare negative and positive quantities, satisfactory and defective products, or passed and failed students. The chart in Figure 8-13 divides

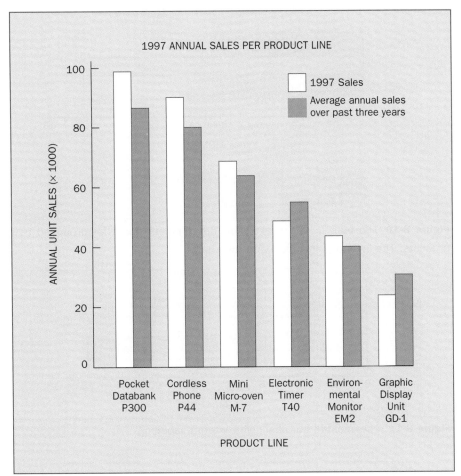

Figure 8-9 A vertical bar chart that lets readers compare current statistics with past statistics, and so determine trends.

products returned for repair into two groups: those that are covered by warranty, and those that are not. Each bar represents 100% of the total number of items repaired in a particular product age group and is positioned about the zero line depending on the percentage of warranty and nonwarranty repairs.

Histograms

A histogram looks like a bar chart, but functionally it is similar to a graph because it deals with two continuous variables (functions that can be shown on a scale to be increasing or decreasing). It is usually plotted like a bar chart because it does not have enough data on which to plot a continuous curve (see Figure 8-14). The chief visible difference between a histogram and a bar chart is that there are no spaces between the bars of a histogram.

Although seen rarely, histograms are useful when there is only limited data

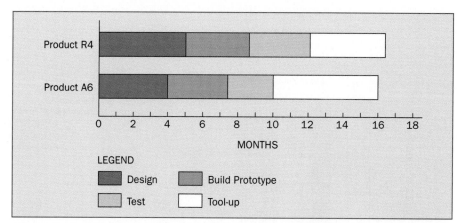

Hand-drawn bar charts offer only limited scope for creative presentations

Figure 8-10 The bars in this chart show development times for proposed new products. The legend is included with the chart.

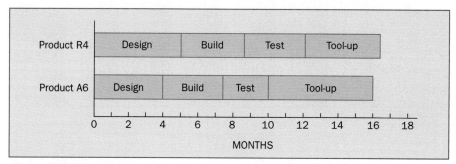

Figure 8-11 A segmented bar chart with internal labelling.

Computer graphics automatically create imaginative presentations

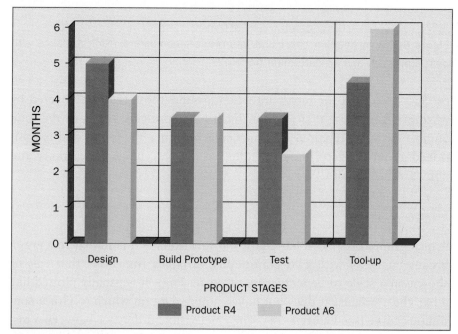

Figure 8-12 A computer-generated 3-D vertical bar graph showing the same information as presented in Figure 8-11.

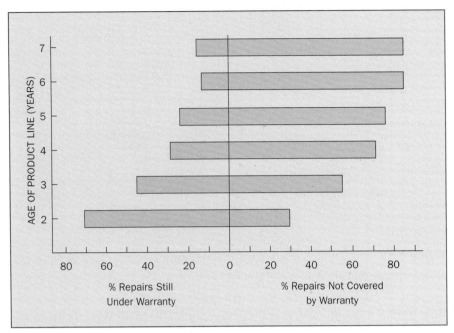

Figure 8-13 A bar chart constructed on both sides of a zero point.

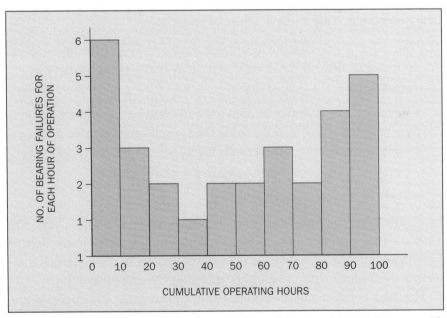

Figure 8-14 This histogram shows the number of bearing failures for every 10 hours of operation. Considerably more data would have been required to construct a curve.

Surface Charts

A surface chart (Figure 8-15) is like a graph, in that it has two continuous variables that form the scales against which the curves are plotted.

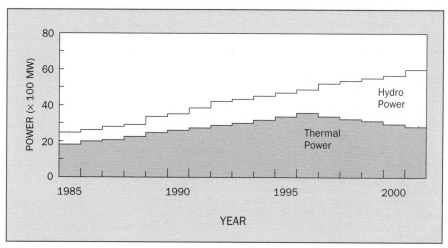

Figure 8-15 This surface chart adds thermal data to hydro data to show projected energy resources of a power utility.

But, unlike a graph, individual curves cannot be read directly from the scales.

The uppermost curve on a surface chart shows the *total* of the data being presented. This curve is achieved as follows:

1. The curve containing the most important or largest quantity of data is drawn in first, in the normal way. This is the Thermal curve in Figure 8-15.

2. The next curve is drawn in above the first curve, using the first curve as a base (i.e. "zero") and adding the second set of data to it. For example, the energy resources shown as being variable in 1995 are:

Thermal Power:	33 000 MW
Hydro Power:	14 500 MW

Difficult to draw by hand, yet useful for depicting a cumulative effect

In Figure 8-15, the lower curve for 1995 is plotted at 33 000 MW. The 1995 data for the next curve is 14 500 MW, which is added to the first set of data so that the second curve indicates a *total* of 47 500 MW. (If there were a third set of data, it would be added in the same way.)

The area between curves is shaded to indicate that the curves represent the boundaries of a cumulative set of data. Normally, the lowest set of data has the darkest shade, and each set above it is progressively lighter.

Pie Charts

Probably the most widely recognized and readily understood

A pie chart is aptly named, because it looks like a whole pie viewed from above with cuts in it ready for people of varying appetites. It is a pictorial

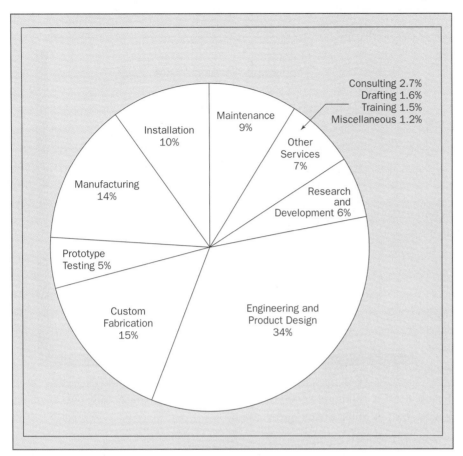

Figure 8-16 Pie chart shows Macro Engineering Inc's products and services. "Slices" add up to 100%.

device for showing approximate divisions of a whole unit. The pie chart in Figure 8-16 depicts the percentage of work done by Macro Engineering Inc in eight major product or service categories. Figure 8-17 is a computer-generated pie chart showing the stages of product development.

If a pie chart has a lot of tiny wedges that would be difficult to draw and hard to read, you may combine some of them into a larger single edge and give it a general heading, such as "miscellaneous expenses," "other uses," or "minor effects." All the wedges must add up to a whole unit, such as 100%, $1.00, or 1 (unity).

Diagrams

Diagrams include any illustration that helps the reader understand the report narrative yet does not fall within the category of graph, chart, or table. It can range from a schematic drawing of a complex circuit to a simple plan of an intersection. There is, however, one restriction: any

A diagram must be readily understood; it should rarely need to be explained

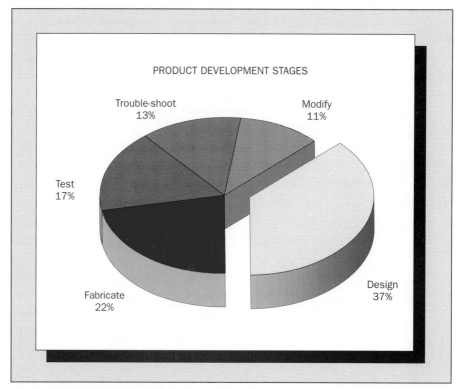

Figure 8-17 A computer-generated exploded view pie chart, with one segment emphasized by being pulled partly away.

illustration included in the narrative part of a report must be clear enough to be understood easily. This means that complex drawings should be placed in an appendix and treated as supporting data.

Diagrams should be simple, easy to follow, and contribute to the narrative. (See Figure 7-3 on page 246 for two good examples.) They can comprise organization charts (see Chapter 2), flow diagrams (Figures 2-4 and 9-1), site plans (Figure 8-18 on page 287), and sketches.

Photographs

A photograph can do much to help a reader visualize shape, appearance, complexity, or size. For example, the photographs of Tina Mactiere and Wayne Robertson add depth to the narrative description of the two engineering companies in Chapter 2. The criterion when selecting a photograph is that it be clear and contain no extraneous information that might distract the reader's attention.

Photographs, however, create difficulties during the printing process. Most photocopiers do not reproduce them well, and to prepare them for offset printing, which permits them to be rendered clearly, is time-consuming and

With digital scanning equipment, it's becoming easier to use photographs

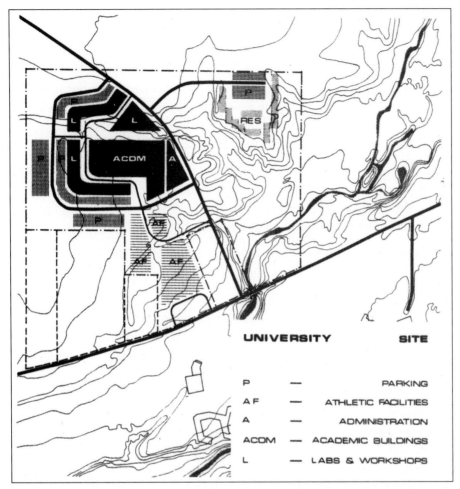

Figure 8-18 A site plan that illustrates where college facilities are to be located. (Courtesy Smith, Carter, Searle—W L Wardrop & Associates Ltd, Winnipeg, Man.)

potentially expensive. Consequently, before deciding to use photographs check whether your printer can reproduce them clearly and economically. If the printer cannot handle them easily, you may have to glue individual copies of the photographs into your report (a time-consuming process), place them in a sleeve or envelope at the back of the report (a cumbersome arrangement), have copies printed professionally (an expensive proposition), or replace the photographs with good sketches.

Tables

A table may be a collection of technical data, as in Tables 8-1 and 8-2, or a series of short narrative statements, as in Tables 8-3 and 11-1. Whether you should insert a table into the report narrative or place it in an attachment or appendix depends on three factors:

1. If the table is short (i.e. less than half a page) and readers need to refer to it as they read the report, then include it as part of the report narrative (i.e. in the discussion), preferably on the same page as the text that refers to it.
2. If readers will be able to understand the discussion without referring to the table as they read the report, but may want to consult the table later, then place the table in an attachment or appendix.
3. If readers will need to refer to a table but the data you have will occupy a full page or more, then

- summarize the table's key points into a short table to be inserted into the report narrative, and
- place the full table in an attachment or appendix.

There are five additional guidelines that apply to tables, particularly if the tables contain columns of numerical data:

1. Keep a table simple by limiting it only to data the readers will *really* need, and create as few columns as possible.
2. Decide whether the table is to be "open" (i.e. without lines separating the columns, as in Tables 8-1 and 8-2) or "closed" (with lines separating the columns, as in Table 8-3).
3. Insert units of measurement, such as decibels, bolts, kilograms, or seconds, at the head of each column rather than after each column entry (see the "min:sec" and "°C" entries at the top of the columns in Table 8-1).
4. Centre a table number and an appropriate title above the table.
5. Draw readers' attention to a table by referring to it in the report narrative and commenting on a specific inference to be drawn from the table. For example:

> The voltage fluctuations were recorded at 10-minute intervals and entered in column 3 of Table 7, which shows that fluctuations were most marked between 8:15 and 11:20 a.m.

Positioning the Illustrations

Whenever possible place each illustration on the same page as or facing the narrative it supports. A reader who has to keep flipping pages back and forth between narrative and illustrations will soon tire, and your reasons for including the illustrations will be defeated.

When reports are printed on only one side of the paper, full-page illustrations can be difficult to position. The only feasible way to place them conveniently near the narrative is to print them on the back of the preceding page, facing the words they support. But this in turn may pose a

Use word-processing software to create tables, for words as well as numbers

Consider illustrations as an integral part of a document, not just an "add on"

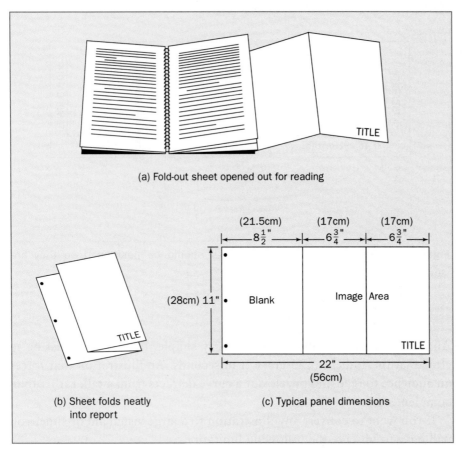

Figure 8-19 Large illustrations can be placed on a fold-out sheet at rear of report.

printing problem. A more logical solution is to limit the size of illustrations so that they can be placed beside, above, or below the words.

When an illustration is too large to fit on a normal page, or is going to be referred to frequently, consider printing it on a foldout sheet and inserting it at the back of the report (see Figure 8-19). If the illustration is printed only on the extension panels of the foldout, the page can be left opened out for continual reference while the report is being read. This technique is particularly suitable for circuit diagrams and flow charts.

Position horizontal full-page illustrations sideways on a page so that they are viewed from the right (see Figure 8-20). This holds true whether they are placed on a left- or right-hand page.

If your reader may want to refer to any figures or tables from your report, we suggest you add a List of Figures and Tables right after your Table of Contents. List the figure or table number, the title or caption, and the page number.

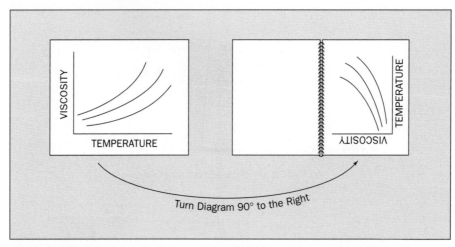

You align horizontal illustrations this way, even if some words appear to be upside down

Figure 8-20 Page-size horizontal drawings should be positioned so they are viewed from the right.

Illustrating a Talk

An illustration for a talk must be utterly simple. Its message must be so clear that the audience can grasp it in seconds. An illustration that forces an audience to read and puzzle out a curve detracts from a talk rather than complements it.

If you want to convert any illustration to a large visual aid or slide, you will have to observe the following limitations:

Never project a visual taken directly from a printed document

- Make each illustration *tell only one story*. Avoid the temptation to save preparation time by inserting too much data on one chart. Be prepared to make two or three simple charts in place of a single complex one.
- Use bold letters large enough to be read easily by the back row of your audience.
- Use very few words, and separate them with plenty of white space.
- Give the illustration a short title.
- Insert only the essential points on a graph. Let the curves tell the story, rather than bury them in construction detail.
- Accentuate key figures and curves with a bold or colored pen (but remember that from a distance some colors look very similar).
- Avoid clutter—a simple illustration will draw attention to important facts, whereas a busy one will hide them.
- Test your visuals to make sure people sitting in the back of the room can read them.

If you are preparing to illustrate an in-plant briefing you may hand-letter your tables, charts, and graphs with a felt pen on large sheets. But if you are preparing to present a technical paper before a conference audience, you must prepare your illustrations more professionally. In the first instance you are expected to produce a standard job at no great expense. In the second you are expected to convey an image of yourself and the company you represent; in effect, the quality of your illustrations will demonstrate the quality of your company's products or services.

If you prepare your own illustrations, you can use one of the currently available computer software programs to produce good quality tables, charts and diagrams. Refer to the program's accompanying documentation of available styles for charts, graphs and tables. Most inkjet and laser printers accept transparencies so you can print your visuals directly onto them. See the section about *Computer-Designed Graphs and Charts* earlier in this chapter for additional guidelines when using a computer to create graphics.

Working with an Illustrator

Although you may prepare your own illustrations, in a large technical organization you may work with a company draftsperson or illustrator who will prepare your drawings, graphs, and charts according to your requirements. Good communication between you and the illustrator is essential if the drawings you want are to appear in the form you visualized when you wrote your report.

An illustrator needs to have much more than a bare, roughed-out sketch to work from. You will have to describe your project and its outcome in detail, so that the illustrator will know

- the background to and purpose of the report or oral presentation,
- who the readers or listeners will be, what their technical knowledge is, and how they will use the information contained in the report or presentation,
- what each illustration is to portray, and what particular aspects are to be emphasized,
- what size each illustration is to be (vertical and horizontal dimensions) and, for a talk, the size of the expected audience and their likely distance from the screen,
- how much the illustrations' size will be reduced when they are printed or converted into slides or transparencies (a drawing that is to be reduced must have lines that are not too fine), and
- when you need the illustrations.

If you're lucky, you will work with an illustrator; more often, you will create your own visuals

You can help an illustrator even more by providing a sketch of each proposed illustration and a copy of the words the illustration is to support. Better still, talk to the illustrator *before* you write your report or make your speaking notes, describe what illustrations you plan to use, and ask for suggestions for their preparation.

Project 8.1: Who Buys "Planit"?

You work for a very successful software company that publishes a monthly user magazine. The company's most successful software has been "Planit," a program for organizing a user's business operations.

Part 1

You have to illustrate a consumer survey

The editor of the user magazine asks you to provide an illustration showing the breakdown of buyers by user groups for last year, when there were 14 236 sales, and suggests you keep the illustration simple. The buyers were:

Hospitals	1588
Small businesses	1011
Public utilities	2165
Consultants	233
Manufacturers	3176
Architects	217
Writers/editors	116
Land surveyors	245
Sales representatives	866
City/town planners	1155
Engineers	1732
Miscellaneous	433
Radio and television stations	1299

Part 2

When you give your illustration to the editor, the response is: "I like that. But I'd also like another one comparing last year's buyers with those of four years ago, which was the first year we marketed Planit."

The sales of Planit four years ago were:

Hospitals	174
Small businesses	1393
Public utilities	1132
Consultants	174
Manufacturers	2351
Sales representatives	958
City/town planners	1306

Engineers	784
Miscellaneous	261
Radio and television stations	174

Project 8.2: Comparing Electricity Costs

Your company markets heat pumps, and wants to demonstrate to electricity users that a heat pump can significantly reduce their electricity bills. Prepare an illustration based on the following actual monthly bills for four different dwellings last year. The residences are identical five-room bungalows built at the same time and in the same block on Margusson Avenue. The only differences are that No. 216 and 234 do not have air-conditioning (marked "No AC" in the table), while No. 227 and 248 do (marked "+AC"), and No. 234 and 248 each have a heat pump.

Monthly Electricity Bills

	Actual Bills No Heat Pump		Amount Saved With Heat Pump	
	216	227	234	248
	No AC	+AC	No AC	+AC
January	$ 48.07	$ 49.23	$ 12.06	$ 11.59
February	45.15	46.01	11.68	11.86
March	43.20	42.86	10.90	11.13
April	41.11	42.20	4.15	3.80
May	38.37	42.15	0.36	3.48
June	35.20	48.16	—	6.10
July	33.06	57.19	—	18.26
August	32.87	56.80	—	17.44
September	34.11	47.10	2.64	7.21
October	38.62	39.20	7.85	8.60
November	41.67	41.10	9.62	10.51
December	43.20	42.89	11.58	11.77

Use these figures to create an easy-to-read illustration

Project 8.3: Assessing Relay Life

Eight years ago the Interprovincial Telephone Company bought and installed 90 000 relays, to be used in a long-range testing program that would assess failure rates. The relays purchased were:

Type	Quantity	Type	Quantity
Nestor 221	40 000	Macro R40	20 000
Vancourt 1200	20 000	Camrose Series 8	10 000

As relays failed they were replaced and the failures were recorded and totalled for each year:

		Number of Failed Relays		
Year	Nestor 221	Vancourt 1200	Macro R40	Camrose 8
1	1 123	901	1 180	105
2	1 080	805	690	124
3	1 007	762	321	131
4	1 076	813	279	306
5	1 140	878	322	402
6	1 656	910	415	545
7	2 210	956	478	609
8	3 303	1 012	584	891
Totals	12 595	7 037	4 269	3 113

It's difficult to predict ahead from a table, but much simpler on a graph or chart

Prepare a graph or chart depicting these failure rates and predicting failures for the next five years.

Project 8.4: Illustrations Related to Other Projects

Prepare a graph, chart, table, or diagram to accompany the report you would write for the following projects in Chapter 6.

1. Project 6.1

Prepare a site map of your city (or part of your city) and on it show the three or four potential office locations for the local branch of H L Winman and Associates. Use the description of each site to select an imaginary location in each case. If you have also identified a fourth (real) location, include it in the illustration.

2. Project 6.2

Prepare a comparison table to accompany the report evaluating highway paints.

3. Project 6.3

Prepare a comparison table to accompany the report describing your research into computers for the Association of Small Business Operators.

4. Project 6.4

Prepare a graph or chart to show how sound levels differ under the four sets of conditions at Mirabel Realty:

Be wary! The positions of the measuring points may skew your illustration

- When only Mirabel Realty is working.
- When both Mirabel Realty and Superior Giftware are working.
- When only Superior Giftware is working.
- When neither office is working.

(Prepare more than one illustration if necessary.)

Include a "normal" office working sound level. Remember that your illustration(s) should *compare* the sound levels, not simply show the levels recorded for the four sets of conditions.

Explain why you have chosen a particular type of graph or chart. Also explain how your graph or chart helps convey the idea to the reader that, when both offices are working or only Superior Giftware is working, the sound level *increases* as you approach the north wall.

WEBLINKS

Technical Illustration
www.arcm.com/illustra.html

This site contains a discussion of the purpose of technical illustrations and has a link to descriptions and samples of product renderings, exploded diagrams, and cutaway views.

Illustration Samples from NASA Lewis Research Center
lerc.nasa.gov/WWW/STI/graphics/samillus.htm

Several examples of detailed illustrations from NASA's Lewis Research Center are available at this site. A contact address is included for readers who want more information about the preparation of technical illustrations.

Graphical Excellence
darkstar.engr.wisc.edu/zwickel/397/graphexc.html

Excellence in statistical graphics consists of complex ideas communicated with clarity, precision, and efficiency. This document gives the technical writer some guidelines for creating useful illustrations. The site is part of the technical writing handbook of the Department of Engineering Professional Development at the University of Wisconsin.

Graphics Guidelines
darkstar.engr.wisc.edu/zwickel/397/graphguide.html

Also from the technical writing handbook of the Department of Engineering Professional Development at the University of Wisconsin, this site describes the characteristics of pie charts, bar and column charts, line charts, schedule charts, flowcharts, and tables.

Chapter 9
Technically-Speak!

This chapter covers two facets of public speaking, both concerned with the oral presentation of technical information. The first is the oral report, sometimes called the technical briefing, delivered to a client or one's colleagues. The second is the technical paper presented before a meeting of scientific or engineering-oriented people. Both depend on carefully honed public speaking skills for their effectiveness, although neither requires vast experience or knowledge in this field. The chapter also describes how to present information at and contribute effectively to office meetings.

The Technical Briefing

Your department head approaches your desk, a letter in hand, and says:

> "Mr Dawes has had a letter from RAFAC Corporation. They're sending in some representatives next Tuesday. I'd like you to give them a rundown on the project you're working on."

Every day visitors are being shown around industrial organizations, and every day engineers and technicians are being called upon to stand up and say a few words about their work. On paper, this sounds straightforward, but to those who have to make the presentation it can be a traumatic experience. Much of their nervousness can be reduced (it can seldom be entirely eliminated, as any experienced speaker will tell you) if they learn a few simple public speaking techniques. The best training, of course, is practical experience, which can be gained only by standing up and doing the job.

Establish the Circumstances

Your first step after being told you have to deliver an oral report is to establish the circumstances affecting your presentation. Go to your

Many people fear hav-
ing to stand up and
speak, more even than
sky diving

department head—or the person who is arranging the event—and ask four questions. The first two will identify your listeners:

1. **Who Will Be in My Audience?**
   If you are to focus your presentation properly, and use appropriate terminology for the people you will be addressing, you need to know whether they will be engineers and technologists knowledgeable in your area of expertise, technical managers with only a general appreciation of the subject, or laypersons with very little or no technical knowledge.

2. **What Will They Know Already?**
   If you are to avoid boring your listeners by repeating information they already know, or confusing them by omitting essential background details, you need to find out how much they know now about your subject, or will have been told before you address them.

It's just like writing a report: first, identify your audience

The second two questions deal with the briefing itself:

3. **How Long Do You Want Me to Talk?**
   Find out if you are to describe the project in detail or simply touch on the highlights. The answer will directly influence how deeply you cover the topic.

4. **Where Is the Presentation to Take Place?**
   Identify whether you are to make your presentation in your company's conference room or training room, or at a client's or some other premises. Within your own company you can easily identify what audio-visual facilities are available, and where your audience will be seated in relation to you. If you will be speaking at another location you should ask for a description of the facilities or, even better, a chance to view them in advance.

When you have firmly established these factors, you can start making notes. Jot down the topics you intend to discuss, and arrange them in an interesting, logical order. But be careful to avoid letting your familiarity with the subject blind you to characteristics that are unimportant to you but might be interesting to an unenlightened listener.

Find a Pattern

A briefing is like a progress report...

The best technical briefings follow an identifiable pattern, just as written formal reports do. You can establish a pattern for your briefing by mentally placing yourself in your listeners' shoes and asking yourself three questions.

1. **What Are You Trying to Do?** The answer will help you build your *Introduction*, as you would for a formal report. Offer your listeners

some background information, which may comprise

- how your company became involved in the project (with, perhaps, a comment on your own involvement, to add a personal touch),
- exactly what you are attempting to do (in more formal terms, your objectives), and
- the extent or depth of the project (i.e. its scope).

2. **What Have You Done So Far?** This is equivalent to the *Discussion* section of a formal report. Your answers should cover

- how you set about tackling the project,
- what you have accomplished to date (work done, objectives achieved, results obtained, and so on), and
- preliminary conclusions you have reached as a result of the work done (if the work is complete, these will be the final conclusions).

...describe what you have done; say what you're doing now; outline what you plan to do next

3. **What Remains to Be Done?** (or **What Do You Plan to Do Next?**) This question is relevant only if the project is still in progress, in which case it is equivalent to the *Future Plans* section of a written progress report. Your answers to this question should cover

- the scope of the planned future work,
- results you hope to achieve, and
- a time schedule for reaching specific targets and final completion.

If the project is complete, this question is not relevant and is replaced by an alternative question: **What Are the Results of Your Project?** The answers are then combined with the final answer to question 2, and as such are equivalent to the *Conclusions* section of a written report.

Now you have a pattern for the main part of your presentation. But, as the flow diagram in Figure 9-1 shows, you still need to start with a quick synopsis of the project in easy-to-understand terms—the equivalent of a report *Summary*—and to end by repeating the key points and the main outcome (this becomes a *Terminal Summary)*. Then invite your listeners to ask questions.

Prepare to Speak

Make Speaking Notes

Prepare your speaking notes on prompt cards no smaller than 150 × 100 mm. Write in large, bold letters that you can see at a glance, using a series of brief headings to develop the information in sufficient detail. A specimen prompt card is shown in Figure 9-2.

The amount of information you include will depend on the complexity of the subject, your familiarity with it, and your previous speaking experience. As a general rule, the notes should not be so detailed that you can-

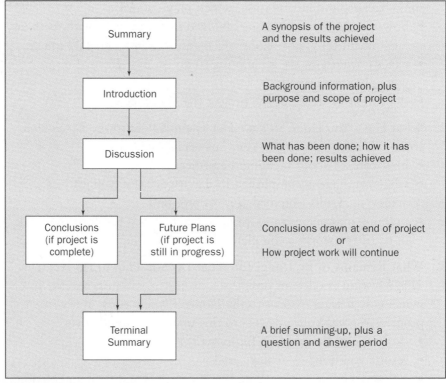

Figure 9-1 Flow diagram for a technical briefing.

Use the pyramid approach to structure your briefing

not extract pertinent points at a glance. Neither should they be so skimpy that you have to rely too much on your memory, which may cause you to stumble haltingly through your presentation.

Prepare Visual Aids

Visual aids can help you give a clearer, more readily understood briefing. They may range from a series of steps listed as headings on a flip chart, through computer-generated slides projected by an LCD panel, to a

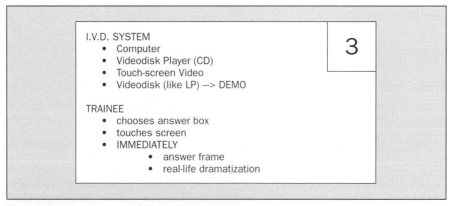

Figure 9-2 Prompt card for an oral report.

Write in bold letters that can be read from 0.7 to 0.9 metres distance working model that demonstrates a complex process. Here are some essential hints for creating effective visuals.

- Strive for simplicity: let each visual make just one point. A visual aid should support your spoken narrative; you should rarely have to explain it.
- Use large, bold letters that will be visible from the back of the room. Ideally, use upper and lower case letters rather than all capitals.
- Use color to accentuate key words or parts, but in moderation. Some colors, such as green and blue or red and orange, are difficult to tell apart from a distance.
- Place a short title above or below each slide.

Let each slide provide just one message

If you will be preparing computer-generated slides using a software program such as *Power Point* or *Astound*, follow these guidelines:

- Select a design that is appealing and provides good contrast between the background color and the color of the lettering you will place in front of it.
- Avoid using dazzling transitions that thrill the audience but may take attention away from the message in the slide. Choose transitions that make a smooth, effective, but conservative change between one slide and the next, and maintain continuity by using similar transitions throughout the presentation.
- Create a natural progression from one slide to the next. If you are presenting, say, four bulleted points on a slide, bring in each point with each successive slide. Also, when you bring in a new point, show it in dark, bold type, but let the previous points be in softer, less bold type.
- Print copies of your slides onto 8 1/2 × 11 inch sheets, four or six slides to a page. You will need these as a prompt when presenting the slides.

Resist the temptation to use fancy transitions!

There are additional suggestions for preparing visual aids such as graphs, charts, and diagrams in Chapter 8.

Practise working with your visual aids, first on their own and then as part of the whole presentation. This gives you a chance to check whether you have keyed them in at the correct places, and whether the entries in your notes are sufficiently clear to permit you to adjust from speech to a visual aid and then back again without losing continuity.

Don't overlook the practical aspects of the briefing. If you have equipment to demonstrate, consider its layout in relation to the sequence of your presentation. Try to arrange the briefing so that you will move progressively from one side of the display area to the other, instead of jumping back and forth in a disorganized way. If the display is large and easy to see, let it remain unobtrusively at the back of the area. If it is small, consider moving it forward and talking from beside or behind it.

Practise, Practise, Practise

Take a leaf from the experienced technical speaker's notebook and practise your briefing. Run through it several times, working entirely from your prompt cards, until you can do so without undue hesitation or stumbling over awkward words. If the cards are too hard to follow, or contain too much detail, amend them. Then ask a colleague to sit through your demonstration and give critical comments.

Practise...practise... practise

Modify your notes after each practice reading. Where necessary, insert more information; in other places, delete unneeded words. As you grow familiar with the notes you will find that your confidence increases and certain sentences and phrases spring readily to mind at the sight of a single word or topic heading; this will help you to maintain oral continuity.

Time yourself each time you rehearse your presentation. Aim to speak for slightly less time than allowed; for example, plan to speak for 17 or 18 minutes for a talk scheduled to last 20 minutes. This will give you time to include some previously unanticipated remarks, should you want to do so at the last minute.

Now Make Your Presentation

Control Your Nervousness

There are very few people who are not at least a little nervous when the time comes to stand up and speak before an audience. Some nervous tension is perfectly normal and can even help a speaker give a better performance. You will find that, once you start speaking, your nervousness will gradually decrease. A lot depends on the quality of your speaking notes: if you have done a thorough job preparing them, know they are reliable, and have practised using them, you will find the familiar phrases and sentences form easily. Then you will begin to relax and so speak with even greater confidence, which in turn will help you relax even more.

Tell Your Story Three Times

Figure 9-1 shows an opening summary, a central discussion, and a closing or terminal summary. We call this the "Tell—Tell—Tell" method of presentation: you tell your story three times:

Start by telling your audience where you plan to take them...

Tell 1:	Tell your readers what they most need to hear: the key points. Then outline very briefly the main topics you will cover.
Tell 2:	Now provide all the details, in the same order you mentioned them in *Tell 1*.
Tell 3:	Sum up by very briefly repeating the key points, and possibly by offering a recommendation.

...end by telling them where they have been

Capture Audience Attention

Grab your listeners' attention by offering an interesting start: tell your audience immediately where you are going to take them, and why you are going to take them there. Here are two examples:

A dull start: "Today, I want to tell you about the effects of poor quality control when manufacturing electrical products. This happened in the fall of 1997, at our plant in Montreal. The problem began when a supplier failed to monitor production quality adequately…"

An exciting start: "Have you, in your company, ever built a better mousetrap? A product that significantly outclasses the competition? Well, we did, in the spring of 1997. It was an immediate success and we had to start a second production line to keep up with the demand. But five months later we had a disaster on our hands: warranty returns were reaching an unprecedented 30%!! The reason: poor quality control at one of our suppliers' plants…"

Get off to a flying start!

Sharpen Your Platform Manner

Knowing some elementary platform techniques can help improve your performance. There are ten:

1. Arrive early and check that the projector, computer, LCD panel, and microphone work. Simultaneously, identify the light switches you will need to control, if you have to dim the lights when presenting video programs or LCD slides.
2. Appear businesslike and cheerful.
3. Speak from notes. You will lose contact with your audience if you read from a prepared speech.
4. Let your enthusiasm *show*. If you present your information vigorously, your audience will see that you really enjoy talking about your subject and will listen more attentively.

Be an energetic, enthusiastic speaker

5. Look at your audience. Try to speak to individuals in turn, rather than the group as a whole. Pick out someone in one part of the room and talk to that person for a few moments, then turn to someone in another part of the room. Let each listener feel he or she is being addressed personally.
6. Use humor sparingly, and only if it fits naturally into your presentation. Never insert a joke to "warm up" an audience. If you do use humor, make sure your audience is laughing *with* you, not *at* you.

7. Speak at a moderate speed. We recommend 120 to 140 words per minute.
8. Speak up. If possible try speaking without a microphone, since this gives you much greater freedom of movement and tonal flexibility. If the room is large and you have to use one, try to obtain a lavalier (travelling) microphone that clips onto your clothing and has a long cord. Better still, ask for a radio microphone.
9. Pause occasionally to study your speaker's notes. Never be afraid to stop speaking for a few moments to consolidate your position and establish that you have covered every major topic.
10. Avoid distracting habits that divert audience attention. For example, avoid pacing back and forth or balancing precariously on the edge of the platform (the audience will be far more interested in seeing whether you fall off than in following your topic). Also avoid nervous behavior, such as jingling keys or coins in your pocket (put them in a back pocket, out of reach), playing with objects on the speaker's table (remove them before you start speaking), or cracking your knuckles.

Watch your body language; make it work for you, not against you

Reach Out to Your Audience

Although good pre-platform preparation and knowledge of platform techniques can give you confidence, they are not sufficient in themselves to break down the initial barrier between speaker and audience. Successful speakers develop a well-rounded personality they use continuously and unconsciously to establish a sound speaker-audience relationship. When you speak to an audience only once, and then only briefly, probably the most important attributes to develop are enthusiasm and sincerity: enthusiasm about your topic, and sincerity in wanting to help your audience learn about it.

The time and effort you invest in preparing for a briefing will depend on your confidence as a speaker and your familiarity with the subject. The more confident you are, the less time you will need to prepare. No one will expect you to give a fully professional briefing at your first attempt, but your listeners (and your employer) will appreciate your efforts when they see that you have prepared your talk carefully and are presenting it in an interesting manner.

The Technical Paper

Chapter 7 discussed the steps Mickey Wendell would have to take to publish a magazine article or technical paper. (He is a senior technologist in H L Winman and Associates' materials testing laboratory, and he has discovered that an additive called Aluminum KL mixed with cement in the right proportions produces a concrete with high salt resistance.) This chapter assumes that the papers committee of the Combined Conference on Concrete liked Mickey's abstract and summary, and the chairperson of the committee has notified him that his paper has been selected for presentation at the forthcoming Toronto conference. Mickey has four months to prepare for it.

Speaking at a conference can create even greater stress

Presenting a paper before a society meeting is more demanding than delivering the same information at a technical briefing. The occasion is more formal, the audience usually is much larger, and the speaker is working in unfamiliar surroundings. Yet the guidelines for preparing and presenting a technical briefing, discussed earlier in this chapter, still apply.

Unfortunately, many experienced engineers and scientists duck their responsibility to the audience and simply read their papers verbatim. The result is a dull, monotonous delivery that can turn even a superior technical paper into a dreary, uninteresting recital. The key is to start preparing early, to make good speaker's notes and to practise speaking from them.

The spoken version of a technical paper does not have to cover every point encompassed by the written version. In the 15 to 20 minutes allotted to speakers at many society meetings, there is time to present only the highlights—to capture listeners' interest, so they will want to read the published version.

Before presenting his paper, Mickey Wendell needs to prepare three versions of his speaking notes. The first will consist mainly of brief topic headings derived from the written paper. He should jot these headings onto a sheet of paper and then study them with four questions in mind:

1. Which points will prove of most interest to the audience?
2. Which are the most important points?
3. How many can I discuss in the limited time available?
4. In what order should I present them?

When Mickey was writing his paper, he was preparing information for a reader. Now he is preparing the same information for a listener, and the rules that guided him before may not apply. The logical and orderly arrangement of material prepared for publication is not necessarily that which an audience will find interesting or easy to digest.

Focus attention on the paper's most interesting features

Mickey may assume that the audience at a society meeting is technically knowledgeable, has some background information in the subject area, and is interested in the topic. Most of his audience will likely be civil engineers and technologists, with a sprinkling of sales, construction, and management people. He must keep this in mind as he examines his list of headings, identifies which points he intends to talk about, and arranges them in the order he feels will most suit his listeners. This version he will use for his practice sessions.

(Mickey should avoid writing speaking notes in the margin of his printed paper. The sequence may seem illogical to his audience and, if he is nervous, he may be tempted to read rather than speak extemporaneously.)

Next, Mickey needs to practise speaking from his notes, just as he would for a technical briefing. It will not be enough to scan or read the notes to himself and assume he is becoming familiar with them. He must speak from them out loud, as though he is presenting the paper to an audience.

Then, when he has practised enough and feels his preliminary notes are satisfactory, Mickey can prepare his final speaking notes. These he should keystroke with a minimum 14 pt typeface, or handletter them (in black ink) in clearly legible letters.

Mickey should plan to have his final notes ready at least three days before leaving for the conference. To a certain extent the headings in the first set of notes have helped to trigger familiar phrases and sentences. Now he has to familiarize himself with new pages. During these last practice sessions, he should hold a full dress rehearsal, presenting his paper before some of his colleagues with, among them, someone qualified to comment on his platform techniques. If this is not possible, he should at least try speaking the paper alone, standing at a rostrum or desk to simulate actual conditions. This "dry run" will also give him the opportunity to check speaking time.

Taking Part in Meetings

We all have occasion to attend meetings. In industry you may be asked to sit on a committee set up for a multitude of reasons, from resolving technical problems that are tying up production to organizing the company's annual picnic. The effectiveness of such meetings is controlled entirely by those taking part. Meetings attended by people *aware of their role* as participants can move quickly and achieve good results; those attended by individuals who seize the opportunity to air personal complaints can be deadly dull and cripple action. Unfortunately, cumbersome, long-winded meetings are much more common than short, efficient ones.

Meetings can be either structured or unstructured, depending on their purpose. A structured meeting follows a predetermined pattern: its chairperson prepares an agenda that defines the purpose and objectives of the meeting and the topics to be covered. The meeting then proceeds logically from point to point. An unstructured meeting uses a conceptual approach to derive new ideas. Only its purpose is defined, since its participants are expected to introduce suggestions and comments that may generate new concepts (this approach is frequently called "brainstorming"). We will discuss the structured meeting here, because you are much more likely to encounter it in industry.

A meeting is composed of a chairperson and two or more meeting participants, one of whom often is appointed to be secretary for that particular meeting. (The secretary makes notes of what transpires during the meeting and, after the meeting, writes the "minutes," or meeting record.) Each person's role is discussed here.

The chairperson may run the meeting, but the participants control its progress

The Chairperson's Role

Good chairpersons are difficult to find. A good chairperson controls the direction of a meeting with a firm hand, yet leaves ample room for the participants to feel they are making the major contribution. The chairperson must be a good organizer, an effective administrator, and a diplomat (to smooth ruffled feathers if opinions differ widely). Much of the success of a meeting will result from the chairperson's preparation before the meeting starts, and ability to maintain control as it proceeds.

Prepare an Agenda

Approximately two days before the meeting the chairperson should prepare an agenda of topics to be discussed and circulate it to all meeting participants. The agenda should identify:

- The date, time, place, and purpose of the meeting.
- The topics that will be discussed (numbered, and in the sequence they will be addressed), divided into two groups:

     Action Items
     Discussion Items

Place the most important points at the top of the agenda

  This arrangement ensures that participants deal with the most important items early in the meeting.

- The person who is delegated to record the meeting's minutes.

A typical agenda is shown in Figure 9-3.

Run the Meeting

The chairperson's first responsibility is to start the meeting on time: a person with a reputation for being slow in getting meetings started will encourage latecomers. The second responsibility is to keep the meeting as short as possible without seeming to "railroad" decisions. The third responsibility is to maintain control.

The meeting should be run roughly according to the rules of parliamentary procedure. (Since most in-plant meetings are relatively informal, full parliamentary procedure would be too cumbersome.) The chairperson should introduce each topic on the agenda in turn, invite the

MACRO
ENGINEERING INC.

600 Deepdale Drive, Toronto ON M5W 4R9

To: Members, Electronic Facsimile Research Committee

The monthly meeting of the Electronic Facsimile Research Committee will be held in conference room B at 3 p.m. on Friday, September 19, 1997. The agenda will be:

Action Items:

1. Accelerated completion of DPS-2A ultra-narrowbeam system installation. *(R Taylor)*
2. Purchase of three portable computers. *(W Frayne)*
3. Proposals: Papers for 1998 International Electronics Conference. *(D Thomashewski)*

Name **the person who will speak about each particular topic**

Discussion Items:

4. Test program report: high speed feed HS-4. *(C Bundt)*
5. Proposal for development of CD-ROM data storage unit. *(R Mohammed)*
6. Plans for annual Research Division Banquet. *(C Tripp; J Kosty)*
7. Other business: (Please send topics to me by 10 a.m. Thursday, September 18.)

This month's meeting secretary: J Kosty.

Daniel H Thomashewski

Daniel K Thomashewski
Chairperson, EFR Committee
September 13, 1997

Figure 9-3 An agenda for a meeting.

person specializing in the topic to present a report, then open the topic for discussion. The discussion offers the greatest challenge for the chairperson, who must

- permit a good debate to generate among the members, yet steer a member who digresses back to the main topic,
- sense when a discussion on a subject has gone on for long enough, then be ready to break in and ask for a decision, and
- know when strong opinions are likely to block resolution of a knotty problem, and assign one person or a small subcommittee to investigate further and present the results at the next meeting.

An effective chairperson encourages reluctant contributors, dampens over-enthusiastic ones

Sum Up

Before proceeding from one topic to the next, the chairperson should summarize the outcome of the discussion on the first topic. The outcome may be a general conclusion, a consensus of members' opinions, a decision, or a statement of action defining who is to do what, and when. In this way all members will be aware of the outcome, and the secretary will know what to enter into the minutes.

Effective "summing up" helps the secretary identify what should be said in the minutes

The chairperson should also sum up at the end of the meeting, this time reviewing major issues that were discussed and the main results. At the same time, the chairperson should point the way forward by mentioning any important actions to be taken and the time, date, and place of the next meeting (presuming there is to be one).

The best way to learn to be a good chairperson is to watch others undertake the role. Study those who seem to get a lot of business done without appearing to intrude too much in the decision making. Learn what you should not do from those whose meetings seem to wander from topic to topic before a decision is made, have many "contributors" all speaking at the same time, and last far too long.

The Participants' Role

You can contribute most to a meeting by arriving prepared, stating clearly your facts, ideas, and opinions when called on, and keeping quiet the remainder of the time. If you observe these three basic rules you will do much to speed up affairs. Let's examine them more closely.

Come Prepared

If a meeting is scheduled to start at 3:00 p.m. do not wait until 2:30 to gather the information you need. Arriving with a sheaf of papers in your hand and shuffling through them for the first 15 minutes creates a disturbance and makes you miss much of what is being said. Start gathering information as soon as the agenda arrives, sort your information to identify the specific items you need, then jot down topic headings and details you may want to quote.

Plan your presentation like a technical briefing

Preparation becomes even more important if you have been researching data on a particular topic and will be expected to present your findings at the meeting. Start by dividing your information into two compartments:

1. Facts your listeners *must* have if they are to fully understand the case you are making and reach a decision (if, for example, you are requesting their approval to take a specific course of action). These become "Need to Know" facts.
2. Details your listeners only *may* be interested in, and do not necessarily need to understand your case or reach a decision. These are "Nice to Know" facts.

Your plan will be to present only the **Need to Know** facts in your prepared presentation, but to have the **Nice to Know** facts ready in case one of the listeners asks questions about them.

The second step is to examine the Need to Know facts to identify the two or three pieces of information your listeners will *most* need to hear. These become your Main Message, or Summary Statement.

The three compartments form a pyramid-style speaking plan, as illustrated in Figure 9-4.

Be Brief

In your opening remarks summarize what your listeners most need to hear from you (your Main Message), and then follow immediately with facts and details from the Need to Know compartment. If you have small amounts of statistical data to offer, either project a transparency onto a

Structure your information like a written report or proposal

Opening remarks to grab listeners' attention:	MAIN MESSAGE	A Summary Statement: what listeners most need to hear
Facts to amplify and consolidate the Main Message:	NEED TO KNOW	Information listeners *must* hear to fully understand the presenter's case
Facts held in reserve:	NICE TO KNOW	Details listeners only might be interested in

Figure 9-4 Plan for presenting information at a meeting.

screen or print copies and distribute them when you begin to speak. (If you have a lot of information to distribute, print copies ahead of the meeting and ask the chairperson to distribute them with the agenda, so that everyone can examine your data before coming to the meeting.)

At the end of your presentation invite questions from the meeting participants. For your answers, draw on the data you have prepared for the Nice to Know compartment, but be ready to analyse your data in depth. If some listeners seem to resist your ideas, try to avoid becoming defensive. Say that you can see their point of view, and then explain why your approach is sound or offers a better alternative than the one they may be suggesting. In particular, avoid getting into a personal confrontation with one or more of the meeting participants. And when the chairperson calls on the members to indicate whether they accept your ideas or will approve your proposal, if they respond negatively be ready to accept their decision gracefully.

Keep your cool; defend your point logically and rationally

Keep Quiet

There are many parts of a meeting when your role is to be only an interested observer. At these times you should keep quiet unless you have a relevant question, an additional piece of evidence, or an educated opinion. Avoid the annoying habit of always having something to add to the information others are presenting (recognize what others already know: you are not an authority on everything). At the same time, do not withhold information if it would be a genuine contribution. Be ready to present an opinion when the chairperson indicates that a topic should be discussed, but only if you have thought it out and are sure of its validity. Recognize too, that a discussion should be a one-to-one conversation between you and the chairperson, or sometimes between you and the topic specialist. It should never become a free-for-all with each person arguing a point with his or her neighbor.

The Secretary's Role

Sometimes a stenographer is brought in to act as secretary and record the minutes of a meeting, but more often the chairperson appoints one of the participants to take minutes. If you happen to be selected, you should know how to go about it.

Recording minutes does not mean writing down everything that is said. Minutes should be brief (otherwise they will not be read), so there is room only to mention the highlights of each topic discussed. Items that must be recorded are: (1) main conclusions reached; (2) decisions made (with, if necessary, the name[s] of the person[s] who made them, or the results of a vote); (3) actions agreed upon, and who is to take the action; (4) what

MACRO
ENGINEERING INC.
600 Deepdale Drive, Toronto ON M5W 4R9

ELECTRONIC FACSIMILE RESEARCH COMMITTEE

Minutes of Meeting

Friday, September 19, 1997, 3:00 p.m.

In Attendance: C Bundt   R Mohammed
W Feldman   R Taylor
W Frayne   D Thomashewski (Chair)
J Kosty (Secretary) C Tripp
Regrets: D Wilton

Minutes	*Action*
Action Items:	
1. The DPS-2A ultra-narrowbeam system is not yet operational because the S-76 interface needs further modification. R Taylor expects modifications to be complete September 24, and a system operational date of September 28.	*R Taylor*
2. The committee approved W Frayne's proposal to purchase three portable computers, two out of this year's budget, one out of next year's. W Frayne and R Mohammed will survey available computers and present a definitive proposal at the October 16 meeting.	*W Frayne* *R Mohammed*
3. Three papers are to be submitted to the program committee for next year's International Electronics Conference. Deadline: October 9.	*E Bundt* *J Tripp* *W Feldman*
Discussion Items:	
4. C Bundt reported that phase 1 of the HS-4 test program was completed August 27, but no further progression	
8. The Research Division banquet will be held in collaboration with the annual Awards Dinner on March 13, 1998. The banquet committee will establish a joint plan with the Awards Committee.	*C Tripp* *J Kosty*

Minutes should reflect key outcomes, not describe what everybody said

J Kosty

J Kosty, Secretary

Figure 9-5 Minutes of a meeting. Note the "Action" column, which draws individuals' attention to their post-meeting responsibilities. (The dotted line indicates a break between two pages and some omitted information.)

is to be done next and who is to do it; and (5) the exact wording of any policy statements derived during the meeting.

The best way to get this information quickly is to write the agenda topics on a lined sheet of paper, spacing them about two inches vertically. In these spaces jot down the highlights in note form, leaving room to write in more information from memory immediately after the meeting.

The completed minutes should be distributed to everyone present, preferably within 24 hours. They should be a permanent record on which the chairperson can base the agenda for the next meeting (if there is to be one), and participants can depend for a reminder of what they are supposed to do. The format shown in Figure 9-5 also provides an "action" column to draw participants' attention to their particular responsibilities.

Write the minutes while the events are still fresh in your mind

ASSIGNMENTS

Speaking situations you are likely to encounter in industry will develop from projects on which you are working. Hence assignments for this chapter are assumed to grow naturally out of the writing assignments presented in other chapters.

Project 9.1: Speaking Situations Evolving from Other Projects

You have to attend a meeting to present the results of a study or investigation you have carried out, where you will brief managers or a client on your findings and recommendations. In each of the following instances, which are drawn from projects in Chapters 4, 5, and 6, you are to list in point form the information you would convey

- as your "Main Message," and
- as your "Need to Know" details.

In some projects—mainly in Chapter 4—you will be able to draw on the details provided in the assignment instructions without first doing the report writing project itself. In others—primarily in Chapters 5 and 6—it will help if you have first completed the study and written the report.

Be ready to present this information orally.

1. Project 4.3, Part 2

Frank Moroni has flown into Regina, Saskatchewan, and is holding a meeting with the owner/manager and operations manager of High Gear Truck and Car Rentals (HGTCR). They are in room 202 of the Ramada Inn. Frank telephones you and asks you to brief him and the two Regina managers about the picketing that is occurring at the car rental outlet, and report on progress concerning the car wash installation. Drive over there and tell them about the problem and its effects.

Prepare to describe the progress and the problems

2. Project 4.3, Part 4

The car wash installation project is now finished. Frank Moroni invites you to fly to Toronto to attend a car wash evaluation meeting to be attended by seven southern Ontario HGTCR branch managers. "During the meeting I'll be calling on you to describe the car wash installation," he says when he telephones you, "and any problems you ran into."

You are now at the meeting, and Frank has just asked you to describe the project.

3. Project 4.7

When you knock on Vern Rogers's office door, your incident report in your hand, you see he is speaking on the telephone. He signals you to come in and sit on his guest chair. When the telephone call ends, he says: "You're looking pretty glum. You have something to tell me?"

You hold up your report.

"No, no," Vern says. "Just tell me about it."

Do so.

4. Project 4.8

Maintenance crew supervisors from microwave sites 1 through 8 are attending a meeting at H L Winman and Associates. Because you have just returned from microwave site 14, where you have been investigating connector problems, Andy Rittman asks you to come in to the meeting and brief the supervisors on your findings.

5. Project 5.1

When you arrive home following the long drive back from Harmonsville (and the helicopter flight to and from Sylvan Lake), you find a message from David Yanchyn, the research engineer at Triton Mining Corporation, on your answering machine:

Describe the results of a field trip

> "Geologist Frances Cheem and I will be in your city tomorrow morning. Can you come to a meeting in the Holiday Inn at 9 a.m., and tell us what condition the access road is in at Sylvan Lake? Thanks."

6. Project 5.2

When you take your finished "computer research" report into Susan Jenkins, she says: "Nice timing! We're having a management meeting tomorrow morning. Bring 10 copies of the report with you and be prepared to describe your findings before you hand it out."

You're now attending the meeting and chairperson Martin Dawes has just said to you: "Right! We're ready to hear your report!"

7. Project 5.3

Robert Delorme telephones you and asks, "Have you finished the Quillicom landfill study?"

You tell him you have, but you have not yet written the report.

"Then I want you to go to the Quillicom Town Council meeting at 7:30 p.m. tonight. The councillors want to hear what recommendations you will be making."

8. Project 5.4
Paulette Machon (vice president of operations at Baldur Agri-Chemicals—BAC) telephones to say she will be coming to your office tomorrow and bringing BAC's manager of human resources with her. Rather than wait for your report to reach them, they want you to brief them on your findings into BAC's power house problems, and then they want to discuss the implications with you.

9. Project 6.1
You have completed the report in which you have investigated new office locations for the local branch of H L Winman and Associates. When you telephone branch manager Vern Rogers to tell him the report is ready, he invites you to come to the office at 2 p.m. the next day, so you can describe your findings to the department managers.

Brief management on the results of your investigation

10. Project 6.2
Highways engineer Claude Auger asks you to brief his highway engineers on the results of your highway paint study.

11. Project 6.3
You are just putting the finishing touches on your personal computer report for the Association of Small Business Operators (ASBO), when ASBO executive director Milton Lajzieks telephones. "We're holding a society administrative committee meeting on Tuesday morning. Could you come in and tell the eight committee members your results? You can have between 10 and 15 minutes."

12. Project 6.4, Part 3
When you deliver your sound level study to Trudy Parsenon, you tell her there is a problem at Mirabel Realty and it's going to cost a fair sum to remedy it.

She telephones you two days later. "Three people from head office will be here next week on a routine visit. I'd appreciate it if you could come in one afternoon to describe your findings to them. You'll do a better job than I will, because you know the study better than I do."

Project 9.2: Informing Technicians of a New Product
You have researched information on a new manufacturing material, method, or process, as described in Project 7.4, and have prepared a written description. Now you have to inform other technicians about the product at a lunchtime briefing organized by your department head.

Make an in-depth presentation to unknowledgeable listeners

Part 1

On a sheet of paper write brief notes in point form identifying what you will say under each of the following topic headings:

- Summary Statement
- Purpose (of product, material, or process)
- Details (what it does and why it is unique)
- Conclusions

Part 2

Make your presentation.

WEBLINKS

Preparing Outstanding Presentations
www.ieee.org/pcs/creimold.html

This series of articles from Cheryl Reimold's "Tools of the Trade" column was published in the Institute of Electrical and Electronics Engineers Professional Communication Society's newletter. Preparing Outstanding Presentations includes sections on understanding your audience; basic presentation structure; the introduction, body and summary of a presentation; effective visuals; and making visuals memorable.

Giving a Scientific Talk: A Guide for Botanists
www.botany.uwc.ac.za/botany/talks.htm

In this article, Derek Keats and Alan Millar argue that oral presentations are an inefficient means of giving information because little of it is retained. Such presentations, however, are effective for stimulating discussion, constructive criticism, and interest among colleagues and students. The authors describe the primary features of a good talk and provide detailed suggestions for creating audio-visual aids suitable for scientific presentations.

The Elements of a Professional Presentation
www.access.digex.net/~nuance/keystep1.html

This site contains detailed information about preparing professional presentations. It includes sections on knowing your subject matter and audience, developing a theme, preparing the script, selecting visual aids, producing visuals, rehearsing, presenting, avoiding deadly sins, and following up.

Visual Aids in Presentations
darkstar.engr.wisc.edu/zwickel/397/presgraf.html

This site, which is part of the technical writing handbook of the Department of Engineering Professional Development at the University of Wisconsin, contains a summary of the use of visual aids in presentations.

Crossing a Bridge of Shyness: Public Speaking for Communicators
www.eeicom.com/eye/shyness.html

Diane Ullius, the author of this article published in *The Editorial Eye*, teaches oral and written communication skills at Editorial Experts Inc. and at Georgetown University. She offers practical advice about getting over the fear of speaking in public.

Chapter 10
Communicating with Prospective Employers

As head of the administration and personnel department of H L Winman and Associates, Tanys Young is responsible for hiring new staff. Recently she advertised for an engineer to coordinate a new project, and received 48 applications from across the country. Since it would have been impractical to interview all the applicants, she narrowed the field down to the nine applicants she felt had the best qualifications, basing her selection on the information contained in the 48 application letters and resumes she received.

One of the applicants was Eugene Koenig of New Westminster, BC, who believed—quite rightly—that he probably was the most qualified person for the job. But his name was not one of the nine on Tanys's short list, and so he was not interviewed. Tanys has since met the nine selected applicants and has offered the job to the most promising person. She will never know that Eugene would have been a better person to hire. And Eugene will never know why he was not considered.

What went wrong? The fault was entirely Eugene's, whose letter and resume failed to persuade Tanys that she should talk to him.

In today's highly competitive employment market, job seekers have to tailor each resume and application letter so that together they capture the interest of a particular employer. (To mail copies of an identical resume and similar letter to every employer is a wasted effort.) They must carefully orchestrate the whole employment-seeking process, from preparing their resumes to presenting themselves personally at an interview.

This chapter describes the various stages of the job-seeking process, and the careful steps that you, as an applicant, must take if you want every employer you contact to consider you seriously as a potential employee. The process starts with the preparation of a personal data record and then proceeds through preparing a resume, writing an application letter, completing an application form, attending interviews, and accepting or declining a job offer.

The Employment-Seeking Process

Figure 10-1 illustrates the five steps a *successful* applicant has to take before being employed. At each step the number of contenders for a particular job is reduced until only one person remains. Tanys Young received letters and resumes from 48 potential employees, but asked only 25 of them—those she felt most nearly met the company's requirements—to complete a company application form. Thus in one step she cut the field in half. Then, from the application forms and resumes, she selected nine people to interview.

The significant factor here is that Tanys narrowed the field down to only 19% of the original applicants *based solely on their written presentations*. Like Eugene Koenig, you cannot afford to be one of the over 80% who were eliminated from the interview stages because they prepared inadequate written credentials.

The five steps of the job-seeking process are outlined below. At each step you have to present a confident, positive image of yourself if you are to proceed to the next step.

1. **Initial Contact.** Your first step as a job seeker is to approach a prospective employer and ask to be considered for employment, either by responding to an advertisement (this may be over the Internet) or by approaching the employer "cold" (in the hope that the employer either has or shortly will have an opening). You may make this initial contact by presenting yourself at the employer's door, by writing a letter or email message, or by telephoning. The personal visit and the letter or email are better, because they let you *place your resume in the employer's hands.*

The image you convey — on paper or online — has a major impact on a potential employer

A resume may be submitted electronically

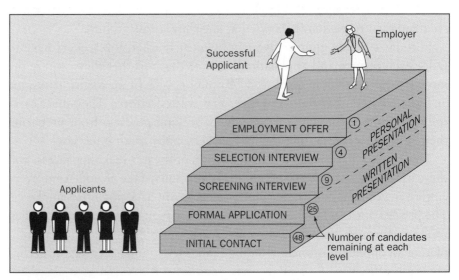

Figure 10-1 The five stages or steps of the job-seeking process.

2. **Formal Application.** In step 2 of the job-seeking process the employer will likely ask you to complete a company application form so that all applicants are documented in the same way.

3. **Screening Interview.** The first interview you attend helps the employer identify which applicants have the strongest potential. In a large firm such as H L Winman and Associates, the screening interview may be conducted by only one person, usually the employment manager or an employment representative.

4. **Selection Interview.** The most promising candidates are asked to attend a second interview. This time the manager of the department where the successful applicant will work also is present, and sometimes is accompanied by technical specialists.

5. **Job Offer.** The employer makes a formal offer of employment to the successful applicant, often first by telephone and then by letter. The applicant responds, also by telephone and letter, to confirm his or her acceptance.

Not all employment-seeking processes develop exactly along these lines. Sometimes an applicant will obtain a company application form beforehand and submit it in step 1, with his or her letter and resume. At other times there may be only one job interview or, in some cases, there may be three or more. With advances in technology and the Internet, the initial contact may be made by the company after retrieving your electronic resume from a database.

Using the Internet in Your Job Search

The Internet has opened up a world of opportunities and possibilities to people seeking employment. It provides a completely new way for employers and potential employees to meet and learn about each other. Employment information and service are just a click away.

You can use the Internet to research information about companies, search for job postings, or enter your electronic resume into a database. Both large and small companies are investing in sites on the World Wide Web and using them to distribute new information to potential customers and to communicate with existing customers. These web sites provide not only product information but also company information, so if you are looking for information about a particular company, its history, its corporate structure, its beliefs and philosophies, check to see if it has a web site. This will help prepare you for an interview or help you decide if you would like to work for that company. Many corporate web sites also list job openings.

You can use the Internet to place an online international classified advertisement. This is a new method for locating a job: online job ad services

You research corporate information online...

...or announce your availability as a potential employee

and electronic employer databases allow job hunters to quickly compile a list of prospects. Gone are the endless days spent reading through newspapers and directories!

Companies from all around the world in all types of industries use the Internet to advertise employment opportunities. Search capabilities allow you to narrow your criteria to specific locations or job requirements. There are also a number of commercially run Internet services that specialize in matching people to positions, and nonprofit Bulletin Board networks that allow anyone to list or look at job openings.

Another way to use the Internet in your job search is to post your resume so potential employers can find you. It not only is a quick and effective way of transmitting your credentials, but also demonstrates your understanding and ability with the new technology. When H L Winman and Associates were looking for a Computer Services Manager, they turned to the Internet and found Susan Jenkins. Since one of the Computer Services Manager's responsibilities would be to give the company a presence on the web, looking for someone presenting themselves on the web was a good place to start the hiring process.

Consequently, we recommend using an Internet resume service as the best way to have your electronic resume seen. These companies specialize in organizing, indexing, and distributing resumes (see the section "The Electronic Resume" later in this chapter for suggestions on preparing this type of resume). Independent database services are companies that match people with jobs.

This technique works particularly well for people with definable education and experience and in-demand skills. For example, engineers, computer scientists, and people in other technical fields fit this category. Conversely, many liberal arts resumes may get left in "cyberspace" never to be matched. In these cases, electronic resume databases may not be the best route for lodging your resume.

Start by contacting a local Internet resume specialist

Developing a Personal Data Record

There are three ways you can go about writing a resume: you can rely solely on your memory; you can dust off and update a previous resume; or you can create a new resume from a permanent personal data record (PDR). Using a PDR is best, because it provides a much broader information base for you to draw on.

A PDR becomes particularly useful in later years, when our ability to recall names, addresses, dates, and specific details of earlier employment diminishes. It can also be invaluable if, when calling initially on a potential employer, you are asked to complete an application form on the premises.

If you do not already have a PDR, prepare one now. You may find it's a pain to start, but is not difficult to update and keep current. There are four topic areas for which you will need to record details (and update them approximately once a year):

Start a databank: store your history online

- Education
- Work Experience
- Extracurricular Activities
- References

You can write them on paper or cards, or store them in a computer or on disk.

Education

List the schools, colleges, and universities you have attended or are attending. Start at junior high school and record the name of each school, the address and telephone number, and the dates you were there, and for high school, your graduation date and your area of specialization. For college and university, particularly note courses taken, special options, and the full name of the degree, diploma, or certificate you were awarded. Include grades or at least a grade point average (GPA) for each year. List any additional courses or seminars you completed and the date attended. Don't forget about courses taken while working for an organization.

Work Experience

For each job you have held in the past—and, if you are currently employed, the job you now hold—list

- the full name, address, and telephone number of the company or organization, and the full name and title of each supervisor you worked for,
- the dates you started and finished employment and, if you held several positions within the company, the name of the position and the date you were appointed to it,
- your job title, or titles if you held several positions,
- your specific responsibilities and duties for each position, paying particular attention to the supervisory aspects and responsibilities of any job that you carried out without supervision,
- any special skills you learned on the job,
- special awards or words of praise you received, or results you achieved, and
- projects you were involved in, including the type of technology you learned or used.

Be as detailed as possible: don't rely on your memory

Extracurricular Activities

List your activities in organizations that were not necessarily part of the job you have held or your education, but which show your participation and leadership qualities. These help identify you as a well-rounded, balanced person. For example:

- Membership in a club, society, or group, particularly noting your responsibilities as an active participator or committee member. (For example, member of sports committee or secretary of administrative committee.)
- Participation in community activities such as the Big Brother or Sister Organizations, YM or YWCA, YMHA, 4-H Club, Parent-Teacher Association (PTA), or local community club. Particularly describe any executive or administrative positions you held, with special responsibilities and dates.
- Involvement in a technical society on a local or national level, with particular mention of any conferences you attended or papers you presented or published.
- Participation on a sports team, with special mention of your role as a team leader or coach.
- Involvement in hobby activities such as stock car racing or rebuilding, a computer club, or dog breeding.
- Awards you have received for any activities you have been involved in.

Don't overlook your personal attributes and life experience

For each activity, include the dates of your involvement and the name, address, and telephone number of a person who can vouch for your participation. Make sure you indicate whether you were elected to the position or are doing the work voluntarily.

References

List the names of people you feel are best fitted to speak on your behalf. They fall into two groups: those who can vouch for your *capabilities* (as an employee, student, or committee member), and those who can speak for your *character*. Always contact these people first and ask if you may list them as a reference. Then, for each person write down

Be thorough: you cannot tell in advance who you may want to use as a reference

- full name, professional title (such as plant engineer), place of employment, and job position,
- employer's address and telephone number,
- home address and telephone number, and
- how long you have known them.

(If a reference has changed jobs, list details of both the previous and current employer.)

For each person you worked with in an extracurricular activity, also list

- the name of the organization you both were involved with, and the reference's position within that organization, and
- whether the person prefers to be called, or written to, at home or at work.

Preparing a Resume

A resume contains key information about yourself, carefully assembled and presented so that prospective employers will be impressed not only by your qualifications but also by your ability to display your wares effectively. (The correct spelling is "résumé," but common usage has made the accentless "resume" acceptable. In some countries it's called a *curriculum vitae*, or CV.)

Technical people tend to be conservative when they write their resumes, yet today's employment environment demands they be *competitive*. If a resume is to capture an employer's attention, it must display its writer's wares to full advantage. The resume is a sales tool and the writer is the product.

The three resumes presented on the following pages range from fairly conservative to clearly provocative. You will have to decide which you want to use, keeping four factors in mind: which will best represent you as an individual; which will best present your qualifications; which will most suit the position you are applying for; and which will most likely appeal to the particular employer.

Your resume style must fit your personality...

For ease of reference, we will refer to the three styles as the traditional resume, the focused resume, and the functional resume. All three have one important feature in common: they open with a summary statement that (1) describes the applicant's strongest qualifications from the *employer's* point of view, and (2) identifies that the writer is seeking work in a particular field. Ideally, there is a logical connection or development between these two pieces of information, and they are presented in a short paragraph of no more than two or three sentences. For example:

OBJECTIVE
I have four years' experience supervising the installation and testing of wire and fibre optic telephone communication systems, and a recent MSc in electronics engineering with a major in fibre optics. I am now seeking employment where I can apply my knowledge and experience in fibre optics engineering.

An assertive statement such as this at the start of a resume draws the employer's attention rapidly to the applicant's primary experience and education, and to the employment direction the applicant wants to pursue. If the resume "hits the mark" successfully, the employer automatically reads further to learn more about the applicant.

To be of most value, the opening statement is focused to suit the needs of a particular employer, or sometimes a group of employers engaged in similar work. The implications for job applicants are far-reaching: now they have to invest much more time, care, and research into resume preparation to ensure their resumes are clearly directed toward a specific audience. However, with the speed and ease of using a word-processor, writing specific resumes for specific jobs is no longer a tedious process.

...and that of the particular employer you are contacting

The Traditional Resume

For decades the most widely recognized approach to resume writing has been to divide a job applicant's information into five parts, each preceded by an appropriate heading:

> Objective
> Education
> Experience
> Extracurricular Activities
> References

The traditional resume is particularly suitable for recent university or college graduates who have limited work experience, or for students who shortly expect to graduate. Alison Witney is a biological sciences undergraduate who has held two previous jobs totalling three years of full-time employment. Her resume is shown in Figure 10-2. Comments on the resume, plus guidelines you can use to write a resume of your own in the traditional format, are presented below and keyed to the circled numbers in the figure.

1 Job applicants with only limited work experience should try to keep their resumes down to one page.

2 Each line of Alison's name, address, and telephone number is centred to give the top of the resume a balanced appearance.

3 There is no need for Alison to list all the primary and secondary schools she attended; it is enough to state the name of her senior high school and the year she graduated. She should then list each college or university she attended, plus the type of course enrolled in, the diploma or degree received, and her year of graduation (or expected graduation). Alison has decided to include her grade point average (GPA) because it is high. This is optional, but if you do include it be sure to include it for all schools.

A short, concise, focused resume is welcomed by employers!

1

2

ALISON V WITNEY
1670 Fulham Boulevard
Truro, NS B2N 6C4
Tel: (902) 474-6318
email: avwitney@nsonline.net

OBJECTIVE
To work in a position related to Animal Biology or Health Science, where I can use to good advantage both my Diploma in Biological Science and my experience as a veterinary assistant.

3

EDUCATION AND TRAINING
• Will graduate with a Diploma in Biological Science from Amiento Technical College, June, 1997 (GPA to date: 3.7).
• Graduate of Morton Stanley High School, Corisand, Nova Scotia, 1993 (avg: 92.3%).

An appealing, uncrowded appearance, coupled with good words, will catch an employer's attention

WORK EXPERIENCE

4

5

1995 to date   **Animal Treatment Centre**, Amiento, NS. Veterinary assistant, responsible for reception, grooming, and exercising of animals, assisting veterinarian during operations, changing dressings, administering injections and anaesthetics, and performing administrative duties such as accounting and ordering of supplies. (One year full-time, two years part-time.)

6

1993 to 1995   **Remick Airlines**, Fredericton, NB. Accounts clerk in air freight department; coordinating billings, preparing invoices, following up lost shipments, assisting clients, and writing monthly reports. For nine months assisted in payroll preparation.

1988 to date   **Bar None Riding Stables**, Corisand, NS. Part-time employment teaching the care and handling of horses, and basic riding techniques, to young riders. Assisted in grooming, cleaning, feeding, and saddling-up.

7

ADDITIONAL INFORMATION
• Winner of two educational awards: Morton Stanley Science Scholarship (1992) and Amiento Technical College Biology Scholarship (1997).
• Member of YWCA since 1989, where I now teach swimming and lifesaving.
• Interests: horseback riding and jumping, swimming, and water skiing.

8

REFERENCES
The following people have agreed to supply references on my behalf:

Dr Alex Gavin
Veterinary Surgeon
Animal Treatment Centre
2230 Wolverine Drive
Amiento, NS B3R 2G1
Tel: 474-1260
Fax: 474-1355
email: a.gavin@atlantic.vet.net

Mr Charles Devereaux
Owner-Manager
Bar None Riding Stables
2881 Westshore Drive
Corisand, NS B4L 3A2
Tel: 632-2292
Fax: 631-3105
email: bar.none@nst.ns.ca

Figure 10-2 A traditional resume or biography of experience.

4 Experience is usually presented in reverse order, with the most recent work experience appearing first and earliest experience last. You should provide more details about recent experience (as Alison has done), and about earlier work that is similar to that of the position you are seeking, than for less-related work. Quote dates as whole years for long periods of employment, but as month and year for short periods; for example, Jun 1991 – Feb 1992.

5 For each employer, state the name of the company or organization first, emphasize it with bold type, and then identify the city and province in which it is located. Then describe the position held (give the job title), and what the work involved. Particularly draw attention to the *responsibilities* of the job rather than merely list the duties you performed. Use words that create strong images of your self-reliance, such as:

coordinated	organized
monitored	implemented
presented	supervised
planned	directed

(Note that Alison uses "responsible for," "administering," "coordinating," and "teaching.")

If you have held several short part-time jobs, describe them together and draw attention to the most important, like this: "Several after-school jobs, primarily as a stock clerk in a groceteria."

6 The two-column arrangement of dates and work experience is important because it gives a less crowded appearance to the page. If the job descriptions were carried to the left—under the dates—the job details would appear as heavy, less visually appealing blocks of information.

7 Employers are particularly interested in an applicant's activities and interests outside normal work. They want to know if the person is more than a routine employee who arrives at 8 a.m., works until 4:30 p.m., then drives home, eats supper, and presumably watches television all evening. Information on your hobbies, interests, and participation in sports and community activities tells prospective employers that you recognize your role in society, are not too rigid or too narrow, and adapt well to your environment. Employers reason that such an applicant will make an interesting, active employee who will not only contribute to the company, but also take part

Let the words you use convey a positive impression

An employer wants to see "the whole picture"

in social and sports functions. Outside activities represent a balanced lifestyle and provide outlets for stress.

8 Try to draw your list of references from a cross section of people you have worked for, been taught by, or served with on committees, and ensure that their relevance is apparent (their connection to one of your previous jobs or activities must be clear). Before including them in your list, check that all are willing to act as references.

Both Alison Witney and Colin Farrow (whose resume appears in Figure 10-3) are well aware of the important role a resume's appearance plays in a prospective employer's readiness to consider an applicant. Submit a carefully arranged and printed resume. (When printing your resume, if possible use a laser jet printer; inkjets can smudge.)

The Focused Resume

Job applicants who have more extensive experience to describe do better if they focus an employer's attention on their particular strengths and aims. This means asking themselves what a prospective employer is *most likely to want to know* after reading their opening statement. (Probably it will be: "What have you done that specifically qualifies you to achieve the objective you have presented?") To answer, applicants must focus on their work experience rather than their education, and particularly on work that is *relevant to the position they are seeking* (which means they must first research information about the company).

If their experience is sufficiently varied, then they can go one step further and divide the Work Experience section of their resume into two parts: (1) work related to the position they are seeking; and (2) work in unrelated areas. They must place all of this information *ahead* of the "Education" section, so that there is a natural flow from their Objective to their Related Experience. Thus, the parts of a focused resume are:

Objective (or Aim)
Related Experience
Other Experience
Education
Extracurricular Activities
References

Colin Farrow's two-page resume in Figure 10-3 adopts this sequence. The circled numbers beside the resume refer to the comments below.

1 Colin has sufficient information to warrant preparing a two-page resume, but he should not run over onto a third page. (A third page

Focus your resume to match the employer's primary interest

can be used, however, if an applicant has published papers and articles or has obtained patents for new inventions, which can be listed on a separate sheet and identified as an attachment.)

② Colin's summary statement clearly shows his thrust toward structural engineering and his desire to obtain employment in that field.

Divide your work experience into "directly related" and "less related" compartments

③ The positions described within each Experience section should be listed in reverse order, the most recent experience being described first and the earliest experience described last. The most recent and most relevant experience should be described in considerably greater depth than early or unrelated experience (compare the descriptions of Colin's Northwestern Steel Constructors' experience with his Bowlands Stores' experience).

④ As in the traditional resume, each employer's name is listed first (in boldface type) and followed by the city and province. The person's position or job title is identified next, and then a description of what the job involved. If several positions have been held within the same firm, each is named and its duration stated so that the applicant's progress within the firm is clear. Each position should draw particular attention to the personal responsibilities and supervisory aspects of the job, rather than just list specific duties. Verbs should be chosen carefully, so they make the position sound as comprehensive and self-directed as possible. If the paragraph grows too long, it can be broken into subparagraphs like these:

...appointed crew chief responsible for
- installing interconnection and distribution systems
- hiring, training, and supervising local labor
- ordering and monitoring delivery of parts and materials
- arranging and supervising subcontract work
- preparing progress and job completion reports.

Economize on space yet appeal to the eye

⑤ Single-spaced typing should be used as much as possible to keep the resume compact. At the same time there should be a reasonable amount of white space on each side and between major paragraphs to avoid a crowded effect. Although we normally recommend setting the right margin "ragged right," for Colin's resume a justified right margin does not seem too rigid.

⑥ Education can be listed either in chronological or reverse sequence. If a resume is to be sent to another province, or if the applicant

COLIN R FARROW, P.Eng

408 Medwin Street

Brandon, Manitoba R7C 0B3

Tel: (204) 548-1612

email: c.farrow@mbonline.com

OBJECTIVE

After four years comprehensive experience as an engineering technologist installing and testing transmission line towers in Northern Canada, I returned to university where I obtained a B.Sc in Structural Engineering. Now I am seeking employment where I can use my experience and education to research and test tower anchors and grouts in permafrost areas.

Immediately announce your strengths and show how they can be used by the employer

RELATED WORK EXPERIENCE

June 1995 to October 1997	**Fairborne and Warren Associates,** Consulting Engineers, Brandon, Manitoba. Project engineer managing construction of microwave transmission towers and associated structures between Brandon and The Pas, Manitoba, for Manitoba Telephone System. Wrote specifications, coordinated and monitored contractors' work, prepared progress reports, and maintained liaison with client.
June 1988 to August 1992	**Northwestern Steel Constructors Ltd,** Winnipeg, Manitoba. Crew chief, supervising team installing high-voltage transmission line towers between Flin Flon, Manitoba, and Minnowin Point, N.W.T. After 30 months was assigned to assist project engineers of Ebby, Little and Company, testing concretes and grouts installed in discontinuous permafrost (10 months). For final year, appointed installation coordinator, responsible for scheduling and supervising installation crews. Resigned to attend university.

OTHER WORK EXPERIENCE

January 1983 to February 1987	**Canadian Forces**, Construction and Maintenance Directorate. For first two years, member of crew installing communication systems (buildings and towers) at CF bases between Armstrong, Ontario, and Prince George, BC. For final two years, antenna installation and maintenance technician at CFB Comox, Vancouver Island. Attained rank of corporal.
September 1980 to December 1982	**Bowlands Stores**. Stock clerk in Store No. 26, St. Boniface, Manitoba. (One year part-time while at high school, 1 1/4 years full-time.)

/2...

Figure 10-3 A focused resume for a job applicant with a varied background.

Colin R Farrow – page 2

EDUCATION

- B.Sc in Structural Engineering, University of Manitoba, 1995.
- Diploma in Civil Engineering Technology, Red River Community College, Winnipeg, Manitoba, June 1988.
- Graduate (Grade 12), Dakota Collegiate, St Vital, Manitoba, 1981.

ADDITIONAL ACTIVITIES/INFORMATION

- Member, Association of Professional Engineers of Manitoba (APEM).
- Member, Certified Technicians and Technologists Association of Manitoba (CTTAM).
- Awarded Orton R Smith Scholarship for proficiency in applied mathematics, Red River Community College, 1987.
- Courses attended in Canadian Forces:
  * Construction Techniques, 1983.
  * Supervisory Skills Development, 1985.
  * First Aid and Safety Methods (various courses), 1984 to 1986.
- Junior Leader, St Vital YMCA, 1979 to 1982, teaching swimming and aquatic activities to boys and girls age 9 to 15. Awarded Red Cross Bronze Medallion, 1980. Lifeguard at Grand Beach, Manitoba, summers of 1980 and 1981.

REFERENCES

The following have agreed to provide information regarding my qualifications and work capabilities:

Martin G Warren, M.Sc
Projects Coordinator
Fairborne and Warren Associates
360 Rosser Avenue
Brandon, Manitoba R7A 0K2
Tel: (204) 544-1687
Fax: (204) 544-1628
email: mgw13@aol.com

Philip G Karlowsky
Contracts Manager
Northwestern Steel Constructors Ltd
3335 Notre Dame Avenue
Winnipeg, Manitoba R3H 2J4
Tel: (204) 632-1450
Fax: (204) 632-2177
email: p.karlowsk@norsteel.mb.ca

If applying to an educational institution, consider placing the Education section ahead of Work Experience

We recommend including two references, rather than writing "References available on request"

was educated in another province, he or she should identify the city and province of each educational institution attended.

7 Employers are *interested* in a job applicant's accomplishments and extracurricular activities, particularly those describing community involvement and awards or commendations. This part of a resume can be preceded by a heading such as "Extracurricular Activities" instead of "Additional Information."

8 Both of the people Colin has chosen as references can be cross-referenced to his previous work experience. Telephone numbers and email addresses are important, because most employers prefer to talk to rather than receive a letter from a reference.

The Functional Resume

Of the three resumes discussed here, the functional resume goes furthest in *marketing* a job applicant's attributes. For some employers its approach may seem too forthright—too blatantly "pushy"; for others, particularly employers seeking someone for a technical sales position, its approach helps demonstrate that the applicant has strong capabilities.

The functional resume is the only one to offer *opinions*: its objective identifies in general terms what the applicant believes he or she can do to improve the quality of the employer's product or service, and then follows immediately with the applicant's key qualifications—the capabilities the applicant believes best demonstrate that he or she is qualified to do what the objective proclaims.

The functional resume is not for everyone, yet for certain people and jobs its direct approach is ideal

To prove that the applicant's opinions are valid, the third section establishes—with clear facts and figures—what he or she has done for previous employers or organizations. This results in a revised arrangement of the resume's parts:

Objective
Qualifications
Major Achievements
Employment Experience
Education
Awards/Other Activities
References

The intent of this arrangement is to target the resume not just for a particular employer but also for a particular position. It is especially useful under two circumstances: for job applicants who have experience in marketing and want to be employed in technical sales; and for applicants who have a lean educational background but have proven and demonstrable practical experience that can be of value to a specific employer.

The resume in Figure 10-4 shows how Reid Qually uses the functional method to capture the attention of the marketing manager of a company engaged in selling cellular telephone services. The circled numbers beside his resume are keyed to the following comments.

Use subtle marketing techniques to promote yourself

1 Reid has positioned his name in the top right corner of the page because, in a pile of resumes, his name will stand out just where the person's hand is placed to flip through the pages. The line underneath his name helps draw the reader's eye to it. Reid has also saved a few lines by putting his contact address centred, all on one line.

2 Reid has written his Objective with a specific employer in mind. He has heard that King Cell—a relatively new West Coast player in the cellular telecommunications field—is planning to expand and hopes to become a major provider of cellular telephone services across the country. By echoing the company's philosophy, he is almost certain to catch management's attention.

3 Reid is aware that, as soon as the personnel manager at King Cell has read his Objective, he or she is likely to think: "You have told me what you want to do. Now tell me *why* you think you can do it." So he immediately offers five reasons, each demonstrating that he can handle the job. Note particularly that

- each is short, so that the reader assimilates the information quickly,
- each starts with a strong "action" verb (i.e. *identify, create, establish*), which creates a strong, definite image, and
- each is an opinion (although not recommended for other types of resumes, opinions can be used here because Reid will follow immediately with *evidence* to support his assertions).

Opinions must be supported by solid evidence

4 Reid's evidence provides *facts*, which demonstrate he has already established a solid track record. Reid keeps each piece of evidence short and offers definitive details (i.e. percentages, names, and dates), which adds credibility to his statements.

5 Reid can keep details of his work experience short because he has already identified his major accomplishments. For each employer he provides

Reid G Qually **❶**

7 - 2617 East 38th Avenue • Vancouver, BC • V5R 2T9 • Tel: 604-263-4250 • email: qually@interex.net

Reid sets the scene with a commanding, professional approach

Objective **❷**

To use my proven skills in marketing to increase market share for a Canadian company providing cross-country cellular telephone services and selling cellular telephone systems.

Qualifications

I have proven capability to **❸**

- Identify special-interest client groups and develop innovative marketing strategies for them.
- Create results-oriented proposals and focus them to meet specific client needs.
- Follow through with clients, both before and after a sale.
- Supervise and coordinate the efforts of small groups.
- Establish strong interpersonal relations with clients, management, and sales staff.

Major Achievements

For previous employers and organizations I have

- Devised an innovative lease/purchase marketing plan for first-time customers, resulting in a 34% increase in lease agreements and a 23% increase in follow-on sales over a 12-month period (for Morton Sales and Leasing, in 1995). **❹**
- Increased sales and leases of facsimile machines by 31%, and answering machines by 26%, over a nine-month period (for Advent Communications Limited, in 1997–1998).
- Received a company-wide "Salesperson of the Year" award (from Provo Department Stores, in 1992).
- Advised and coordinated Electronic/Computer Technology students who won a nationwide IEEE "Carillon Communication Award" (for Pacific Rim Community College, 1997).

Achievements must be stated strongly and supported with specific details

/2...

Figure 10-4 A functional resume identifies in detail what an applicant feels he or she can do for a particular employer

Employment Experience

(5)

June 1997
to the present

Advent Communications Limited, Vancouver, BC.
Assistant Marketing Manager, responsible for coordinating four representatives selling facsimile transmission (fax) and telephone answering equipment to commercial customers.

November 1993
to June 1996

Morton Sales and Leasing, New Westminster, BC.
Sales representative marketing fax machines and cordless telephones to business accounts and private customers.

July 1990
to October 1993

Provo Department Stores, Store No. 17, Calgary, Alberta.
Sales representative in Home Electronics Department. Responsible for over-the-counter sales of stereos, video-cassette recorders, and portable radios.

Education

(6)

June 1997

Certificate in Commercial and Industrial Sales, Pacific Rim Community College, Vancouver, BC (placed 2nd in course with GPA of 3.84).

1992 to 1996

Various courses in theoretical and applied electronics, at Pacific Rim Community College, Extension Division (partial credit toward electronics technician certificate).

June 1990

Graduated from Rosemount High School, Port Coquitlam, BC.

Awards and Other Activities

(7)

October 1996 and
November 1997

Coordinator, IEEE "Papers Night," Pacific Rim Community College, at which students of Electronics and Computer Technology presented term projects.

1994 to present

Associate Member, Institute of Electrical and Electronics Engineers Inc (IEEE).

1993 to present

Member, British Columbia Sales and Advertising Association; currently vice-president.

Information in a functional resume must be easy to find

References

Two people will provide immediate references; other names are available.

(8)

James B Morton
President, Morton Sales & Leasing
330 Pruden Avenue
New Westminster, BC V3J 1J5
Tel: (604) 475-3166
Fax: (604) 475-2807
email: j.morton@bconline.com

Dr Fergus Radji
(Chairman, Burnaby Section, IEEE)
Pacific West HV Power Consultants
1920 – 784 Thurlow Street
Vancouver, BC V6E 1V9
Tel: (604) 488-1066
Fax: (604) 489-2722
email: radji@pacwest.bc.ca

- start and finish dates (by month),
- employer's name (emphasized, in bold or italic letters),
- employer's location (city and province), and
- his job title and major responsibilities.

To maintain continuity, he lists his employment experience in reverse sequence.

6 Reid has only limited formal education, so he draws attention to his high grade point average (GPA) on returning to school after a long absence.

7 In a functional resume, the "other activities" section provides additional information to support statements in the Qualifications and Major Achievements sections.

8 Reid has asked several people to act as references but lists only two, partly because they are best able to speak about his qualifications, and partly to keep his resume down to two pages.

Reid's use of bullets on page 1 and a two-column format with dates on the left of page 2 provide variety in his resume's overall layout yet continuity within each page. The bulleted items on page 1 can be read easily—Reid wants his readers to learn quickly about him—while the facts on page 2 can be examined in more detail.

The functional resume is an effective way for a job applicant like Reid to present himself to a particular employer, but it *must* be done well if it is to create the right impact. Ideally, an applicant should use it only if he or she is confident that the employer will not be "turned off" by its non-traditional approach.

Never be afraid to use a display technique for your resume that will enhance its professional quality and make it stand out among other resumes. (We do not mean you should make your resume "flashy," because an overdone appearance can evoke a negative reaction from a reader.) An engineer with technical editing experience recently prepared a two-page resume that he had printed side-by-side on 11×17 inch (280×420 mm) paper, and then folded the sheet so that the resume was inside. On the outside front he printed only his name and the single word "Resume" (see Figure 10-5). On the back he created a table in which he listed the major projects he had worked on and, for each, itemized his degree of involvement. When employment managers placed his resume among other resumes submitted for a particular job opening, its professional appearance captured their interest and resulted in his being called in for more interviews than he had anticipated.

An unusual yet conservative appearance can help "sell" you as a strong, imaginative applicant

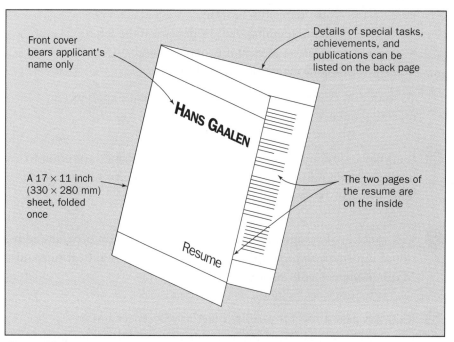

Figure 10-5 An imaginatively prepared resume.

The Electronic Resume

In today's competitive work place there are more people looking for fewer jobs, which means that Human Resources departments are often flooded with far too many resumes for the jobs they have available. (The Human Resources staff may even hesitate to advertise a position because they fear the overwhelming response they might get!) Instead they tend to maximize their hiring time by using automated computer systems that narrow the search for them. In effect, instead of posting a job and seeing who applies, they first go into the databases and see who is out there.

Resume submission methods step boldly into the millenium!

The key is to register or submit your resume to a number of independent databases so you are getting the maximum coverage. An electronic resume can be scanned, coded (for example in HTML), or both, so the computer can read and categorize it. Susan Jenkins developed two resumes. Figure 10-6 on pages 338 and 339 shows the resume she sent to an independent database service and Figure 10-7 on page 340 shows the HTML version she used on her personal World Wide Web Home Page. For the latter, she used a basic menu structure with links to additional information. She also made arrangements with her Internet Service Provider to collect data on who accessed her page, so she could follow up with an email message to make contact with the organization. H L Winman and Associates found her resume through a request to the database and were emailed the scanned version of the resume shown in Figure 10-6.

Usually, when you submit your resume to an independent database service it is scanned in and categorized, labelled, and keyworded in several variations. For example, it might be categorized by **education**, and then by **experience** and yet again by **location** or a certain **skill**. An employer enters a request into the system and provides keywords to describe the position, such as "Engineer," "B.Sc EE," "microwave technology," "telecommunications," "development." The system searches through resumes to locate any that match and presents these resumes to the employer. See how the process gets narrowed down?

The traditional paper resume is *not* obsolete: you just have more options. Often a resume provider service will take your paper resume and scan it into their system. You will still, however, need to carry a resume to the interview. Even a scanned resume that has been keyworded and indexed for a computerized retrieval system may end up being viewed or downloaded, once it has been flagged as a possible match. With this in mind, formatting and word choice are still critical since they present an image of you.

All three types of resumes described earlier will work as an electronic resume. There are, however, some additional factors you need to consider when preparing an electronic resume.

A smart job seeker uses every avenue to "record a hit"

Keywords

The resumes that get listed first are the ones who have matched the most keywords. So, when developing an electronic resume, you need to think like a Human Resources Manager and include as many keywords as possible.

Effective keywords are the key to getting noticed!

A keyword is usually a noun, not an action verb. This is a change from how we recommend you write paper resumes, using strong action verbs like *managing, implementing, installing*. The Human Resources Manager would search for words that describe the qualities or skills needed for a particular position, words like *account manager, CAD skills, member IEEE*. The search often includes other company names, particular tools or technologies, and even hobbies.

H L Winman and Associates entered the keywords *Computer Specialist, three years experience, web pages, web site design* and *manager* when they were looking for someone to fill the new Computer Services Manager position. And that's how they found Susan.

Readability

When a resume is scanned the quality is often compromised. We suggest using simple fonts and formatting and recommend a sans-serif font like Arial. When a serif font is scanned the letters tend to run together, making the

Susan R Jenkins
517 – 210 Olivia Crescent
London, Ontario N5Z 3E9
Tel/Fax: 519-438-0761
s.jenkins@interact.on.ca

Objective

To obtain a position as a computer services consultant/coordinator for a major engineering firm, so I may use my expertise in computer system maintenance and upgrading to optimum effect.

Experience

May 1995 to date

Superior Manufacturing Systems, Cambridge, Ontario
Information Technologist
Responsible for upgrading software, configuring new systems and managing computer accounts and server space for a research and development lab of 138 employees.

April 1993 to date

Jenkins Communication Services, London, Ontario
Independent Consultant
Part-time business designing and developing web sites for small businesses and organizations.

June 1990 to
April 1993

Woolland Computer Services, London, Ontario
Computer Specialist
One year full time, after high school graduation; two years part-time, while attending university. Duties included
- direct sales of computers and software
- on-site servicing of computers and training of users
- training new staff
- demonstrating new technology

Education

B.Sc in Computer Science, University of Western Ontario, London, Ontario, 1995 GPA 3.8
Senior Project: Developed a hypertext information system for athletic department.

East Elms High School, Stratford, Ontario, 1990 84.6%

Extracurricular Activities

Westferry Ski Club, London, Ontario, 1990 to date
Received ski instructor certification in 1990, served as club secretary 1994–1996.

"Theatre for Youth" Stratford, Ontario, 1988–1994
Actor-in-training for one year, then as electrical/computer technician for five years, responsible for designing and implementing computer-generated dramatic effects.

Susan Jenkins's resume may look routine, yet it has been designed for electronic scanning and coding

Note the sans serif type and minimum punctuation

Figure 10-6 Electronic resume prepared by Susan Jenkins, which she submitted to an independent database service.

Special Awards   Recipient of Miller Foundation scholarship in Computer
Science, University of Western Ontario, 1994

Awarded Maitland Trophy for best overall performance,
East Elms High School, 1989

References

The following people will provide more information about my work and
involvement in company activities. Additional references can be provided
on request:

Margaret Ferbrache
Owner/Manager
Woolland Computer Services
313 Oak Street
London ON N2R B6J
Tel: 519-323-6972
Fax: 519-323-6647
email: ferbrach@woollcom.on.ca

David Singh
Program Director
Theatre for Youth
PO Box 2120
Stratford ON N6A 7M3
Tel: 519-717-6690
email: d.singh@players.net

words difficult to read. The same is true for small-sized fonts. Consequently we recommend using a 12 pt font. Bold fonts should be used for emphasis, rather than italics and underlining, since the latter do not scan well.

When formatting your resume avoid using graphics, shading or horizontal or vertical lines, since a scanner may not be able to handle them or may not interpret them correctly. This includes commas, parentheses, and extra periods. Keep the formatting as simple as possible, as in the resume Susan Jenkins submitted to an independent resume database service (in Figure 10-6).

Ensure your name is at the top!

Make sure that your name is the first thing listed on your resume since that is how the computer will index you. You don't want to be known as "Resume" or "Education"! Although Reid Qually's name is positioned first, the underlining would make his resume unsuitable as an electronic resume, because the address may not be read correctly.

Submitting an Original

Before sending your resume to an independent database service, ask them for specific guidelines on preparing and submitting it. Then we suggest you mail your resume (not send it by email or fax), because that eliminates another electronic conversion it has to go through. When mailing it,

An employer accessing this home page clicks on a particular bulleted heading to bring specific details on-screen

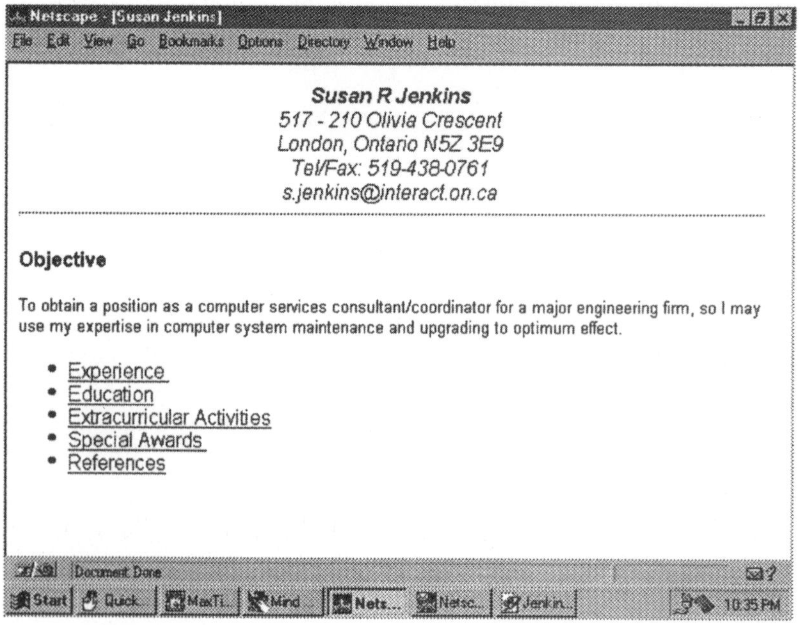

Figure 10-7 An HTML coded resume prepared by Susan Jenkins, which she used on her World Wide Web personal Home Page.

include a thin slice of cardboard to ensure it won't be folded or crumpled. Never staple your resume because the staples will be removed to scan it and the holes left behind will end up as black blobs on the electronic version. Use white paper and print the resume with a LaserJet for the clearest, cleanest original.

Writing a Letter of Application

Although some resumes may be delivered personally, the majority are mailed or submitted electronically, both with a covering letter. Because potential employers will probably read the letter first, it must do much more than simply introduce the resume. The letter needs to state your purpose for writing (that you are applying for a job) and demonstrate that you have some very useful qualifications that the reader should take the time to consider.

An assertive, interesting, and well-planned application letter can prompt employers to place your letter and resume with those whose authors they want to interview. Conversely, a dull, unemphatic letter may cause the same employers to drop your application on a pile of "also rans," because its approach and style seem to imply you are a dull, unemphatic person.

Write using a firm yet personable style

A letter of application should adopt the pyramid method of writing: it should open with a brief summary that defines the purpose of the letter, continue with strong, positive details to support the opening statement, and close with a brief remark that identifies what action the writer wants the reader to take. These three parts are illustrated in Figure 10-8 on page 342.

There are two types of application letter. Those written in response to an advertisement for a job that is known to be open, or at the employer's specific invitation, are known as solicited letters. Those written without an advertisement or invitation, on the chance that the employer might be interested in your background and experience even though no job is known to be open, are referred to as unsolicited letters. The overall approach and shape of both letters are similar, but the unsolicited letter generally is more difficult to write.

The Solicited Application Letter

The main advantage in responding to an advertisement, or applying for a position that you know to be open, is that you can focus your application letter on facts that specifically meet the employer's requirements. This has been done by Alison Witney in Figure 10-9, which responds to an advertisement in a Nova Scotia local newspaper.

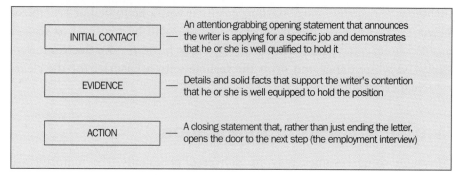

Figure 10-8 Writing plan for a job application letter.

The following comments and guidelines are keyed to the circled numbers beside Alison's letter.

1 For a letter that will have a personal address at the top, you would be wiser to use the modified block style shown here rather than the full block style in which every line starts at the left margin. (See Figure 3-11 of Chapter 3 for information on the modified block letter.) Because the appearance of a modified block letter is better balanced on the page, it creates a more pleasant initial impression. Each line of the applicant's name, address, and telephone number is centred at the top of the page; the date and the signature block at the end of the letter start at the page centreline.

2 Whenever possible, personalize an application letter by addressing it by name to the personnel manager or the person named in the advertisement. This gives you an edge over applicants who address theirs impersonally to the "Personnel Manager" or "Chief Engineer." If the job advertisement does not give the person's name, invest in a telephone call to the advertiser and ask the receptionist for the person's name and complete title. (You may have to decide whether to send your letter and resume to someone in the personnel department or to a technical manager who is more likely to be aware of the quality of your qualifications and how you could fit into his or her organization.)

3 This is the **Initial Contact,** in which Alison summarizes her key points about herself that she believes will most interest her reader and states that she is applying for the advertised position. Note particularly that she creates a purposeful image by stating confidently "I am applying...." This is much better than writing "I wish to

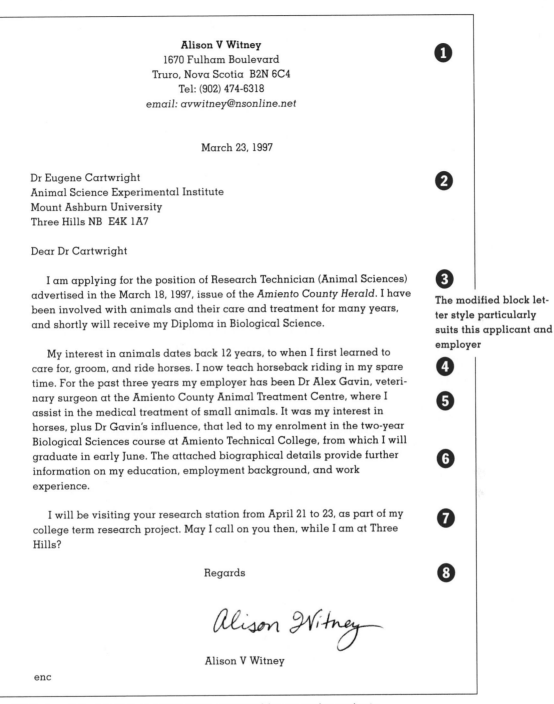

Alison V Witney
1670 Fulham Boulevard
Truro, Nova Scotia B2N 6C4
Tel: (902) 474-6318
email: avwitney@nsonline.net

❶

March 23, 1997

Dr Eugene Cartwright
Animal Science Experimental Institute
Mount Ashburn University
Three Hills NB E4K 1A7

❷

Dear Dr Cartwright

I am applying for the position of Research Technician (Animal Sciences) advertised in the March 18, 1997, issue of the *Amiento County Herald*. I have been involved with animals and their care and treatment for many years, and shortly will receive my Diploma in Biological Science.

❸

The modified block letter style particularly suits this applicant and employer

My interest in animals dates back 12 years, to when I first learned to care for, groom, and ride horses. I now teach horseback riding in my spare time. For the past three years my employer has been Dr Alex Gavin, veterinary surgeon at the Amiento County Animal Treatment Centre, where I assist in the medical treatment of small animals. It was my interest in horses, plus Dr Gavin's influence, that led to my enrolment in the two-year Biological Sciences course at Amiento Technical College, from which I will graduate in early June. The attached biographical details provide further information on my education, employment background, and work experience.

❹

❺

❻

I will be visiting your research station from April 21 to 23, as part of my college term research project. May I call on you then, while I am at Three Hills?

❼

Regards

❽

Alison V Witney

Alison V Witney

enc

Figure 10-9 A solicited letter of application prepared by an undergraduate.

apply...," "I would like to apply...," or "I am interested in applying...," all of which create weak, wishy-washy images because they only imply interest rather than purposefully apply for a job. An equally confident opening is "Please accept my application for...."

④ The **Evidence** section starts here. It should offer facts drawn from the resume and expand on the statements made in the first paragraph. Avoid broad generalizations such as "I have 13 years' experience in a metrology laboratory," replacing them with shorter descriptions that describe your exact role and responsibilities, and stress the supervisory aspects of each position. The name of a person for whom you worked on a particular project can be usefully inserted here because it adds credibility to the role and responsibilities you are describing.

⑤ You may indent the first line of each paragraph about 1 cm, as Alison has done, or start each line flush with the margin, as Colin Farrow has done in Figure 10-10.

Draw on key information in your resume to support your opening statement

⑥ The **Evidence** section covers the key points an employer is likely to be interested in and draws the reader's attention to the attached resume. If the paragraph grows too long, divide it into two shorter paragraphs (as Colin Farrow has done).

⑦ This paragraph is Alison Witney's **Action Statement,** in which she effectively opens the door to an interview by drawing attention to her imminent visit to the advertiser's premises. She avoids using dull, routine remarks such as "I look forward to hearing from you at your earliest convenience" or "I would appreciate an interview in the near future," both of which tend to close rather than open the door to the next step.

⑧ Contemporary usage suggests that most business letters should end with a single-word complimentary close such as "Regards" or "Sincerely," rather than the more formal but less meaningful "Yours very truly."

The Unsolicited Application Letter

An unsolicited application letter has the same three main parts as a solicited letter and looks very much the same to the reader. To the writer, however, there is a subtle but important difference: it cannot be focused to fit the requirements of a particular position an employer needs to fill. This means the job applicant has to take particular care to make the letter sound both positive and directed. Here are some guidelines to help you shape such a letter.

- Make a particular point of addressing your letter to the person, by name and title, who will most likely be interested in you. This may mean selecting a particular department or project head, who will immediately recognize the quality of your qualifications and how you would fit into the organization, rather than applying to the personnel manager. Never address an unsolicited letter to a general title such as "Manager, Human Resources," because, if the company does not use such a title and you have not used a personal name, it will likely be the mail clerk who decides who should receive your letter.

- Try to find out enough information about a firm so that you can visualize the type of work it does and how you and your qualifications would fit the company's needs. This will enable you to focus your letter on factors likely to be of most interest to the employer.

<aside>Start by researching the company's home page</aside>

- Try to make your initial contact positive and interesting even though you are not applying for a particular position, as Colin Farrow has done in his unsolicited letter in Figure 10-10.

Like Alison Witney, Colin has used the modified block format for his letter. It is longer than Alison's because he has more information to present, and to do so he has created two **Evidence** paragraphs.

Completing a Company Application Form

Filling in company application forms can become a boring and repetitive task, yet any carelessness on an applicant's part can draw a negative reaction from readers. Each company or organization usually uses its own specially designed form that, although it asks for generally the same basic information, may vary in detail. Consequently the suggestions below apply primarily to the *approach* you should take rather than suggest what you should write:

- When visiting prospective employers, always carry your personal data record with you so you can readily search for details such as dates, telephone numbers, and names of supervisors.
- Treat every application form as though it is the *first* one you are completing—write carefully, neatly, and legibly. Never let an untidy application form subconsciously prepare an employer to meet an untidy worker.

The letter's appearance
and sans-serif type cre-
ate a crisp, professional
impression

December 15, 1997

Vern A Rogers, P.Eng
Branch Manager
H L Winman and Associates
574 Reston Avenue
Winnipeg MB R3C 4G2

Dear Mr Rogers

(A) As a Structural Engineer who has specialized in constructing and maintaining
transmission line towers and associated buildings for 10 years, and who has particular experience working in permafrost, I am applying for a position with H L
Winman and Associates so I can use my expertise to good effect.

(B) My experience evolves from three periods of employment. For four years I
installed and maintained communications systems with the Canadian Forces.
Subsequently I became a crew chief and installation coordinator with the
Northwestern Steel Construction Company, where for four years I was responsible for erecting and testing high-voltage transmission line towers between Flin
Flon, Manitoba, and the Northwest Territories. For the past two years I have been
a project engineer supervising the construction and installation of microwave towers on Manitoba Telephone System's Brandon–The Pas extension.

The enclosed resume describes my responsibilities in greater detail and my particular involvement in testing structures erected on discontinuous permafrost. I
hold a Diploma in Civil Engineering Technology from Red River Community College
in Winnipeg, and a B.Sc in Structural Engineering from the University of
Manitoba. I am keen to return to the north and the challenge of building on
unstable soil.

(C) I would welcome the opportunity to meet you and learn more about your project
at Winterton Lake. As I travel frequently between Brandon and Winnipeg, I will
call you when I next expect to be in your city.

Sincerely

Colin R. Farrow

Colin R Farrow
enc

Figure 10-10 An unsolicited letter of application prepared by an experienced engineer. A is
the initial contact, B is the evidence, and C is the action statement.

- Complete *every* space on the form, entering N/A (not applicable), Not Known, or a short horizontal line in spaces that do not apply to you or for which you genuinely do not have information. This will prevent an employer from thinking you carelessly (or, worse, intentionally) omitted answering the question.

Take care: every word you write conveys an image of how you approach a task

- Take care that your familiarity with your city and street names does not cause you to abbreviate or omit them. If you write "Wpg" for Winnipeg or omit the "St.," "Ave.," or "Crescent" from your street name (because you *know* it is a street, avenue, or crescent), you may create the impression that your approach to work is to take shortcuts whenever possible.
- Use words that describe the responsibility and supervisory aspects of each job you have held (as you would for a resume) rather than list only the duties you performed.
- Particularly describe extracurricular activities that show your involvement in the community, or activities in which you held a teaching or coaching role.
- Pay particular attention if there is a section on the form that asks you to comment on how your education and past experience have prepared you for the position. Think this through very carefully before you write so that what you say shows a natural progression from past experience to the job you are applying for. If you can, and if they fit naturally, add a few words to demonstrate how the position fits your overall career plan. This can be a particularly difficult section to write so do not be afraid to obtain an opinion of its effectiveness from another person.

The most difficult part to write! You should prepare in advance for such a question

Attending an Interview

This is the third step in the job application process and the first occasion when you meet a prospective employer (or, more often, the employer's representative) face to face.

Prepare for the Interview

The key to a good interview is thorough preparation. If you have prepared yourself well, the interview will likely run smoothly and you will present yourself confidently.

As soon as you are invited to attend an interview—or, better still, before you are called—start researching facts about the company (or organization, if it is a government establishment). Presumably, you will have done some research before submitting your letter of application. Now you need to identify additional information, such as the number of persons the company

Attending an interview can create as much anxiety as speaking before an audience employs, specific fields in which it is involved, work for which it is particularly well known, its major products and services, important contracts it has received (news of which has been released to the media), locations of branch offices, and the company's involvement in community activities. (An ideal way to do this is to carry out a search on the World Wide Web.) Such knowledge can be extremely useful during the interview, because it permits you to ask intelligent questions at appropriate places—questions that indicate to the interviewer that you have done your homework.

You also need to prepare for difficult questions an interviewer may pose to test your readiness for the interview and the sincerity of your application. You may be asked:

- *Why do you want to join our organization?*
- *How do you think you can contribute to our company?*
- *Why do you want to leave your present employer?* (Asked only of persons who are already employed.)
- *Why did you leave such-and-such company on such-and-such date?* (Asked of persons whose resumes show no explanation for a previous employment termination.)
- *What do you expect to be doing in five years? Ten years?*
- *What salary do you expect?*

Lack of preparation will show up in your body language and how you answer questions

If you have not prepared for such questions, and so hesitate before answering, an interviewer may interpret your hesitation to mean you find a question difficult to answer or there are factors you would rather conceal. In either case, you may inadvertently provide an entirely misleading impression of yourself.

An interviewer who asks what salary you expect is partly testing your preparation for the interview and partly assessing how accurately you value yourself. For an undergraduate at a university or college, the question is largely academic: undergraduates compare notes and quickly learn what starting salaries are being offered. But for a person who recently has been or currently is employed, the question is important and must be anticipated. Always know the salary you would like to receive and think you are worth. Avoid quoting a salary range, such as "between 22 and 24 thousand dollars," because it indicates unsureness. Quote a definite figure, such as $23 000, and you will sound much more confident. If you fear that the salary you want to quote may be too high, you can always add the qualification "...depending, of course, on the opportunities for advancement and the fringe benefits your company offers."

An interview is like a two-way street: traffic should flow in both directions

You should be ready to ask questions during the interview. The interviewer wants to acquire information about you, but you should also learn things about the company and the opportunities it can offer. Consider what questions you would like answered, jot them onto a small card, and store the card in a pocket or purse. Then when the interviewer asks, "Now, do you have any questions?" you can pull out the card.

Make the entries on your card brief and clearly legible, and keep the list short so you can scan it quickly. Remember, too, that the quality of your questions will demonstrate how carefully you have given thought to the interview.

Create a Good Initial Impression

Remember that you are being evaluated from the moment you step into the interview room. Consequently,

- walk in briskly and cheerfully,
- shake hands firmly, because a limp handshake creates an image of a limp, indefinite applicant,
- repeat the person's name as you are introduced and look him or her directly in the eye, and
- sit when invited to do so, pushing yourself well back in the chair, making yourself comfortable, and avoiding folding your arms across your chest (which psychologically suggests you are resisting questions).

Participate Throughout the Interview

An interview normally falls into three fairly easy-to-distinguish parts. The initial part is an exchange of pleasantries between yourself and the interviewer, who wants you to be at ease. To help you adjust to the interview environment, he or she may ask questions on topics you can answer confidently, such as a major news item or something from the hobbies and interests section in your resume. This initial part of the interview normally is short.

An experienced interviewer will try to ease your nervousness

In comparison, the middle part of the interview is quite long. So that the interviewer can find out as much as possible about you, he or she will want to hear your opinions and have you demonstrate your knowledge of certain topics. Although the interviewer will want to control the direction the interview takes, you will be expected to develop your answers and to comment on each topic in sufficient depth to establish that you have real knowledge and experience, backed up by well-thought-out opinions.

The closing portion of the interview also is short. The interviewer will ask if you have any questions and will discuss details about the company and employment with it. By this stage the interviewer should have a good impression of you, and you should know whether you want to be employed by the company.

An effective interviewer will pose questions and subsequent prompts in such a way that you are carried easily from one discussion point to the next and are automatically encouraged to provide comprehensive answers. If, however, you face an inexperienced or inadequately prepared interviewer, the responsibility becomes yours to develop your answers in greater depth than the questions seem to call for.

For example, the interviewer may ask, "How long did you work in a mobile calibration lab?"

You might be tempted to reply "Three years," and then sit back and wait for the next question. You would do much better to reply: "For three years total. The first year and a half I was one of four technicians on the North Bay to Sioux Lookout circuit. And then for the next year and a half I was the lab supervisor on the Dana, Saskatchewan, to Prince George route."

An answer developed in this depth often provides the prompt (piece of information) from which the interviewer can frame the next question.

Sometimes you will face a single interviewer, while at other times you may face an interview board of two to five people. In a single-interviewer situation you will naturally direct your replies to the interviewer and should make a point of establishing eye contact from time to time. (To maintain continuous eye contact would be uncomfortable for both you and the interviewer.) In a multiple-interviewer situation

- direct most of your questions, and your responses to general questions, to the chairperson (but if an answer is long, occasionally look briefly at and talk momentarily to other board members),
- if a particular board member asks you a specific question, address your response to that person, and
- if a board member has been identified as a specialist in a particular discipline, direct questions to that board member if they especially apply to that field.

In certain interviews—often when applicants are being interviewed for a high-stress position—you may be presented with a stress question. A stress question is designed to place you in a predicament to which there may be two or even more answers or courses of action that could be taken. You are expected to think *briefly* about the situation presented to you and then to select what you believe is the best answer or course of action. Often you will be challenged and expected to defend the position you have taken.

The secret is not to let yourself be rattled and to defend your answer rationally and reasonably even though the questioner's challenging may seem harsh or unreasonable. Remember that the interviewer is probably more interested in seeing how you cope in the stress situation than in hearing you identify the correct answer.

Here are seven additional factors to consider:

- Use your voice to good effect; make sure everyone can hear you, speak at a moderate speed (think out your answers before speaking) and, where appropriate, let your enthusiasm *show*.
- Be ready to ask questions, but have a clear idea of what you want to ask before you pose them. The interviewer will recognize a good question and the clarity of thought behind it.
- If you do not know the answer to a question, say you don't know rather than try bluffing your way through it.
- If you do not understand the question, again don't bluff. Either say you do not understand or, if you think you know what the interviewer is driving at, rephrase the question and ask if you have interpreted it correctly. (Never imply that the interviewer posed the question poorly.)
- Use humor with great care. What to you may be extremely funny may not match the interviewer's sense of humor.
- Bring demonstration materials to the interview if you wish (such as a technical proposal or report you have authored, or a drawing of a complex circuit you designed) but be aware that you may not have an opportunity to display them. If the topic they support comes up during the interview, introduce them naturally into your response to a question. But remember that the interviewer does not have time to read your work, so the point you are trying to make must be readily identifiable. Never force demonstration materials on an interviewer.
- Do not smoke, even if the interviewer smokes and invites you to do so.

Finally, try to be yourself. Remember that interviewers want to see the kind of person you really are. If you relax and answer questions comfortably and purposefully, they will gain a good impression of you. If you try too hard to be the kind of person you think the interviewers want you to be, or to give the kind of answers you think they want rather than the answers you really believe in, they may detect it and judge you accordingly.

Above all, present an image of "the real you"

Accepting a Job Offer

The telephone rings and the personnel representative you met during your interview tells you that the company is offering you employment at a salary of $xxxx. You accept the offer! And then she asks when you can start work. (Employers recognize that if you are attending college there will be a waiting period until your course is finished and you have graduated; similarly, an employed engineer or engineering technologist has to resign from his or her present position, normally giving either two weeks' or one month's notice.) You quote a starting date to the personnel representative, which she agrees to, and then she says she will confirm the offer in writing. She also asks you to write a letter confirming your acceptance of the position.

The two letters become, in effect, a contractual agreement: the employer offers you work under certain conditions, which you agree to. The letters can also prevent any misunderstandings from developing, which can occur if arrangements are made only by telephone. Consequently your acceptance letter should:

Accept a job offer *in writing*, like sealing a contract

- Announce that you are accepting the offer of employment.
- Repeat any important details, such as the agreed salary and starting date.
- Thank the employer for considering you.

The following acceptance letter conforms to this pattern:

Dear Ms Tataryn

I am confirming my telephone acceptance of your May 19 offer of employment as an engineering technologist in the controls department. I understand that I am to join the company on June 15 and that my salary will be $26 500 annually.

Thank you for considering me for this position. I very much look forward to working for Magnum Electronics.

Sincerely

Sometimes an applicant may receive two offers of employment at the same time and will have to decline one. The letter declining employment should follow roughly the same pattern:

- Decline the offer.
- Briefly explain why.
- Thank the employer.

Here is an example:

Also decline a job offer in writing, with a smile on your face

Dear Mr Genser

I very much regret I will be unable to accept your offer of employment. Since my interview with you I have been offered employment elsewhere and now must honor my commitment to the other company.

Thank you for considering me for this position.

Regards

Declining a job offer pleasantly and formally in a carefully worded letter like this is insurance for the future: one day you may want to work for that employer!

Project 10.1: Preparing a Resume

You are to prepare a resume describing your background, education, work experience, extracurricular activities, and other interests. Do it in three parts.

Part 1

If you do not already have one, prepare a personal data record (PDR), using 200 × 125 mm (8 × 5 inch) file cards.

Part 2

Write down the following information:

- The name of a real employer for whom you would like to work at the end of your course.
- The type of position you would be qualified to hold with that particular employer.
- The type of resume that would be most effective to use.

Part 3

Prepare the resume (assume that you will be graduating from your course in two months).

Create a resume

Project 10.2: Applying for a Locally Advertised Position

From your campus student employment centre or your local newspaper, identify a company currently advertising a position you could apply for at the end of your course.

Part 1

Write a letter applying for the position (assume that you will be graduating in six weeks). Also assume that you are attaching a resume to your letter. If you are replying to a newspaper advertiser, attach a copy of the advertisement to your letter.

Part 2

Assume that the company you wrote to in Part 1 sends you an application form. Obtain a standard application form from your campus student employment centre and complete it as though it is the advertiser's form.

Rehearse applying for a real position

Part 3

Now assume that the company has telephoned and asked you to attend an interview next Tuesday. On a sheet of paper write down five questions you would ask during the interview. After each question explain why the question is important and what answer you hope it will elicit from the interviewer.

Project 10.3: Unsolicited Application for Employment

This project assumes that you are seeking permanent employment at the end of a technical training program, but few job openings have been advertised in your field. Write to Macro Engineering Inc in Toronto, or to H L Winman and Associates (to the attention of one of the department heads in Calgary, or local branch manager Vern Rogers), applying for employment. Use your knowledge of the company and your real background. If it is still early in your training program, you may update the time and assume that it is now two months before graduation.

Project 10.4: Replying to Other Advertisers

Apply for one of these positions

This project assumes that you are seeking permanent employment at the end of a technical or engineering-oriented educational program. You are to reply to any one of the following advertisements, using your present situation and actual background. If it is still early in your training program, you may update the time and assume that it is now two months before your graduation date. In each case enclose a resume with your application letter.

We require a
CIVIL ENGINEERING TECHNOLOGIST

with an interest in building construction to prepare estimates, do quantity takeoffs, and assume responsibility for company sales and promotion activities.

Apply in writing to:
Personnel Department
PRECASCON CONCRETE COMPANY
227 Dryden Avenue
(your city)

(Advertisement in your local newspaper, February 26)

ROPER CORPORATION (ONTARIO DIVISION)
requires a
CHEMICAL TECHNOLOGIST

to join a project group conducting research and development into the organic polymers associated with the coating industry. The successful applicant will also assist in the development of control techniques for producing automated color tinting.

Apply in writing, stating salary expected, to:

> Phyllis Cairns
> Personnel Manager
> P.O. Box 1728
> Cambridge, Ontario
> N3L 2B6

(Advertisement in Engineering Times, April 17)

H L WINMAN AND ASSOCIATES
475 Lethbridge Trail
Calgary, Alberta
T3M 5G1

CIVIL ENGINEERS AND TECHNOLOGISTS

This established firm of consulting engineers needs three Engineers or Technologists to assist in the construction supervision of several grade separation structures to be built over a two-year period. Graduation from a recognized Civil Engineering or Technology course is a prerequisite. Successful applicants will be hired as term employees. Opportunities are excellent for transfer to permanent employment before the project ends.

Apply in writing to:

> A Rittman, Head
> Special Projects Department

(Advertisement on college notice board)

ARCHITECTURAL DRAFTSPERSON
required by
a well-established firm of architects and planners

Applicants should have completed a Design and Drafting course at a community college or similar training institution, in which strength of materials and structural design were a curriculum requirement. Experience in preparation of reports and proposals will be considered in selecting the successful applicant.

Write to:

The Corydon Agency
Room 604 – 300 Main St.

(Advertisement of March 28 on college notice board)

MACRO ENGINEERING INC
invites applications from
ENGINEERS, TECHNOLOGISTS, AND TECHNICIANS
interested in Metrology

We operate a first-class Standards Laboratory that is a calibration centre for precision electrical, electronic, and mechanical measuring instruments used in our military and commercial equipment maintenance programs.

Applicants for senior positions should hold a B.Sc. in Electrical Engineering. Technologists and Laboratory Technicians should be graduates of a recognized electronics or mechanical technology course who have specialized in precision measurement.

Applications are invited from forthcoming graduates, who will work under the direction of our Standards Engineer. All applicants must be able to start work on July 1.

Apply in writing to:

F Stokes
Chief Engineer
Macro Engineering Inc
600 Deepdale Drive
Toronto, Ontario
M5W 4R9

(Advertisement in your local newspaper, March 18)

INTER-MOUNTAIN PAPER COMPANY

Offers excellent opportunities for recent graduates to join an expanding manufacturing organization in the pulp and paper industry.

ENGINEERS AND ENGINEERING ASSISTANTS

Positions are available for mechanical engineers and technologists to assist in the design, installation, and testing of prototype production equipment. Previous experience in a manufacturing plant would be helpful. Innovative ability will be a decided asset.

ELECTRICAL ENGINEERING TECHNOLOGIST

This person will assist the Plant Engineer in the maintenance of power distribution systems. Applicants should be graduates of a two-year course in Electrical Technology with good knowledge of automatic controls and machine application. Ability to read blueprints and working drawings is essential.

ELECTRONICS ENGINEER OR TECHNOLOGIST

Two positions are open for electronics specialists who will maintain and troubleshoot microprocessor-controlled production equipment.

COMPUTER SPECIALIST

This position will suit either a graduate of a Computer Engineering course or an Electronics Technologist who has specialized in Computer Electronics. Duties will consist of installation, maintenance, and troubleshooting of mainframe and personal computers.

ENVIRONMENT SPECIALISTS

Persons selected will test air pollutants and water effluents from our paper mill and production plant, and assess their environmental impact. Applicants should be graduates of a recognized course in the environmental or biological sciences.

Salaries for the above positions will be commensurate with experience and qualifications. Excellent fringe benefit program available. Write in confidence to:

Manager of Industrial Relations
INTER-MOUNTAIN PAPER COMPANY
Kamloops, BC
V2D 2R3

(Advertisement in your local newspaper, March 10)

(Advertisement in last month's issue of North American Wildlife)

(Advertisement in your local newspaper, February 26)

TECHNOLOGIST OR TECHNICIAN
required by manufacturer of
automated car wash equipment

Duties: Assist plant engineer design and test mechanical and electrical components; supervise installations in cities across US and Canada; troubleshoot problems at existing installations.

Applicants must be free to travel extensively. Excellent salary and promotion possibilities.

Apply in writing to:

Personnel Manager
DIAL-A-WASH Incorporated
2020 Waskeka Drive
Halifax, Nova Scotia
B3M 2R7

(Advertisement in your local newspaper, April 23)

CELLFAX
requires recently graduated
ELECTRONICS ENGINEERS AND TECHNOLOGISTS

We are a well-established manufacturer of a wide range of cellular telephones and facsimile equipment.

We plan to augment our sales force by hiring electronics specialists who can assess customers' needs and recommend suitable equipment. If you have strong communication skills, apply in writing (and complete confidence) to:

Reg Drabik
Marketing Manager
Cellfax Inc
P.O. Box 7816
(your city)

(Advertisement in last Saturday's local newspaper)

Assume that you are looking for summer employment. Reply to one of the following advertisements:

REMICK AIRLINES
offers
SUMMER EMPLOYMENT
to a limited number of Canadian students.

Duties: To work as part-time dispatchers, baggage handlers, aircraft cleaners, cafeteria assistants, etc, during the peak summer period (June 15–September 10) at Remick Airlines terminals at:

> Muskwega, Ontario
> Weekaskasing Falls, Quebec
> Lake Lawlong, Alberta

Good pay; ample free time; free transportation to destination and return. Write, describing previous work experience (if any), to:

> Sarah Wiebe
> Remick Airlines
> Winnipeg International Airport
> Winnipeg, MB
> R2Y 3X7

(Notice on college notice board)

CITY OF MONTROSE, ALBERTA
TRAFFIC DIVISION

The city of Montrose is carrying out an Origination–Destination (OD) Traffic Study throughout June and July, with computation and analysis to be carried out in August.

Approximately 40 students will be required for the OD Study, and 20 for the computation and analysis. For the latter work, some knowledge of statistical analysis will be an advantage.

Students interested in this work are to apply in writing to V L Sinjun, Room 310, Civic Centre, City of Montrose, Alberta, T7W 6L4.

(Notice on college notice board)

Box 2980, Yellowknife, NWT, X1A 2R4

Applications are invited from students in two-, three-, or four-year technical programs who are seeking three months of well-paid summer employment. The work will include construction of living quarters and associated facilities for construction camps in the Northwest Territories and Yukon. Terms of employment will be:

13 full weeks—June 10 to September 10

Excellent pay: $475 per week

Transportation: Paid

Accommodation: Paid

Health: Applicants should be in excellent physical condition.

Students interested in this work are to write to Ms Rheena O'Connell at the above address. Although previous experience in construction work is not essential, knowledge of general construction methods will be considered an asset for some positions.

(From Construction—North, February 20)

Project 10.6: Unsolicited Application for Summer Employment

Assume that you are looking for summer employment but few summer jobs have been advertised. Write to H L Winman and Associates or Macro Engineering Inc, asking for a summer job. Use your knowledge of the companies, plus your actual background, to write an interesting letter.

Address your letter to the attention of Tanys Young in Calgary or George Dunn in Toronto. If there is an H L Winman and Associates branch in your area, you may address your letter to branch manager Vern Rogers.

Archeus Online: Resume Writing Resources
www.golden.net/~archeus/reswri.htm

Archeus Online is an extensive resume writing resource with links to dozens of sites about this and other job search skills. Archeus also has links to information about cover letters, interviewing, job search books and articles, and related topics.

Writing in the Job Search
owl.english.purdue.edu/by-topic.html#job

The Online Writing Lab at Purdue University has a section devoted to writing job search materials, including applications, resumes, cover letters, acceptance letters, references, and personal statements.

Career Mosaic: Resume Writing Center
www.careermosaic.com/cm/crc/crc15.html

The Career Mosaic resume resource page includes sample resumes, ASCII text and electronic resume information, cover letters, thank-you letters, a proofreading checklist, and resume tips and links.

Translating Resumes for the Internet
www.nytimes.com/library/jobmarket/0107sabra.html

In this "Careers" article from *The New York Times*, Sabra Chartrand gives practical advice about how to design an electronic resume. Links to other job search sites are included.

Canadian Jobs Catalogue
www.kenevacorp.mb.ca/

Canadian Jobs Catalogue contains links to more than a thousand Canadian employment-related sites, including job and resume banks, employer sites, and related Internet sites.

What Careers Can Technical Writing Offer?
www.cas.ilstu.edu/English/Hypertext96/Gleason/finalprojpage/careers.html

This site provides a short description of technical writing as a career and has links to related technical communications career sites such as the Tech-Web Multimedia Job Hotline.

Chapter 11

The Technique of Technical Writing

This chapter concentrates on a few writing techniques that will enable you to convey information both quickly and efficiently. It considers technical writing from five points of view: how to create the whole document, how to structure paragraphs, how to write individual sentences, how to use specific words, and how to create a good technical style. At the end of the chapter you will find several pages of exercises that test your ability to adopt an effective writing style and, in some cases, establish a suitable tone.

As we wrote this chapter, we assumed you are already proficient in grammar and can recognize and correct basic writing problems. If you need practice in basic writing, we suggest you refer to a textbook such as *The Canadian Writer's Handbook*.[1] You can also refer to the Glossary of Technical Usage (see page 423) for information on how to form abbreviations and compound adjectives, spell problem words, and use numerals or spell out numbers in narrative.

The Whole Document

Three factors affect the whole document: the tone you set, the writing style you adopt, and the arrangement of information on the page. Of the three, tone is by far the least tangible: a reader is less likely to be aware of the tone you establish than the writing techniques you use and the arrangement of paragraphs and headings.

Tone

Whether your writing should be formal or informal will depend on the situation and your familiarity with the reader. Formal reports should adopt

---

[1] William E Messenger and Jan de Bruyn, *The Canadian Writer's Handbook*, 3rd ed (Scarborough, Ontario: Prentice Hall Canada, 1995).

You need to know your
reader if you are to set
the right tone

a formal tone. (Note, however, that a formal tone is neither stiff nor pompous; there is no room for writing that makes readers feel uncomfortable because they are not as knowledgeable as you are.) Business correspondence is generally less formal, depending on its importance. For example, a management-level letter proposing a joint venture on a major defence project would be formal, whereas letters between engineers discussing mutual technical problems would be informal, and email would be very informal. A memorandum report also can be informal, since normally it would be an in-plant document written between people who know each other.

Varying levels of tone are evident in the following extracts from three separate H L Winman and Associates' documents, all written on the same subject.

1. **Extract from a Memorandum.** John Wood's materials testing laboratory has compression-tested samples of concrete for Karen Woodford of the civil engineering department. In his memorandum reporting the test results, John writes:

   Informal Tone I have tested the samples of concrete you took from the sixth floor of Tarryton House and none of them meets the 33.25 MPa you specified. The first failed at 28.08 MPa; the second at 26.84 MPa; and the third at 27.95 MPa. Do you want me to send these figures over to the architect, or will you?

2. **Extract from a Letter Report.** Karen Woodford conveys this information to the architect in a brief letter report:

   Semiformal Tone Our tests of three samples taken from the sixth floor of Tarryton House show that the concrete at 52 days still was 5.63 MPa below your specification of 33.25 MPa. We doubt whether further curing will increase the strength of this concrete more than another 1.10 MPa. We suggest, however, that you examine the design specifications before embarking on an expensive and time-consuming remedy.

3. **Extract from a Formal Report.** The architect rechecked the design specifications and decided that 28.50 to 29.00 MPa still would not satisfy the design requirements. He then requested that H L Winman and Associates prepare a formal report he could present to the general contractor and the concrete supplier. Karen Woodford's report said, in part:

Formal Tone At the request of the architect we cut three 0.3 × 0.15 metre diameter cores from the sixth floor of Tarryton House 52 days after the floor had been poured. These cores were subjected to a standard compression test with the following results (detailed calculations are attached at Appendix A):

Imagine you are speaking personally to the reader and adjust your tone accordingly

Core No.	Location	Failed at:
1	0.46 m W of col 18S	28.08 MPa
2	0.84 m N of col 22E	26.84 MPa
3	1.42 m N of col 46E	27.95 MPa

The average of 27.62 MPa for the three cores is 5.63 MPa below the design specification of 33.25 MPa. Since further curing will increase the strength of the concrete by no more than 1.10 MPa, we recommend rejecting this concrete pour.

Although the information conveyed by these three examples is similar, the tone the writer adopts varies in response to each situation.

The sequence in which you write longer reports also affects tone. To set the right tone throughout, write in reverse order, starting with the full development. Writing a report in the order in which it will be read is difficult, if not impossible. An engineering technologist who writes the summary before the full development will use too many adjectives and adverbs, big words when shorter words would be more effective, and dull opening statements such as "This report has been written to describe the investigation into defective RL-80 video terminals carried out by H L Winman and Associates." He or she will be writing without having established exactly what to say in the full development.

Deal with all the details before you start writing

The comments on Karen Woodhouse's formal report in Chapter 6 explain how Karen set about writing her report in reverse order (see pages 190 and 192). In brief, they identify the following writing sequence:

Step 1   Assemble and document all the details and technical data. These will become the appendices to your report.

Step 2   Write the Discussion, or full development. Direct it to the type of technical reader who will use or analyse your report in depth.

Step 3   Write the Introduction, Conclusions, and Recommendations. Keep them brief and direct them to a semitechnical reader or person in a supervisory or managerial position.

Step 4   Write the Summary. Direct it to a nontechnical reader who has absolutely no knowledge of the project or the contents of your report.

Style

Style is affected by the complexity of the subject you are describing, and the technical level of the reader(s) to whom you are writing. Consequently you need to "tailor" your writing style to suit each situation, following these guidelines:

Adjust sentence length to suit the subject and the reader's familiarity with it

1. When presenting low-complexity background information, and descriptions of nontechnical or easy-to-understand processes, write in an easygoing style that tells readers they are encountering information that does not require total concentration. Use slightly longer paragraphs and sentences, and insert a few adjectives and adverbs to color the descriptions and make them more interesting.

2. For important or complex data, use short paragraphs and sentences. Present one item of information at a time. Develop it carefully to make sure it will be fully understood before proceeding to the next item. Use simple words. The more punchy style will warn readers that the information demands their full attention.

3. When describing a step-by-step process, start with a narrative-type opening paragraph that introduces the topic and presents any information that the readers should know or would find interesting. Then follow it with a series of subparagraphs, each describing a separate step, that
   - develop only one item or aspect of the process,
   - are short, and
   - are parallel in construction (the importance of parallelism is discussed later in this chapter).

For example:

Precede subparagraphs with bullets or sequential numbers...

    If you use subparagraphs or a bulleted list to present a series of points, follow these guidelines:
   - Indent each subparagraph as a complete unit of information, to show readers how you are subordinating your ideas.
   - Precede each subparagraph with either a bullet (as has been done here) or a sequential number (1, 2, 3, etc).
   - Number the subparagraphs if you want to identify that the information is presented in a prescribed sequence or in decreasing order of importance, or if you will want to refer to the subparagraph later in your report. At all other times use bullets.

Notice the difference between the two bulleted lists on this page. The list at the end of paragraph 3 is introduced by 14 lead-in words (*Then follow...separate step, that*) which do not form a complete sentence and do not end with a colon. Consequently the bulleted items each start with a lower case letter and end with a comma (except the last item), because the lead-in words and the bulleted items really make one long, complete sentence. Conversely, the list in the example immediately preceding this paragraph is introduced by 17 lead-in words *(If you use...follow these guidelines:)*, which *do* create a complete sentence. Consequently the lead-in words end with a colon, and the bulleted items each start with a capitalized letter and end with a period. They are complete thoughts in themselves.

Appearance

In some industrial documents—particularly specifications, technical instructions, and military reports—the paragraphs and subparagraphs are numbered. The simplest paragraph numbering system starts at 1 and numbers the paragraphs consecutively to the end of the document. More complex systems combine numbers, letters, and decimals to allow for subparagraphing (but note that roman numerals are *not* used). Three possible arrangements are shown in Figure 11-1.

Method (A) is a numerals-only decimal system, simple and unambiguous, often used by the military and for specifications. However, there can be problems when a document has several levels of subparagraphs. If the subparagraphs are indented as full blocks, farther to the right at each sublevel, the extended decimal numbers (e.g. 3.4.1.5) can take up a lot of space and so create very narrow subsubsubparagraphs on the right side of the page. Yet if the sentences of all subparagraphs are run back to the left

Use a simple numbering system

```
1.                A.                1.
2.                1.                2.
3.                2.                3.
3.1                   (a)               3.1
3.1.1                 (b)               3.2
3.1.2                     (1)               a)
3.1.2.1                   (2)               b)
                  B.                    (1)
etc, up to        1.                    (2)
6 digits          2.                        etc.
                      etc.

METHOD (A)        METHOD (B)        METHOD (C)
```

Devise a paragraph numbering system to suit the situation

Figure 11-1 Suggested paragraph and subparagraph numbering systems.

margin, the level of subparagraphing can become difficult for the reader to identify.

Method (B) uses a combination of letters and numerals that more readily permits indenting of full subparagraphs. Extremely simple, it is popular with many report writers even though it can be ambiguous (there can be several paragraph 1's, 2's, etc., one set under A, another under B, and so on).

The third method, (C), combines the decimal and letter-number arrangements for a system with no ambiguity between main paragraphs and subparagraphs. This is the method chosen by Anna King for H L Winman and Associates' reports that carry paragraph numbers, and she suggests that engineers use normal-size paragraphs for the first two levels, but only short paragraphs or single sentences for the lower-level subparagraphs:

1. 2.	} main paragraphs
2.1 2.2	} full subparagraphs
a) b)	} short subparagraphs
(1) (2)	} very short subparagraphs (i.e. one sentence only)

This paragraph numbering system combines well with the heading arrangements illustrated in Figure 11-2.

Longer reports need headings to introduce main sections and become dividers between subsections. With the wide range of fonts and type sizes available with most word processing systems and printers, you can create an almost limitless variety of centre headings, side headings and paragraph headings for your reports. However, certain rules apply, as illustrated in Figure 11-2:

- Use upper and lower case letters for almost all headings. All block capitals are much harder to read.
- Use boldface type rather than underline the headings.
- Keep headings in the same font as the main text.
- Use larger point sizes for the principal headings, and a slightly smaller point size for subsidiary headings.

Make your headings contribute to the overall appearance

Paragraphs

The role of the paragraph is complex. It should be able to stand alone but normally is not expected to. It must contribute to the whole document, yet

Guidelines for Integrating Paragraphs and Headings

The main centre heading (above) is set in a larger boldface type than all other headings. Here, it is set in 14 pt Times New Roman, which is the same font we have used for the all the text and headings in this example.

Subparagraphing Without Paragraph Numbering

If a subsidiary centre heading is used, it is set in 12 pt type (see immediately above), while the side headings and the text in this example are all set in 11 pt type.

For headings, vary font size and use boldface type rather than underline the words

Side Heading

A side heading introduces a new section of text and is set flush against the left margin. Paragraphs following the side heading are also typed with all lines flush against the left margin. In technical writing, the first line of each paragraph is seldom indented.

Subparagraph Headings and Subparagraphing

Subparagraph headings are indented about 1 centimetre in from the left margin, as is any text that follows the subparagraph heading.

Each subparagraph is typed as a solid indented block, so that readers can *see* the subordination of ideas.

Secondary Subparagraphing

If further subparagraphing is necessary, the headings and subparagraphs are indented a further 1 cm (i.e. a total of 2 cm) from the left margin.

Headings Built into the Paragraph. In this lesser-used arrangement, the text continues immediately after the heading. Usually, a paragraph heading applies only to one paragraph of text.

2/...

Figure 11-2 Guidelines for integrating paragraphs and headings.

Subparagraphing Combined with Paragraph Numbering

1. Side Headings

1.1 When paragraph numbers are used, side headings normally are assigned simple consecutive paragraph numbers, as has been done here.

1.2 Where only one paragraph follows a side heading, it is not assigned a separate paragraph number and is typed with its left margin level with the side heading, as has been done in the paragraph immediately below heading 1.3.

1.3 **Subparagraph Headings and Subparagraphing**

Ensure your paragraph numbering system is unobtrusive

If more than one paragraph follows a subparagraph heading, each is assigned an identification number or letter:

a) This would be the first subparagraph.

b) This would be the second subparagraph.

c) Each subparagraph can be further subdivided into a series of very short secondary subparagraphs:

   (1) Here is a secondary subparagraph.

   (2) Ideally, each secondary subparagraph should contain no more than one sentence.

d) Secondary subparagraphs may also be assigned subparagraph headings.

2

it must not be obtrusive (except when called on to emphasize a specific point). And it should convey only one idea, although made up of several sentences each containing a separate thought.

Experienced writers construct effective paragraphs almost subconsciously. They adjust length, tone, and emphasis to suit their topic and the atmosphere they want to create, and sometimes even stretch or bend the rules to obtain exactly the right impact. But even they once had to master the techniques of good paragraph writing, although now they let the rhythm of the words guide them far more than the rules.

A topic sentence at the start of the paragraph

We are not so fortunate. We have to learn the rules and apply them consciously. Yet we must not let our approach become too pedantic, or become so bound by the rules that we write in a stilted manner that lacks interest and rhythm. We should consider the rules as building blocks that help create good writing, not as bars that imprison our creativity.

A topic sentence at the end of the paragraph

Good paragraph writing depends on three elements:

Unity
Coherence
Adequate Development

These elements cannot stand alone. All three must be present if a paragraph is to be useful to its reader.

Unity

For a paragraph to have unity, it must be built entirely around a central idea. This idea is expressed in a topic sentence—often the first sentence—and developed in supporting sentences. In effect, this permits us to construct paragraphs using the pyramid technique, with the topic sentence taking the place of the summary and the supporting sentences representing the full development (see Figure 11-3 on page 372).

The topic sentence does not always have to be the first sentence in a paragraph; there are even occasions when it does not appear at all and its presence is only implied. But until you are a proficient writer you would be wise to place your topic sentences right up front, where both you and your readers can see them. Then, when you have experience to support your actions, you can experiment and try placing topic sentences in alternative positions.

The following paragraph has the topic sentence right up front. It has strong unity because its topic is clearly expressed and the supporting sentences develop it fully.

A fax modem marries the essential features of a fax machine with those of an electronic modem. It takes a document that has been

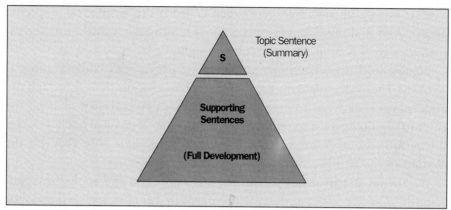

Figure 11-3 Pyramid technique applied to the paragraph.

Every supporting sentence must amplify or evolve from the topic sentence

keyed into the sending computer and transmits it to the receiving computer, where it is immediately viewed on screen or, if the computer is in use, stored for future reading. A fax modem may be a separate unit about the same size as a regular modem, or an integral unit about the size of a credit card that is inserted into a slot in the computer. Its transmission speed is the same as for a regular modem.[2]

Coherence

Coherence is the ability of a paragraph to hold together as a solid, logical, well-organized block of information. A coherent paragraph is abundantly clear to its readers; they can easily follow the writer's line of reasoning and have no problem in progressing from one thought to the next.

Good sequencing and effective transitions help create coherent paragraphs

Most technical people are logical thinkers and should be able to write logical, well-organized paragraphs. But the organization must not be kept a secret; it must be apparent to every reader who encounters their work. Simply summarizing a paragraph in the topic sentence and then following it with a series of supporting sentences does not make a coherent paragraph. The sentences must be arranged in an identifiable order, following a pattern that assists the reader to understand what is being said.

This pattern will depend on the topic and the type of document. Paragraphs describing an event or a process most likely will adopt a sequential pattern; those describing a piece of equipment will probably be patterned on the shape of the equipment or the arrangement of its features. (See Table 11-1.)

---

[2] Angie Morton, *Proposal for Improved Field Communication Capabilities*, Electro-Mechanical Engineering Services Inc, Westholm, Ontario, March 22, 1995, p 3.

Table 11-1 Paragraph patterns used in technical writing.

Topics or Subjects		Writing Patterns	Definitions and Examples
Generally Narrative	Event	Chronological order	The sequence in which events occurred; e.g., the steps taken to control flooding, how a new product was developed, how an accident happened
	Occurrence / Situation	Logical order	Used when chronological order would be too confusing (when describing a multiple-activity process, or several events that occurred concurrently); requires careful analysis to achieve a clear narrative
	Process	Cause to effect	The factors (causes) leading up to the result (effect) they produce; e.g. dirty air vents and failure to lubricate bearings cause overheating and eventual failure of equipment
	Procedure	Evidence to conclusion	The factors, data, or information (evidence) from which a conclusion can be drawn; e.g. how results obtained from a series of tests help identify the source of a problem
	Method	Comparison	A comparison of factors (price, size, weight, convenience) to show the differences between products, processes, or methods
Generally Descriptive — By Shape	Equipment	Vertical	From top to bottom* of a tall, narrow subject
	Scene	Horizontal	From left to right* of a shallow, wide subject
		Diagonal	From corner to corner (for items arranged across a subject)
	Appearance	Circular	From centre to circumference* (for items arranged concentrically) Clockwise* (for items arranged around a subject)
Generally Descriptive — By Features	Building	In order of: Size	From smallest item to largest item*
	Location	Importance	From most important to least important item*
	Features	Operation	The sequence in which the items are used when the equipment is operated

* or vice versa

Narrative Patterns

You can write narrative-type paragraphs to describe a sequence of steps or events. The past-present-future pattern of a progress report or occurrence report, such as Bob Walton's accident report in Figure 4-3, is a typical example. The pattern should be clearly evident, as in two of the following three paragraphs:

<table>
<tr>
<td>

A coherent paragraph (in chronological order)

</td>
<td>

The accident occurred when Dennis Friesen was checking in at the Remick Airlines counter. He placed the company Polaroid camera on the counter while he completed flight boarding procedure. When the passenger ahead of him lifted a carry-on bag from the counter, its shoulder strap tangled with the carrying strap of the camera and pulled the camera to the floor. Dennis examined the camera and discovered a 40 mm crack across its back. Remick Airlines' representative Kathy Trane took details of the incident and will be calling you to discuss compensation.

</td>
</tr>
<tr>
<td>

Two treatments of the same information

A much less coherent paragraph (containing the same information but not presented in an identifiable pattern)

</td>
<td>

The accident occurred when Dennis Friesen was checking in at the Remick Airlines counter. Kathy Trane, a Remick Airlines representative, took details of the incident and will be calling you to discuss compensation. The damaged camera received a 40 mm crack across the back. When the passenger ahead of Dennis removed a carry-on bag from the counter, its shoulder strap tangled with the carrying strap of the company Polaroid camera and pulled it to the floor. Dennis had placed the camera on the counter while he completed flight boarding procedure.

</td>
</tr>
<tr>
<td>

A coherent paragraph (tracing events from evidence to conclusion)

</td>
<td>

We noticed a mild shimmy at speeds above 80 km/h about 10 days after the new tires had been installed. A visual check of all four wheels revealed no obvious defects, so we rotated the four wheels to different positions on the vehicle. This did not eliminate the shimmy but did seem to change its point of origin. To pin down the cause we replaced each wheel in turn with the spare wheel, and found that the shimmy disappeared when the spare was in the left front position. We removed the wheel from that position,

</td>
</tr>
</table>

tested it, and found that it had been incorrectly balanced.

Descriptive Patterns

You can write descriptive paragraphs to describe scenes, buildings, equipment, and any subject having physical features. This pattern can be defined by the shape of the subject, the order in which parts are operated, the arrangement of parts from smallest to largest, or the importance of the various parts. For example:

Excerpt from paragraph (features in order of importance)

The most important control on the bomb aimer's panel is the firing button, which when not in use is held in the black retaining clip at the bottom left-hand corner. Next in importance is the fusing switch at the top right of the panel; when in the "OFF" position it prevents the bombs from being dropped live. Two safety switches, one immediately above the firing button retaining clip and the other to the right of the bank of selector switches, prevent the firing button from being withdrawn from its clip unless both are in the "LIVE" (up) position.

The topic sentence sets the scene

Continuity

A fully coherent paragraph must also have smooth transitions between its sentences. Smooth transitions give a sense of continuity that makes readers feel comfortable. As they finish one sentence, there is a logical bridge to the next. This can be accomplished by using linking words and by referring back to what has already been said. In the example just quoted there is a natural flow from "The most important..." in the first sentence to "Next in importance..." in the second. The third sentence then refers back to the firing button and so relates the newly introduced safety switches to the previous information. The transitions are equally good in the first paragraph describing damage to a Polaroid camera, each sentence containing a component that is a development from one of the previous sentences. This is not true of the second camera-damage paragraph, in which each new sentence introduces a new subject with no reference to what has already been said.

Adequate Development

Paragraph development demands good judgment. You must identify your readers clearly enough so that you can look at each paragraph from their

You want to say enough, but not rabbit on, and on, and on...

point of view. Only then can you establish whether your supporting sentences amplify the topic sentence in sufficient detail to satisfy their interest.

Simple insertion of additional supporting sentences does not necessarily meet the requirements for adequate development. The supporting sentences must contain just the right amount of pertinent information, all directed to a particular reader. There must never be too little or too much. Too little results in fragmented paragraphs that offer snippets of information that arouse readers' interest but do not satisfy their needs. On the other hand, too much information can lead to long, repetitious paragraphs that annoy readers. Compare the following paragraphs, all describing the result of exploration crews' first venture with machinery across the Peel Plateau in the Yukon, intended for a reader who is interested in the problems of working in northern Canada, but who has never seen what the terrain is like.

Each paragraph has a good topic sentence...	**Inadequate development**	Trails left by tractors look like long narrow scars cut in the plateau. Many of them have been there for years. All have been caused by permafrost melting. They will stay like this until the vegetation grows in again.
	Adequate development	To the visitor viewing this far northern terrain from the air, the trails left by tractors clearing undergrowth for roads across the plateau look like long, narrow scars. Even those that have been there for as long as 28 years are still clearly defined. All have been caused by melting of the permafrost, which started when the surface moss and vegetation were removed and will continue until the vegetation grows in again—perhaps in another 30 years.
...but the development is erratic	**Over-development**	To conservationists, whose main interest is the protection of the environment, viewing this far northern terrain from the air is a heartrending sight. To them, the trails left by tractors clearing undergrowth for roads across the plateau look like long, narrow scars. The tractors were making way for the first roads to be built by man over an area that until now had been trodden only by Inuit indigenous to the area, and the occasional trapper. Some of these trappers had journeyed from Quebec to seek new sources of revenue for their trade. But now, in the very short time span of 28

years, man has defiled the terrain. With his
machines he has cut and gouged his way, thought-
lessly creating havoc that will be visible to those
that follow for many decades. Those that preceded
him for centuries had trodden carefully on the permafrost, leaving no trace of their presence. The
new trails, even those that have been there for 28
years, are visible almost as though they had been
cut yesterday. And all were caused by melting of
the permafrost...

In these examples the descriptive pendulum has swung from one
extreme to the other. The first paragraph leaves the reader with questions:
What were the tractors doing? How many years? How soon will the vegetation grow in again? The second paragraph develops the topic sentence
in just enough detail; it explains why the tractors left semipermanent scars
and predicts how long they will remain. The third paragraph, though
interesting, is filled with irrelevant information (e.g. where the trappers
come from) and repetitive statements that detract from the main theme. It
might be suitable for a novel but not for a technical report or description.

Correct Length

We recommend that you try to limit paragraph length to no more than 10
printed lines. But also keep these points in mind:

- You can adjust paragraph length to suit the complexity of the topic
  and the technical level of the reader. Generally, complex topics
  demand short paragraphs containing small portions of information,
  while general topics can be covered in longer paragraphs. However,
  even a complex topic can be covered in long paragraphs for readers
  who are technically able to deal with the topic.

- Variety in paragraph length has a lively visual effect. Conversely, if
  you write a series of equal-length paragraphs you may create an
  impression of dullness.

- Readers attach importance to a paragraph that is clearly longer or
  shorter than those surrounding it. A very short paragraph among several longer paragraphs particularly attracts attention.

- If you write too many short paragraphs close together, your readers
  may feel you are providing them with incomplete snippets of information. Conversely, if you write a succession of very long paragraphs,
  your readers may feel they are facing "heavy going" reading.

*If your work will be
printed in narrow,
newspaper-like
columns, write shorter
paragraphs*

Sentences

Although sentences normally form an integral part of a larger unit—the paragraph—they still must be able to stand alone. While helping to develop the whole paragraph, each has to carry a separate thought. In doing so each plays an important part in placing emphasis—in stressing points that are important and playing down those that are not.

Just as experienced writers first had to learn the elements of good paragraph writing, so they also had to learn the elements of good sentence construction. These are:

Unity
Coherence
Emphasis

Unity and coherence perform a function in the sentence similar to their role in a well-written paragraph.

Unity

The pyramid again!

Although the comparison is not quite so clearly defined, we can still apply the pyramid technique to the sentence in the same way it is applied to the paragraph and the whole document. In this case the summary is replaced by a primary clause that presents *only one thought*, and the full development by subsidiary clauses and phrases that develop or condition that thought (see Figure 11-4). Thus a unified sentence, as its name implies, presents and develops only a single thought.

Compare these two sentences:

A unified sentence that expresses one main thought	The Amron Building will make an ideal manufacturing plant because of its convenient location, single-level floor, good access roads, and low rent.
A complicated sentence that tries to express two thoughts	The copier should never have been placed in the general office, where those using it interrupt the work being done by the administrative staff, who have been consistently overworked since the beginning of the year.

Try writing short, simple, uncomplicated sentences, then later link those that seem to belong together

The first sentence has unity because everything it says relates to only one topic: that the Amron Building will make a good manufacturing plant. The second sentence fails to have unity because readers cannot tell whether they are supposed to be agreeing that the copier should have been placed elsewhere, or sympathizing that the administrative staff have been overworked. No matter how complex a sentence may be, or how many

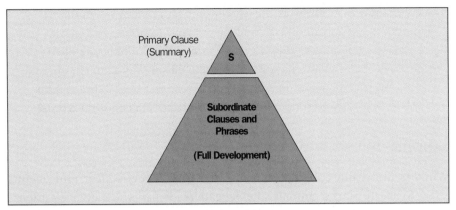

Figure 11-4 Pyramid technique applied to the sentence.

subordinate clauses and phrases it may have, every clause must either be a development of or actively support only one thought, which is expressed in the primary clause.

Coherence

Coherence in the sentence is very similar to coherence in the paragraph. Coherent sentences are continuously clear, even though they may have numerous subordinate clauses, so that the message they convey is apparent throughout. Like the paragraph, they need good continuity, which can be obtained by arranging the clauses in logical sequence, by linking them through direct or indirect reference to the primary clause, and by writing them in the same grammatical form. Parallelism, discussed at the end of this chapter, plays an important role here.

Read examples of good writing in technical journals and magazines (e.g. Scientific American; IEEE Spectrum)

The unified sentence discussing the Amron Building has good coherence because its purpose is continuously clear and each of its subordinate clauses links comfortably back to the lead-in statement (through the phrase "because of its..."). The clauses are written in parallel form (i.e. they have the same grammatical pattern), which is the best way to carry the reader smoothly from point to point.

The first sentence below lacks coherence because there is no logic to the arrangement or form of the subordinate clauses. Compare it with the second sentence, which despite its greater length is still coherent because it continuously develops the thought of "late" and "damaged" expressed in the primary clause.

An incoherent sentence

The Amron Building will make an ideal manufacturing plant because of its convenient location, which also should have good access roads, the advantage of its one-level floor, and it commands a low rent. (34 incoherent words)

A coherent sentence	Many of the 61 samples shipped in December either arrived late or were damaged in transit, even though they were shipped one week earlier than usual to avoid the Christmas mail tie-up, and were packed in polyurethane as an extra precaution against rough handling. (45 coherent words)

Emphasis

Whereas unity and coherence ensure that a sentence is clear and uncluttered, properly placed emphasis helps readers identify the sentence's important parts. By arranging a sentence effectively you can attach importance to the whole sentence, to a clause or phrase, or even to a single word.

Emphasis on the Whole Sentence

You can give a sentence more emphasis than sentences that precede and follow it by manipulating its length or by stressing or repeating certain words. You can also imply to the reader that all the clauses within a sentence are equally important (without affecting its relationship with surrounding sentences) by balancing its parts.

Too many sentences all the same length imply the information is dull!

Although you should aim for variety in sentence length, there may be occasions when you will want to adjust the length of a particular sentence to give it greater emphasis. Readers will attach importance to a short sentence placed among several longer sentences, or to a long sentence among predominantly short ones. They will also detect a sense of urgency in a series of short sentences that carry them quickly from point to point. This technique is used effectively by story tellers:

> The prisoner huddled against the wall, alone in the dark. He listened intently. He could hear the guards, stomping and muttering. Cursing the cold, probably.

In technical writing we have little occasion to write very short sentences, except perhaps to impart urgency to a warning of a potentially dangerous situation:

> Dangerously high voltages are present on exposed terminals. Before opening the doors,
> 1. set the master control switch to "OFF," and
> 2. hang the red "NO" flag on the operator's panel.
> Never cheat the interlocks.

Neither must we err in the opposite direction and write overly long sentences that are confusing. The rule that applies to paragraph writing—adjust the length to suit the complexity of topic and technical level of readers—applies equally to sentence writing. If, on average, your sentences tend to exceed 22 words, they probably are too long.

Similarity of shape can signify that all parts of a sentence are of equal importance. Clauses separated by a coordinate conjunction (mainly *and*, *or*, *but*, and sometimes a comma) tell a reader that they have equal emphasis. The following sentences are "balanced" in this way:

> Eight test instruments were used for the rehabilitation project, *and* were supplied free of charge by the Dere Instrument Company.

> The gas pipeline will be 420 kilometres shorter than the oil pipeline, *but* will have to cross much more difficult terrain.

Keep sentence parts parallel

> The upper knob adjusts the instrument in the vertical plane, *and* the lower knob adjusts it in the horizontal plane.

Emphasis on Part of a Sentence

While coordination means giving equal weight to all parts of a sentence, subordination means emphasizing a specific part and deemphasizing all other parts. It is effected by placing the most important information in the primary clause and placing less important information in subordinate clauses. In each of the following sentences the main thought is italicized to identify the primary clause:

> When the technician momentarily released his grip, *the control slipped out of reach.*

> *The bridge over the underpass was built on a compacted gravel base*, partly to save time, partly to save money, and partly because materials were available on site.

> *He lost control of the vehicle* when the wasp stung him.

> When the wasp stung him, *he lost control of the vehicle.*

Emphasis on Specific Words

Where we place individual words in a sentence has a direct bearing on their emphasis. Readers automatically tend to place emphasis on the first and last words in a sentence. If we place unimportant words in either of these impact-bearing positions, they can rob a sentence of its emphasis:

Emphasis misplaced	Such matters as equipment calibration will be handled by the standards laboratory however.
Emphasis restored	Equipment calibration, however, will be handled by the standards laboratory.
Emphasis misplaced	Without exception change the oil every six days at least.

Let a noun have the last word!

| Emphasis restored | Change the oil at least every six days. |

The verbs we use have a powerful influence on emphasis. Strong verbs attract the reader's attention, whereas weak verbs tend to divert it. Verbs in the active voice are strong because they tell *who did what*. Verbs in the passive voice are weak because they merely pass along information; they describe *what was done by whom*. The sentences below are written using both active and passive voice. Note that the versions written in the active voice are consistently shorter and more direct:

Passive Voice	Active Voice
Elapsed time is indicated by a pointer.	A pointer indicates elapsed time.
The project was completed by the installation crew on May 2.	The installation crew completed the project on May 2.
It is suggested that meter readings be recorded hourly.	I suggest you record meter readings hourly.
The samples were passed first to quality control for inspection, and then to the shipping department where they were packed in polyurethane.	Quality control inspected the samples and then the shipping department packed them in polyurethane.

The passive voice is less preferred in technical writing

We recommend you use the active voice in your technical writing

There are occasions when you will have to use the passive voice because you are reporting an event without knowing who took the action, or prefer not to name a person. For instance, you may prefer to write:

The strain gauge was read at 10-minute intervals.

rather than:

Kevin McCaughan read the strain gauge at 10-minute intervals.

Unfortunately, many scientists and engineers still write in the passive voice. Some have even been advised—wrongly—to shun the active voice because it is too strong and does not fit their "professional" character. Perhaps with the passage of time this outdated belief will disappear.

Completeness

Every day we see examples of incomplete sentences—on television, in magazines, and especially in advertising. Media writers use them to create a crisp, intentionally choppy effect:

SHEER COMFORT!

8200 metres high. Wide seats, just like your living room.
Tempting meals. Complimentary refreshments.
Only on Remick Airlines. Our Business Class. Try us!

But if we do the same in our business letters and reports, our sentences are likely to be read with raised eyebrows.

Forming Sentence Fragments

A common and very easy error to make, particularly when a piece of writing is going well and you do not want to interrupt your enthusiasm, is to inadvertently form a sentence fragment. Normally you correct it later, when you are checking what you have written, but sometimes your familiarity with media writing can cause you not to notice it. These are examples of sentence fragments:

- The meeting achieved its objective. Even though three members were absent.
- The staff were allowed to leave at 3 p.m. Seeing the air-conditioning had failed.

All right for a first rough draft...

The first sentence in each of these examples is complete (it has a subject-verb-object construction), but the two second sentences are incomplete because, to be understood, each depends on information in the first sentence.

A useful way to check whether or not a sentence is complete is to read it aloud entirely on its own. If it contains a complete thought it will be understood just as it stands. For example, *The meeting achieved its objective* is a complete thought because you do not need additional information to understand it. *The staff were allowed to leave at 3 p.m.* is also complete. But you cannot say the same when you read these aloud:

- *Even though three members were absent.*
- *Seeing the air-conditioning had failed.*

In most cases a sentence fragment can be corrected by removing the period that separates it from the sentence it depends on, inserting a comma in place of the period, and adding a conjunction or connecting word such as *and, but, which, who,* or *because*:

- The meeting achieved its objective, even though three members were absent.
- The staff were allowed to leave at 3 p.m. *because* the air-conditioning had failed. (*"Seeing"* has been changed to *"because."*)

...but it should be corrected by the second draft

Here are two others:

- Staff will have to bring bag lunches or go out for lunch from October 6 to 10. While the lunchroom is being renovated. (*Change the period to a comma.*)
- All 20-year employees are to be presented with long-service awards. Including three who retired earlier in the year. At the company's annual banquet. (*Change both periods to commas.*)

In particular check sentences that start with a word that ends in "-ing" (e.g. refer*ring*, answer*ing*, be*ing*) or an expression that ends in "to" (e.g. with reference *to*):

- With reference to your letter of June 6. We have considered your request and will be sending you a cheque. (*Change the period to a comma.*)
- Referring to the problem of vandalism to employees' automobiles in the parking lot. We will be hiring a security guard to patrol the area from 8 a.m. to 6 p.m., Monday through Friday. (*Although the period could be changed to a comma, a better sentence could be formed by reconstructing the fragment:*)
- To resolve the problem of vandalism to employees' automobiles in the parking lot, we will be hiring...

Forming Run-on Sentences

A similar sentence error occurs if you link two separate thoughts in a single sentence, joining them with only a comma or even no punctuation. The effect can jar a reader uncomfortably. For example:

- Ms Solvason has been selected for the word-processing seminar on March 11, she is not eager to attend.

This awkward construction is known as a run-on sentence. It can be corrected by

- replacing the comma with a period, to form two complete sentences:

    ...on March 11. She is not eager to attend.

- or retaining the comma and following it with *which* or *but:*

    ...on March 11, which she is...
    ...on March 11, but she is...

A run-on sentence with no punctuation is even more noticeable:

    The customer said he never received an invoice I made up a new one.

Here, either a period or a comma and a linking word can be inserted between *invoice* and *I*:

Your reader will know what you are saying, but will feel uncomfortable reading your words

...an invoice. I made up...

...invoice, so I made up...

Positioning End Punctuation Correctly

There are particular rules for inserting punctuation after quotation marks and closing brackets:

- If a sentence ends with a quotation mark, place the period *inside* the quotation mark:

  "That's the information we need," the chief engineer remarked. "Now we can start the project."

(The above example shows that when a comma ends an introductory statement, it also is placed *inside* the closing quotation mark.)

These are North American rules; in the UK, they differ

- If a sentence ends with a closing bracket, place the period *outside* the bracket if the words within the brackets do not form a complete sentence:

  Sound levels measured in the laboratory exceeded the tolerance specification (as shown in Table 2).

- However, if the opening bracket is preceded by a period, and the words within the brackets create a complete sentence, then place the period *inside* the bracket:

  Sound levels measured in the laboratory exceeded the tolerance specification. (They were above 82 dBA for more than two hours per day.)

Words

The right words in the right place at the right moment can greatly influence your readers. A heavy, ponderous word will slow them down; an overused expression will make them doubt your sincerity; a complex word they do not recognize will annoy them; and a weak or vague word will make them think of you as indefinite. But the right word—short, clear, specific, and necessary—will help them understand your message quickly and easily.

The keyword here is "specific"...

Words That Tell a Story

Words should convey images. We have many strong, descriptive words in our individual vocabularies, but most of the time we are too lazy to employ them because the same old routine words spring easily to mind. We write "put" when we would do better to write "position," "insert," "drop," "slide," or any one of the numerous descriptive verbs that better describe the action. Compare these examples of vague and descriptive words:

Whenever possible, insert specific words rather than generalizations

Vague Words	Descriptive Words
While the crew was in the town they *got* some spare parts.	they *bought*
	they *purchased*
	they *borrowed*
	they *requisitioned*
We have *contacted* the site.	We have *telephoned...*
	We have *visited...*
	We have *written to...*
	We have *spoken to...*
	We have *faxed...*
	We have *emailed...*
The project will *take a long time.*	will *last four months*
	will *require 300 work hours*
	will *employ two installers for three weeks*

Story writers use descriptive words to convey active images to their readers. Because we are concerned with *technical* writing does not mean we should avoid seeking colorful words. One descriptive word that defines size, shape, color, smell, texture, or taste is much more valuable than a dozen words that only generalize. (But a word of warning: reserve most of these colorful, descriptive words for your verbs and nouns rather than for adverbs and adjectives.)

Analogies can offer a useful means for describing an unfamiliar item in terms a nontechnical reader will recognize:

Relate a complex concept to a well-known idea or fact

A resistor is a piece of ceramic-covered carbon about the size of a cribbage peg, with a 50 mm length of wire protruding from each end.

Specific words tell the reader that you are a definite, purposeful individual. Vague generalities imply that you are unsure of yourself. If you write

It is considered that a fair percentage of the samples received from one of our suppliers during the preceding months contained a contaminant.

you give your reader four opportunities to wonder whether you really know much about the topic:

1. "It is considered"   Who has voiced this opinion?
2. "a fair percentage"   How many?
3. "one of our suppliers"   Who? One in how many?
4. "contained a contaminant"   What contaminant? In how strong a concentration?

All these generalities can be avoided in a shorter, more specific sentence:

> We estimate that 60% of the samples received from RamSort
> Chemicals last June were contaminated with 0.5% to 0.8% mercuric
> chloride.

This statement tells the reader that you know exactly what you are talking about. As a technical person, you should never create any other impression.

Combining Words into Compound Terms

One of the biggest problems for technical writers is knowing whether multiword expressions should be compounded fully, joined by hyphens, or allowed to stand as two or more separate words. For example, should you write:

> cross check, cross-check, or crosscheck?
> counter clockwise, counter-clockwise, or counterclockwise?
> change over, change-over, or changeover?

The tendency today is to compound a multiword expression into a single term. But this bare statement cannot be applied as a general rule because there are too many variations, some of which appear in the glossary.

Most multiword expressions are compound adjectives. When two words combine to form an adjective they are either joined by a hyphen or compounded to form one word. They are usually joined by a hyphen if they are formed from an adjective-noun expression:

Adjective + Noun	As a Compound Adjective
heavy water	heavy-water production
four channels	four-channel receiver
high frequency	high-frequency oscillator

But when one of the combining words is a verb, they often combine into a one-word adjective. Under these conditions they normally will compound into a single-word noun:

Two Words	As a Noun	As an Adjective
lock out	lockout	lockout voltage
shake down	shakedown	shakedown test
cross over	crossover	crossover network

Three or more words that combine to form an adjective in most cases are joined by hyphens. For example, *lock test pulse* becomes *lock-test-pulse generator*. Occasionally, however, they are compounded into a single term, as in *counterelectromotive force*. Specific examples are in the glossary.

The trend is to compound multiword expressions into a single word

Refer to the Glossary for the more common multiword expressions

The Technique of Technical Writing   **387**

Obviously, these "rules" cannot be taken at full face value because there are occasions when they do not apply. Useful guides for doubtful combinations are contained in many dictionaries, and in *The Canadian Writer's Handbook* referenced earlier.

Spelling "Canadian Style"

English-speaking Canadians have inherited a spelling problem. Over the years our two official languages have developed characteristics of their own, which distinguish them from the English spoken in Britain and the French spoken in France. Of the two, French seems to have evolved into the more "pure" Canadian language, simply because it has been less influenced by external forces. In contrast, Canadian English has been strongly and unavoidably influenced by American English.

Do you write "metre" or "meter"?; or "harbor" or "harbour"?

Those of us who write primarily in English are confronted with a dilemma: Should we continue to spell words such as "theatre" and "labour" as they are traditionally spelled in Britain? Or should we spell them "theater" and "labor," as they are spelled by our neighbours (or *neighbors*) in the United States?

Unfortunately there are no clearcut guidelines for the correct spelling of such words in Canada, which means that each of us must decide individually whether we want to spell according to contemporary British usage, or align ourselves with the standards current in the United States.

If almost all of your business correspondence is to people and organizations in Canada, and if your reports are seldom read in the US, then you can safely adopt the British spellings. But if you correspond frequently with people in the US, and write reports that will be read on both sides of the border, then you probably should adopt US spelling as your "standard." The choice, however, is entirely yours.

The better dictionaries offer alternatives

You need a good dictionary to guide you, and preferably one that has been published within the past six to eight years. We recommend *The Concise Oxford Dictionary*[3] if you prefer the traditional British spelling style, and the *Webster's New Collegiate Dictionary*[4] if you choose the American spelling style. These dictionaries show both the preferred and the alternative spellings of words that can be spelled more than one way. For example, the *Oxford* dictionary lists

---

[3] *The Concise Oxford Dictionary of Current English*, 8th ed, ed Richard E Allen (Oxford, England: The Oxford University Press, 1990).

[4] *Webster's New Collegiate Dictionary*, 9th ed (Springfield, Massachusetts: G & C Merriam Company, 1989).

theatre, *theater (the asterisk means "chiefly US")

whereas the *Webster's* dictionary lists

theater or theatre.

The first word in each case is the preferred spelling for the country in which the dictionary has been published, and the second spelling is an alternative, must less common, spelling for that country. (Few people in Britain would write *theater*, and even fewer Americans would write *theatre*.)

A third choice is *Funk & Wagnalls Canadian College Dictionary*,[5] which tends to lean toward American spelling style.

We have chosen to use a widely respected Canadian guideline as our authority for most spellings, both throughout previous chapters and in the glossary. This is *The Canadian Press Stylebook*[6] and its companion booklet *Caps and Spelling*.[7]

Two features of the *Stylebook* are its preference for the "re" spelling for words such as "centre" and "theatre," and the "or" spelling for words such as "honor" and "donor." This is reflected throughout the Glossary of Technical Usage. Additionally, whenever there is a spelling choice, the glossary describes the alternatives and then identifies which spelling we recommend for use in Canada, like this:

> *We use the CP Stylebook as our style guide*

favo(u)r   *favor* pref in US and rec in Can.; *favour* pref in Br
(pref = preferred; rec = recommended)

Note that the emphasis is on the word "recommend": we only *recommend* which spelling we believe to be preferable; the ultimate choice is yours. Remember, however, that you must be consistent in your choice of spellings, particularly within each document you write.

Long Versus Short Words

Big words create a barrier between writer and reader. Some writers use big words to hide their lack of knowledge or because they think it makes them sound important, others because they start writing without first defining clearly what they want to say. There are many long scientific words that we have to use in technical writing; we should surround them with short words whenever possible so our writing will not become ponderous and overly complex.

> *Avoid using an 89 cent word when an equally suitable 25 cent word is available*

---

[5] *Funk & Wagnalls Canadian College Dictionary* (Markham, Ontario: Fitzhenry & Whiteside, 1986).

[6] *The Canadian Press Stylebook* (Toronto: The Canadian Press, 1993).

[7] *Caps and Spelling* (Toronto: The Canadian Press, 1992).

Low-Information-Content Expressions

Words and expressions of low information content (LIC) contribute little or nothing to the facts conveyed by a sentence. They are untidy lodgers. Remove them and the sentence appears neater and says just as much. The problem is that practically everyone inserts LIC words into sentences, and we become so accustomed to them that we do not notice how they fill up space without adding any information. For example:

Vague and Wordy	Pursuant to the client's original suggestion, Mr Richards is of the opinion that the structure planned for the client would be most suitable for erection on the site until recently occupied by the old established costume manufacturer known as Garrick Garments. In accordance with the client's anticipated approval of this site, Mr Richards has taken great pains to design a multi-level building that can be considered to use the property to an optimum extent.

"Waffle" is the word technical editor Anna King uses to describe such cumbersome writing. By eliminating unnecessary expressions (such as *pursuant to*; *of the opinion that*; *for immediate erection on*; *in accordance with*; *can be considered to use*; *an optimum extent*), Anna cut the original 74 words to a much more effective 40 words:

Clear and Direct	Mr Richards believes the building planned for the client should be erected on the site previously occupied by Garrick Garments. He has assumed the client will approve this site, and has designed a multi-level building that fully develops the property.

Table 11-2 contains some of the words and phrases we must try to delete from our writing. They are difficult to identify because they often sound like good but rather vague prose. In the following sentences the LIC words have been italicized; they should be either deleted or replaced, as indicated by the notes in brackets.

- The control is actuated by *means of* No. 3 valve. (**delete**)
- Adjust the control *as necessary* to obtain maximum deflection. (**delete**)
- Tests were run for *a period of* three weeks. (**delete**)
- If the project drops behind schedule *it will be necessary to* bring in extra help. (*I, you, he, we,* or *they will* bring in extra help)
- By Wednesday we had a backlog of 632 units, *and for this reason* we adopted a two-shift operation. (**replace with** *so we adopted...*)

Low-information-content words are untidy lodgers...

- A new store will be opened *in an area* where market research has *given an indication that there actually is* a need for more retail outlets. (delete both expressions; replace the second one with *research has indicated that...*)

Clichés and hackneyed expressions are similar to LIC words and phrases, except that their presence is more obvious and their effect can be more damaging. Whereas LIC words are intangible in that they impart a sense of vagueness, hackneyed expressions can make a writer sound pompous,

...as are clichés and hackneyed expressions

Table 11-2 Examples of low-information-content (LIC) words and phrases.

The LIC words and phrases in this partial list are followed by an expression in brackets (to illustrate a better way to write the phrase) or by an (X), which means that it should be dropped entirely.

actually (X)
a majority of (most)
a number of (many; several)
as a means of (for; to)
as a result (so)
as necessary (X)
at present (X)
at the rate of (at)
at the same time as (while)
at this time (X)
bring to a conclusion (conclude)
by means of (by)
by use of (by)
communicate with (talk to; telephone; write to)
connected together (connected)
contact (talk to; telephone; write to)
due to the fact that (because)
during the course of, during the time that (while)
end result (result)
exhibit a tendency (tend)
for a period of (for)
for the purpose of (for; to)
for this reason (because)
in all probability (probably)
in an area where (where)
in an effort to (to)
in close proximity to (close to; near)

in color, in length, in number, in size (X)
in connection with (about)
in fact, in point of fact (X)
in order to (to)
in such a manner as to (to)
in terms of (in; for)
in the course of (during)
in the direction of (toward)
in the event that (if)
in the form of (as)
in the light of (X)
in the neighborhood of; in the vicinity of (about; approximately; near)
involves the use of (employs; uses)
involves the necessity of (demands; requires)
is a person who (X)
is designed to be (is)
it can be seen that (thus; so)
it is considered desirable (I or we want to)
it will be necessary to (I, you, or we must)
of considerable magnitude (large)
on account of (because)
previous to, prior to (before)
subsequent to (after)
with the aid of (with; assisted by)
with the result that (so, therefore)

Note: Many of these phrases start and end with words such as *as, at, by, for, in, is, it, of, to,* and *with*. This knowledge can help you identify LIC words and phrases in your writing.

These are only partial lists: there are many more LIC expressions

garrulous, insincere, or artificial. Referring to oneself as "the writer," starting and ending letters with such overworked phrases as "We are in acknowledgment of" and "please feel free to call me," and using semi-legal jargon ("the aforementioned correspondent," "in the matter of," "having reference to") are all typical hackneyed expressions. Further examples are listed in Table 11-3.

Some Fine Points

Using Parallelism to Good Effect

Parallelism in writing means "similarity of shape." It is applied loosely to whole documents, more firmly to paragraphs, sentences, and lists, and tightly to grammatical form. Readers generally do not notice that parallelism is present in a good piece of writing—they only know that the sentences read smoothly. But they are certainly aware that something is wrong when parallelism is lacking. Good parallelism makes readers feel comfortable, so that even in long or complex sentences they never lose their way. Its ability to help readers through difficult passages makes parallelism particularly applicable to technical writing.

Good parallelism has a subtle, mostly hidden effect on readers.

The Grammatical Aspects

If you keep your verb forms similar throughout a sentence in which all parts are of equal importance (i.e. in which the sentence has coordination, or is balanced), you will have taken a major step toward preserving parallelism. For example, if I write "unable to predict" in the early part of a balanced sentence, I should write "able to convince" rather than "successful in convincing" in a latter part:

Readers feel uncomfortable when parallelism is violated

Parallelism violated	Mr Johnson was *unable to predict* the job completion date, but was *successful in convincing* management that the job was under control.
Parallelism restored (A)	Mr Johnson was *unable to predict* the job completion date, but was *able to convince* management that the job was under control.
Parallelism restored (B) (more direct alternative)	Mr Johnson *predicted* no completion date, but *convinced* management that the job was under control.

The rhythm of the words is much more evident in the latter two sentences. In sentence (B) particularly, the parallelism has been stressed by

Table 11-3 Typical clichés and hackneyed expressions.

all things being equal	in the foreseeable future
a matter of concern	in the long run
and/or	in the matter of
as a matter of fact	last but not least
as per	many and diverse
attached hereto	needless to say
at this point in time	please feel free to
enclosed herewith	pursuant to your request
for your information (as an intro-	regarding the matter of
ductory phrase)	this will acknowledge
if and when	we are pleased to advise
in our opinion	we wish to state
in reference to	with reference to
in short supply	you are hereby advised

converting the still awkward "...unable...able..." of sentence (A) to the positively parallel "predicted...convinced...."

Two other examples follow:

Parallelism violated	Pete Hansk likes surveying airports and to study new construction techniques.
Parallelism restored	Pete Hansk likes to survey airports and to study new construction techniques. or Pete Hansk likes surveying airports and studying new construction techniques.
Parallelism violated	His hobbies are developing new software programs and stereo component construction.
Parallelism restored	His hobbies are developing new software programs and constructing stereo components. or His hobbies are new software development and stereo component construction.

Parallelism is particularly important when you are joining sentence parts with the coordinating conjunctions *and, or,* and *but* (as in the examples above), with a comma, or with correlatives such as

either...or
neither...nor

Be particularly careful when using the expression *Not only...but also*

Each part of a correlative must be followed by an expression in the same grammatical form. That is, if *either* is followed by a verb, then *or* must also be followed by the same form of verb:

Parallelism violated	You may either repair the test set or it may be replaced under the warranty agreement.
Parallelism restored	You may either *repair* the test set or *replace* it under the warranty agreement.

Application to Technical Writing

Although parallelism is a useful means for maintaining continuity in general writing, in technical writing it has special application. Parallelism can clarify difficult passages and give rhythm to what otherwise might be dull material. When building sentences that have a series of clauses, you can help the reader see the connection between elements by moulding them in the same shape throughout the sentence. There is complete loss of continuity in the following sentence because it lacks parallelism:

Parallelism violated	In our first list we inadvertently omitted the 7 lathes in room B101, 5 milling machines in room B117, and from the next room, B118, we also forgot to include 16 shapers.

When the sentence is written so that all the item descriptions have a similar shape, the clarity is restored:

Maintain similarity of shape within each sentence...

Parallelism restored	In our first list we inadvertently omitted 7 lathes in room B101, 5 milling machines in room B117, and 16 shapers in room B118.

Emphasis can also be achieved by presenting clauses in logical order. If they are arranged climatically (in ascending order of importance), or in order of increasing length, the reader's interest will gain momentum right through the sentence:

...and within each paragraph

No one ever offered to help the navigator struggle out to the aircraft, a parachute hunched over his shoulder, a roll of plotting charts tucked under his arm, a sextant and an astrocompass dangling from one hand, and his precious navigation bag heavy with instruments, computer, radar charts, and the route plan for the seven-hour flight clenched firmly in the other.

Within the paragraph parallelism has to be applied more subtly. If it is too obvious, the similarity in construction can be dull and repetitive. The verbs often are the key: keep them generally in the same mood and they will help bind the paragraph into one cohesive unit (this is closely tied in with coherence, discussed under "Paragraphs"). This paragraph has good parallelism:

The effects of sound are difficult to measure. What to some people is simply background noise, to others may be ear-shattering, peace-destroying drumming. The roar of a jet engine, the squeal of tires, the clatter of machinery, the hiss of air-conditioning, the chatter of children, and even the repetitive squeak of an unoiled door hinge, can seriously affect them and create a distinct feeling of uneasiness.

Similarity of shape is most obvious in the third sentence, with its rhythmic use of "sound" words:

roar of a jet engine	*hiss* of air-conditioning
squeal of tires	*chatter* of children
clatter of machinery	*squeak* of an unoiled door hinge

Here we have words that paint strong images. They not only have the same grammatical form but also are bound together because they relate to the same sense: hearing.

Even if the subject is more mundane, you should still try to use parallelism to carry your reader smoothly through your description, as in this description of a surveyor's transit:

> Two sets of clamps and tangent screws are used to adjust the levelling head. The upper clamp fastens the upper and lower plates together, while the upper tangent screw permits a small differential movement between them. The lower clamp fastens the lower plate to the socket, while the lower tangent screw turns the plate through a small angle. When the upper and lower plates are clamped together they can be moved freely as a unit; but when both the upper and lower clamps are tightened the plates cannot be moved in any plane.

Very technical, yet the rhythm binds the parts together

Application to Subparagraphing

Subparagraphing seldom occurs in literature, but is used frequently in technical writing to separate events or steps, describe an operation or procedure, or list parts or components. Subparagraphing always demands good parallelism.

In the following example, a description of three tests has been divided into subparagraphs (for clarity, only the initial words of each test are shown here):

> Three tests were conducted to isolate the fault:
> - In the first test a matrix was imposed upon the video screen and...
> - For the second test, voltage measurements were taken at...
> - A continuity tester was connected to the unit for test 3 and...

The last subparagraph is not parallel with the first two. To be parallel it must adopt the same approach as the others (i.e. it should first mention the test number and then say what was done):

- For the third test, a continuity tester was connected to the unit and...

To be truly parallel the subparagraphs should start *exactly* the same way:

- In the first test a matrix was...
- In the second test voltage measurements were...
- In the third test a continuity tester was...

But this would be too repetitive. The slight variations in the original version make the parallelism more palatable.

If more than three tests have to be described, a different approach is necessary. To continue with "A fourth test showed...," "For the fifth test...," and so on, would be dull and unimaginative. A better method is to insert a number in front of each paragraph:

Seven tests were conducted to isolate the fault:
1. A matrix was imposed upon the video screen and...
2. Voltage measurements were taken at...
3. A continuity tester was connected to the unit and...

Parallelism has been retained and we now have a more emphatic tone that lends itself to technical reporting.

A third version employs active verbs together with parallelism to build a strong, emphatic description that can be written in either the first or third person:

In laboratory tests conducted to isolate the fault we
1. *imposed* a matrix upon the video screen and...
2. *measured* voltage at...
3. *connected* a continuity tester to the unit and...

To change from the first person to the third person, only the lead-in sentence has to be rewritten:

In tests conducted to isolate the fault, the laboratory
1. *imposed*...
2. *measured*...
3. *connected*...

Abbreviating Technical and Nontechnical Terms

You may abbreviate any term you like, and in any form you like, providing you indicate clearly to the reader how you intend to abbreviate it. This can be done by stating the term in full, then showing the abbreviation in brackets to indicate that from now on you intend to use the abbreviation. Here is an example:

Always spell out single-digit numbers (sdn). The only time sdn are not spelled out is when they are being inserted in a series of numbers.

When forming abbreviations, observe these three basic rules:

1. **Use lower case letters,** unless the abbreviation is formed from a person's name:

centimetre	cm
kilogram	kg
approximately	approx
decibel	dB (the *B* represents *Bell* [Alexander Graham Bell])

2. **Omit all periods,** unless the abbreviation forms another word:

horsepower	hp
cubic centimetre	cm^3
cathode-ray tube	crt
metre	m
pascal	Pa
singular	sing.

These guidelines are recognized worldwide

3. **Write plural abbreviations in the same form as the singular abbreviation:**

metres	m
pascals	Pa
kilograms	kg
hours	h or hr

There are, however, exceptions, which have grown as part of our language. Through continued use these unnatural abbreviations have been generally accepted as the correct form. A few examples follow:

for example (exempli gratia)	e.g.	*(There is a slowly growing trend to write these as eg and ie)*
that is (id est)	i.e.	
morning (ante meridiem)	a.m.	
afternoon (post meridiem)	p.m.	
inside diameter	ID	
number(s)	No.	

The Glossary is there to help you

The Glossary of Technical Usage offers a reasonably comprehensive list of standard abbreviations and some technical abbreviations. For specific technical terms, you may need to refer to a list of abbreviations compiled by one of the technical societies in your discipline.

Writing Numbers in Narrative

The conventions that dictate whether a number should be written out or expressed in figures differ between ordinary writing and technical writing. In technical writing you are much more likely to express numbers in figures.

The rules listed below are intended mainly as a guide. They will apply most of the time, but there will be occasions when you will have to make a decision between two rules that conflict. Your decision should then be based on three criteria:

- which method will be most readable;
- which method will be simplest to type; and
- which method you used previously, under similar circumstances.

Good judgment and a desire to be consistent will help you select the best method each time.

The basic rule for writing numbers in technical narrative is

- spell out single-digit numbers (one to nine inclusive)
- use figures for multiple-digit numbers (10 and above).

However, there are exceptions to this rule:

Always use figures

- when writing specific technical information, such as test results, dimensions, tolerances, temperatures, statistics, and quotations from tabular data,
- when writing any number that precedes a unit of measurement: 3 mm; 7 kg; 121.5 MHz,
- when writing a series of both large and small numbers in one passage: During the week ending May 27 we tested 7 transmitters, 49 receivers, and 38 power supplies,
- when referring to section, chapter, page, figure (illustration), and table numbers: Chapter 7; Figure 4,
- for numbers that contain fractions or decimals: 7 1/4, 7.25,
- for percentages: 3% gain; 11% sales tax,
- for years, dates, and times: At 3 p.m. on January 9, 1997; 08:17, 20 Feb 97,

Numerals are more common than spelled-out numbers in most technical narrative

- for sums of money: $2000; $28.50; $20; 27 cents, or $0.27 (preferred), and
- for ages of persons.

Always spell out

- round numbers that are generalizations: about five hundred; approximately forty thousand,
- fractions that stand alone: repairs were made in less than three-quarters of an hour, and
- numbers that start a sentence. (Better still, rewrite the sentence so that the number is not at the beginning.)

There are five additional rules:

- Spell out one of the numbers when two numbers are written consecutively and are not separated by punctuation: 36 fifty-watt amplifiers or thirty-six 50-watt amplifiers. (Generally, spell out whichever number will result in the simplest or shortest expression.)
- Insert a zero before the decimal point of numbers less than one: 0.75; 0.0037.

Always place a "0" in front of an open decimal

- Use decimals rather than fractions (they are easier to type), except when writing numbers that are customarily written as fractions.
- Insert spaces in large numbers containing five or more digits: 1 275 000; 27 291; 4056. (Insert a space in four-digit numbers only when they appear as part of a column of numbers.)
- Write numbers that denote position in a sequence as 1st, 2nd, 3rd, 4th...31st...42nd...103rd...124th...

When writing and abbreviating numerical prefixes such as "giga" and "kilo," follow the guidelines in Table 11-4.

Table 11-4 Numerical prefixes and abbreviations.

Multiple/ Submultiple	Prefix	Symbol	Multiple/ Submultiple	Prefix	Symbol
10^{18}	exa	E	10^{-1}	deci	d
10^{15}	peta	P	10^{-2}	centi	c
10^{12}	tera	T	10^{-3}	milli	m
10^{9}	giga	G	10^{-6}	micro	μ
10^{6}	mega	M	10^{-9}	nano	n
10^{3}	kilo	k	10^{-12}	pico	p
10^{2}	hecto	h	10^{-15}	femto	f
10	deca	da	10^{-18}	atto	a

This guideline also is recognized worldwide

Writing Metric Units and Symbols (SI)

The Glossary of Technical Usage includes terms and symbols prescribed by the International System of Units (SI). The trend toward worldwide adoption of metric units of measurement means that for some time both the imperial inch/pound system and the metric (SI) system will be in use concurrently. The terms and symbols introduced here are those you are most likely to encounter in your technical reading or may use in your technical writing.

The acronym "SI" represents the name "Système International d'Unités." Both the acronym and the name were adopted for universal usage in 1960 by the eleventh Conférence Générale des Poids et Mesures (CGPM), which is the international authority on metrication. Since then, many of the metric terms the conference established have crept into our language. For example, *Hertz*, the unit of frequency measurement, was first introduced as a replacement for *cycles per second* in the early 1960s; now it is used universally, both in the technical world and by the general public. Other terms already in use are:

The "-re" spelling is correct for *litre* and *metre*, although only the "-er" spelling is recognized in the US

	non-SI	*SI*
Temperature:	degrees Fahrenheit	degrees Celsius
Length:	miles, yards, feet, inches	kilometres, metres, millimetres
Weight:	tons, pounds, ounces	tonnes, kilograms, grams, milligrams
Liquid volume:	gallons, quarts	kilolitres, litres

There are nine general guidelines for writing SI symbols, and these are applied consistently to the entries in the Glossary of Technical Usage. SI symbols must be written, typed, or printed

The abbreviation for *litre* is L, because a lower case letter l looks like the numeral 1

1. in upright type, even if the surrounding type slopes or is in italic letters,
2. in lower case letters, except when the name of the unit is derived from a person's name (e.g. the symbol **F** for *farad* is derived from *Faraday*), in which case the first letter of the symbol is capitalized (e.g. **Wb** for *weber*),
3. with a space between the last numeral and the first letter of the symbol: **355 V, 27 km** (not 355V, 27km),
4. with no "s" added to a plural: **1 g, 236 kg,**
5. with no period after the symbol, unless it forms the last word in a sentence,
6. with no space between the multiple or submultiple symbol and the SI symbol: **3.6 kg, 150 mm, 960 kHz,**
7. with a solidus (oblique stroke: /) to represent the word *per*: **m/s** (metres per second), and with only one solidus used in each expression,

8. with a dot at midletter height (·) to represent that symbols are multiplied: **1m·s** (lumen second),
9. always as a symbol, when a number is used with the SI unit (e.g. "the tank holds 400 L"), but spelled out when no number is used with the unit (e.g. "capacity is measured in litres" [*not* "capacity is measured in L"]).

Writing Non-Gender-Specific Language

History has provided us with a scenario in which men were the warriors and hunters, and subsequently the breadwinners, and women were the home bodies who cooked and reared children and catered to their men's needs. Today, all that has changed and it is universally recognized in developed countries that women and men are equal and can for the most part have equal occupations and equal roles. Consequently, we now see men as secretaries, nurses, and child care workers, and women as airline captains, engineers, truck drivers, and backhoe operators.

Unfortunately, our language hasn't kept pace and we still see some people who write like this:

A secretary will be brought in to record the minutes of the client/contractor project meeting. *She* will be responsible for making travel arrangements for all meeting participants.

After much deliberation, the committee decided to hire an engineer to look into the problem. *He* will evaluate the extent of erosion that occurred when the river overflowed its banks.

Awareness is the key to writing non-gender-specific language

Neither of these writers knew whether the secretary and the engineer were going to be male or female, yet *they automatically assumed* that the secretary would be a woman and the engineer would be a man. Traditionally, that was what they and their parents, and their grandparents, and their parents before them, were accustomed to.

It's our job to eradicate gender-specific references like these from our writing, until it becomes *automatic* always to write non-gender-specific references. For example:

A secretary will be brought in to record the minutes of the client/contractor project meeting, and to make travel arrangements for all meeting participants.

Interestingly, these examples are shorter than the original sentences

After much deliberation, the committee decided to hire an engineer to evaluate the extent of erosion that occurred when the river overflowed its banks.

Eliminate Masculine Pronouns

When describing engineers, scientists, architects, managers, supervisors, technical people, and even accountants and lawyers, historically our language has abounded with masculine pronouns. The engineer described earlier is a typical example. Here is another, this time an excerpt from a company's operating procedures:

> **4.3 Senior Systems Engineer.** *His* primary role is to plan, schedule, manage, and coordinate the activities of the engineers within the Systems Engineering Department. *He* also is responsible for preparing budgets and maintaining fiscal control of operations performed by the department, and for maintaining liaison with and reporting progress to clients.

There are several ways you can remove the male pronouns:

1. Repeat the job title, and abbreviate it:

> **4.3 Senior Systems Engineer (SSE).** The SSE's primary role is to plan, schedule, manage, and coordinate the activities of the engineers within the Systems Engineering Department. The SSE also is responsible for preparing budgets and maintaining fiscal control of operations performed by the department, and for maintaining liaison with and reporting progress to clients.

2. Use a bulleted list:

> **4.3 Senior Systems Engineer.** The Senior Systems Engineer is responsible for
> - planning, scheduling, managing, and coordinating the activities of the engineers within the Systems Engineering Department,
> - preparing budgets and maintaining fiscal control of operations performed by the department, and
> - maintaining liaison with and reporting progress to clients.

This is probably the most clear and comfortable revision

3. Create a table:

> **4.3 Senior Systems Engineer**

Primary Responsibility	Secondary Responsibilities
To plan, schedule, manage, and coordinate the activities of the engineers within the Systems Engineering Department.	To prepare budgets and maintain fiscal control of operations performed by the department. To maintain liaison with and report progress to clients.

This revision employs good information design principles

4. Replace the male pronoun with "you" and "your":

> **4.3 Senior Systems Engineer.** *Your* primary role is to plan, schedule, manage, and coordinate the activities of the engineers within the Systems Engineering Department. *You* also are responsible for preparing budgets and maintaining fiscal control of operations performed by the department, and for maintaining liaison with and reporting progress to clients.

(Note: If you use "you" in one part of a document, be consistent and use it throughout the document. Avoid bouncing back and forth between "you" and "he" or "she".)

5. Replace the male pronoun with "he or she":

> **4.3 Senior Systems Engineer.** *His* or *her* primary role is to plan, schedule, manage, and coordinate the activities of the engineers within the Systems Engineering Department. *He or she* also is responsible for preparing budgets and maintaining fiscal control of operations performed by the department, and for maintaining liaison with and reporting progress to clients.

(Note: This is the least recommended method.)

6. Change singular pronouns to plural pronouns:

> **4.3 Senior Systems Engineers.** *Their* primary role is to plan, schedule, manage, and coordinate the activities of the engineers within the Systems Engineering Department. *They* also are responsible for preparing budgets and maintaining fiscal control of operations performed by the department, and for maintaining liaison with and reporting progress to clients.

Suggestions 4, 5, and 6 are less comfortable revisions

(Note: This method can be used only when the description lends itself to using plural nouns and pronouns; i.e. there must be more than one Senior Systems Engineer.)

Replace Gender-Specific Nouns

Each province in Canada has its Workers Compensation Board: an organization that provides financial help to employees who are injured at work. Yet, not many years ago, all Workers Compensation Boards in Canada were known as *Workman's* Compensation Boards. The previous title seemed to imply that the Board provided help *only* to male workers, which was not true. Similarly, until about 12 years ago, flight attendants on airlines were known as stewardesses, implying that the job was held only by females. Again, particularly today, this is plainly inaccurate.

There are many other job titles that are equally gender-specific and predominantly male-oriented. These have been changed in recent years to reflect that the title refers to both male and female employees. Table 11-5 lists gender-specific titles and suggests better alternatives.

The term "man-hours" previously was used to define the time that would be expended on a particular job. Today, we write *work-hours* or *staff hours*.

Be Consistent When Referring to Men and Women

Men throughout recent history have been given the courtesy title *Mr* before their names. Until 20 years ago, women had two courtesy titles, to denote whether they were married or single: *Mrs* and *Miss*. Today, a woman's marital status is *never* implied in the title: all women should be referred to as *Ms*. (English is not the only language to have created this anomaly. For example, in France men are referred to as *Monsieur* and women as *Madame* or *Mademoiselle*. In Russia, a sex-identifying title is not placed before a person's name, but a woman's family name has an "a" added to the end to denote the person is female: for example, Boris Serov; Svetlana Serova.)

Never address a letter to "Dear Sir or Madam." If you are replying to a letter signed by A J Winters, and you don't know if A J Winters is male or female, then in the salutation write "Dear A J Winters:". If you don't know the person's name, then address the letter to *the position the person*

Be particularly careful when addressing letters and writing the salutation

Table 11-5 Preferred names for gender-specific titles.

Titles such as *fireman*, *foreman*, and *salesman* are deeply entrenched, and so are difficult to eradicate

If you are tempted to write:	Consider replacing it with:
actor; actress	actor (for both sexes)
chairman	chairperson (*or* chair)
cowboy	cattle rancher
fireman	firefighter
foreman	supervisor
newsman	reporter
policeman; policewoman	police officer
postman	letter (*or* mail) carrier
repairman	service technician
salesman	sales representative
spokesman	spokesperson
workman	worker; employee
waiter; waitress	server (or "waiter" for both sexes)

holds: "Dear Customer Services Manager:" or "Dear Manager, Human Resources:".

To sum up: we need to be continually vigilant and check that we never use gender-specific expressions in our writing.

Writing for an International Audience

Three years ago, when Macro Engineering Inc started doing business with companies in Eastern Europe, Senior Engineer Sharleen Boullé had to learn a different set of rules for communicating with the company's new customers and business associates. She found it strange that the way she wrote her letters and memos had to be changed.

"I took it for granted that everyone used the pyramid method to construct their letters," she explained. "But then I found that in many Eastern European and Asian countries, starting immediately with the main message—getting right down to business—is considered downright rude."

In a growing global economy, we have to be aware of and adapt to cultural differences

Writing Business Correspondence

At first, Sharleen thought it would be just a matter of explaining to people she corresponded with in Russia, Lithuania, and the Ukraine, that she wrote this way because in the west we had discovered it was much more efficient. She thought it would help their economy if they adopted our methods. But she had forgotten the hundreds of years of history that has fashioned the Eastern and Asian cultures, and found that she had to adapt her methods to suit them. "You have to understand the culture prevalent in each society and adjust to it," Sharleen says now.

She didn't entirely discard the pyramid method. She decided she could still use the pyramid for the central part of her letters, but that she would have to precede it with a personal greeting and polite remarks concerning the health and happiness of her reader (and, often, also of her reader's family). And she would have to follow the pyramid with a polite closing remark, such as wishing the reader continuing good health and prosperity in the months and years ahead.

So at first she constructed her letters to Eastern Europe as shown in Figure 11-5. "Yet even then I had to be cautious," Sharleen continued. "I learned fairly quickly that this arrangement was fine for readers I have corresponded with before, but to new readers—particularly older, more traditional readers—it still seemed too abrupt. For them, I had to move the Summary Statement to further down in the letter."

Adapt the pyramid to suit the reader's culture

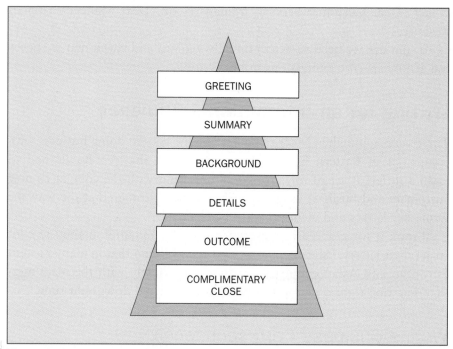

Figure 11-5 "Adapted" pyramid for a letter to an Eastern European or Asian reader.

Revising the Writing Plan

Sharleen's revised writing plan looked like this:

1. GREETING
2. BACKGROUND
3. DETAILS
4. OUTCOME and SUMMARY STATEMENT (combined)
5. COMPLIMENTARY CLOSE

This, of course, is the inverted pyramid, which is the reverse of the sequence promoted in previous chapters. And it applies not only to formal business letters but also to memos and electronic means of communication: faxes and particularly email.

However, changing the focus is not the only part of a message to an Eastern or Asian country that requires attention. When you write in English to readers who normally speak another language—German, French, Italian, Spanish, Malayan, or Chinese, for example—you have to choose words that will be clearly understood. (This also holds true for different cultures who speak the same language. In Great Britain, for instance, the word "fortnight" is commonly used to mean "two weeks," yet in the United States and even in parts of Canada it would not be understood. Similarly, the word "presently" has different meanings in the US and Great Britain.)

Avoid jargon: choose words that will be understood by both writer and reader

Writing Guidelines

Here are some guidelines to follow when writing to an international audience:

- Avoid long, complex sentences.
- Avoid long, complex words. If you have a choice between two or more words or expressions that have roughly the same meaning, choose the simpler one. For example, write "pay" rather than "salary" or "remuneration."
- Use the same word to describe the same action or product consistently throughout your letter. Decide, for example, whether you will refer to money in the bank as *funds, currency, deposits, capital,* or *money.*
- Use a word always in the same sense. You would confuse a foreign-language reader if you were to write "It would not be *appropriate* to transfer funds from Account A to Account B" (meaning it would not be suitable to do it), and then in another sentence you were to write "We had insufficient capital to *appropriate* Company A" (meaning to take over Company A, or buy it out).

Meeting and Speaking

Julian Tarquist was one of Macro Engineering's first computer science specialists to visit clients in Russia and Estonia, and while he was there he discovered some major but subtle differences between western and eastern perceptions of good manners and effective communication techniques. When he went to a meeting, he learned that it was inappropriate to start the meeting by immediately describing why he was there and what he wanted to accomplish.

First, members of both companies would sit around a table, be served coffee, soft drinks, and small cakes or biscuits, be introduced to each other, and engage in polite conversation about travel, where they lived, the weather, the health of each other's families, but *never* about business.

The business discussion did not start until the topic was eventually introduced by a member of the host country. Only then could Julian start describing his plans and ideas, and even then he discovered it had to be done in a roundabout way. Where Julian was accustomed to putting the main message right up front, using the Tell-Tell-Tell method, he found that in the host country it was more customary to start with some history about the subject and then lead gently toward the point he wanted to make. He also discovered that the words he used had to be chosen carefully, because he and his hosts interpreted some words differently.

Now Julian and Sharleen have gained considerable experience communicating internationally. Yet both are ready to admit that, as world markets

Cultural differences particularly influence how you conduct yourself at meetings

Take time to read about doing business in other countries

Talk to people who
have lived there or who
have extensive experi-
ence working there

expand and boundaries between countries become less apparent, they will have to become even more sensitive to the cultural differences that affect how people in other countries act and react to the way we conduct ourselves, and particularly to the way we present information. As Sharleen explained: "I've learned to respect the customs and writing modes in other cultures."

We have room in this book to offer only a few examples and suggestions for communicating to an international audience; just enough to create an awareness of the need to adapt your methods to suit each audience. If you are going to be doing business with other nations, then we recommend you visit your local library and research information on communicating with business people in other cultures. Additionally, within the Society for Technical Communication (STC), a Special Interest Group on International Technical Communication publishes a newsletter titled *Global Talk* that discusses worldwide communication practices. The STC is at 901 N Stuart Street, Suite 904, Arlington, Virginia, 22203-1854, USA.

ASSIGNMENTS

Exercise 11.1

The following sentences and short passages lack compactness, simplicity, or clarity, and particularly contain low-information-content (LIC) expressions. Improve them by deleting unnecessary words, or by partial or complete rewriting.

1. A power outage disrupted work for a period of 4 hours and 13 minutes on January 7, 1997.
2. The company has installed a microprocessor-controlled scanner in close proximity to No. 2 production line.
3. It was installed in an effort to increase quality control of assembled components.
4. If you experience any further problems with the aforementioned scanner, please feel free to call me at any time.
5. We have come to the conclusion that there will be a need for the hiring of two additional product assembly technicians from 1 May to June 30.
6. The *4Tell* software will be initiated as a means of preventing slippage of the schedule.
7. The Freeling Lake Mine survey project has, at this point in time, reached the point of being 3.5 days behind schedule.

Simple ideas confused
by wordy, rambling
sentences

8. During the course of our investigation, it came to our attention that someone has absconded with in the region of 6 to 8 Nabuchi 2001 portable computers.

9. For your information, it was decided by the executive committee that a sum in the amount of $180 000 should be set aside for the purpose of implementing building improvements in 1999.

10. The end result was a 22% decrease in acidity following the introduction of Limasol Plus into the solution.

11. It is with considerable concern that we have noticed a large number of discrepancies between the projected and actual expenditures for 1997.

12. Any attempt to operate the engine in excess of 1500 rpm is likely to result in and be the cause of accelerated bearing failure.

13. The drawings attached hereto are in reference to the May 12 proposed tentative plan for modification of the MicroStat in accordance with directive TP-7.

14. If it is your intention to effect repairs to your model 2120 printer, it would be in your best interests first to assure the availability of parts for this older type of model before starting work.

15. By working an additional 45 minutes per day for the month of January, we will in all probability be back on schedule on or about February 1.

16. Check for grain temperature with the use of a probe 1.5 metre in length inserted vertically downward into the depths of the grain.

17. Registration for next year's Institute in Technical Management should be made before or no later than December 15.

18. It is our estimation that the cost for the maintenance of the Abacus control system for a period of three years (July 1, 1997, to June 30, 2000) will be in the region of $270 000 to $310 000.

19. Between a speed of 86 and 92 km/h the wheel at the left front exhibits a tendency to shimmy.

20. With the completion of this sentence you will bring to a conclusion this exercise on the identification of LIC expressions.

Exercise 11.2

The following sentences offer choices between words that sound similar or are frequently misused. Select the correct word in each case.

1. The test documentation (appears/seems) to be complete.
2. Signals from space probe 811 travelled 86 billion km (farther/further) than signals from space probe 260.
3. Three attempts were made to (elicit/illicit) a response from station K-2.
4. Before clamping the scale onto its base, (orient/orientate) the 0° point on the scale so that it is opposite the "N" mark.
5. Our vehicle was (stationary/stationery) when truck T48106 skidded into it.

These word choices offer unexpected surprises!

6. The inspection team calibrated 38 of the 47 test instruments between April 28 (and/to) May 6. The (balance/remainder) were to be calibrated on May 13 and 14.

7. Fuel consumption is (affected/effected) by driving speed.

8. The quality control inspection team recommended instituting a (preventive/preventative) maintenance program to reduce system outages.

9. We could not complete the modifications on schedule (as/for/because) the monitor had not yet been shipped to the site.

10. Although the report documented the investigation results in full detail, its executive summary (implied/inferred) there were hidden costs.

11. When compared (to/with) the previous period, the (amount/number) of products rejected by the quality control inspectors decreased from 237 to 189.

Some words are so similar it can be difficult to choose between them

12. Under current human rights legislation, employers have to be particularly (discreet/discrete) when making personal enquiries about potential employees.

13. Analyses of the data (is/are) being delayed until all the data (has/have) been received.

14. The inspection team will submit (its/it's/their) final report on October 26.

15. Improved maintenance in 1997 resulted in 27 (fewer/less) service interruptions than in 1996.

16. When the temperature within the cabinet rises above 28.5°C, the blower motor cuts in automatically and operates (continually/continuously) until the cabinet temperature drops to 25°C.

17. Before accepting each job lot of steel from the foundry, samples are tested mechanically and electronically to ensure the steel is (free from/free of) flaws.

18. The chief engineer agreed that the agenda (was/were) too comprehensive for a one-hour meeting.

19. When the secretary telephoned to say that Mr Humphries was too ill to travel, the program committee had to find an (alternate/alternative) keynote speaker for the conference.

20. The (principal/principle) reason for including this exercise in *Technically-Write!* is to draw your attention to the glossary of technical usage.

Exercise 11.3

Rewrite the following sentences to make them more emphatic (in many cases, change them from the passive to active voice). Create a "doer" if one is not identified.

1. The minutes of the April 20 project meeting were recorded by technician Dan Wolski.

2. It was recommended by the quality control inspector that production lots 214 and 307 be reworked.

3. Yesterday's power failure was caused by an explosion at Winfield power station.

4. The fracture of the rotor arm was first noticed by Jan Wright.

5. The *Amaze* software was shipped by express post on Thursday, but it was not received by the customer until Tuesday.

6. Although Whistler Mountain's Olympic Station was obscured by fog for 24 hours, work by the crew repairing the chairlift continued throughout the night.

7. Safety boots and hardhats are to be worn by all technicians working within 100 metres of the construction tower.

8. The 57 personal computers at Multiple Industries will be linked by a local area network (LAN) and central file server that has been designed by Vancourt Business Systems.

9. Due to poor visibility and drifting snow caused by a blizzard on January 15, the highway to Montrose International Airport was closed by the RCMP from 4 p.m. to 9 a.m. the following day.

10. As a result of the high sound levels recorded in the machine shop by Murray Walsh, it was decided by management that the shop would be operated on a reduced scale until sound-reduction measures were implemented by the safety committee.

11. When toxicity levels in No. 3 tank were measured, it was evident from the results that the requirements of specification DZ0286 had been exceeded by 22%.

12. For international travel, all requests are to be approved by the manager of administration, and all travel arrangements are to be made by Columbia Travel at 3030 Windmark Drive.

13. Although delays in the delivery of spare parts had been identified by Cal Poloskiew in two separate letters to the supplier on September 10 and 28, no action was taken by the supplier until he had received four more complaints from site engineers in October.

14. When the *4Tell* software was evaluated, it was discovered by Sheila Fieldstone that the program was only 80% compatible with the operating system used by the company for its mainframe computer.

15. In a memo dated February 10 from Pierre St Germaine, Fern Wilshareen was asked to recommend an engineering technologist from her department to attend the Engineering Technology Symposium in Montreal on April 3 and 4 as the company's representative.

Changing from passive to active voice will shorten these sentences

A passive-voice sentence is longer than an active-voice sentence

Exercise 11.4

Improve the parallelism in the following sentences.

Keep the ideas—like trains—on parallel tracks

1. A survey was carried out to define property boundaries and as a means for settling disputes between property owners.

2. Write your name, address, postal code, and telephone number into spaces 1, 2, 4, and 6 of the application form, and in space 8 list your email address.

3. Agreeing to service Multiple Industries' metal shears on Saturday was easy, but it was much more difficult to find a technician willing to do the work on a Saturday!

4. Getting approval to purchase a replacement control unit will not be a problem, but there will be difficulty in finding a local supplier who can deliver it by June 16.

5. If you decide to order three Nabuchi 310 portable computers, the price will be $1995 each, but the price will be $2250 each if you order only one or two.

6. When the Montrose branch office closed, two staff members were transferred to the Edmonton office and a generous severance package was arranged for five who were laid off.

7. We selected the Nabuchi 700 laserjet printer because it is fast, moderately priced, readily available, and its scalable fonts can be used directly with our word-processing software.

8. The accident delayed installation work for three days and was instrumental in increasing operating costs by 4.6%.

9. Inspection of the air extraction unit revealed
   • rust on the elements,
   • a crack in the inlet bellows,
   • someone had deposited chewing gum on the fan blades, and
   • loose clamps on the fresh air inlet and the outlet pipes.

10. Management conducted an internal audit for three reasons: as a means for demonstrating management's concern for staff working in areas affected by last week's fire; to assess the extent of damage caused by the fire; and to identify areas needing priority repair work.

11. Version 8.1 of the *Arbutus for Windows 95* software has not only made the program more user-friendly, but also the tendency for the version 7.0 software to cause "mouse freeze" has been corrected.

12. The following procedure is to be followed for submitting expense accounts when on protracted field assignments:
    1. They are to be submitted weekly.
    2. Complete them on Monday for the preceding week.
    3. They should be completed in quadruplicate.
    4. Mail copies 1, 2, and 3 to head office, retaining copy 4 for your files.
    5. Be sure to mail them no later than noon on Tuesday.

Exercise 11.5

Abbreviate the terms shown in italics in the following sentences. In some cases you will also have to express numerals in the proper form. (Guidelines for forming abbreviations and writing numbers in narrative are on pages 396 to 399. The Glossary of Technical Usage also contains many technical abbreviations.)

1. John McCormack, at age twenty-three, is the youngest *laboratory technician.*
2. A 24-*inch diameter* culvert, *approximately* 330 *metres* long, will connect the proposed development to the main drainage channel.
3. The sound level recorded near the south wall of the office was 58.6 *decibels*, which was .8 *decibels* higher than the level recorded beside the north wall.
4. 8,000 of the 68 656 cylinders for the Norland contract were manufactured with a *22.5 millimetre inside diameter.*
5. Rotate the antenna in a *counterclockwise* direction until the needle on the dial dips to between the .6 and 1.4 markings.
6. Vibration became severe at speeds above 84 *kilometres per hour*, which forced us to drive more slowly than anticipated. However, this enabled us to achieve a fuel consumption of 6.4 *litres per 100 kilometres.*
7. Radio station DMON's two operating frequencies are 88.1 *kilohertz* and 103.6 *megahertz.*
8. Paragraph three on page seven of the proposal quoted an additional six hundred and fifty-seven dollars ($657.00), tax *included*, for repairing the damage to the roof caused by Hurricane Hazel.
9. One degree of latitude is the equivalent of sixty *nautical miles*, so that each minute of latitude equals one *nautical mile* or 6080 feet (or 1.609 *kilometres*).
10. The new furnace produces 56 000 *British thermal units*, which is seven *per cent* more than the unit it replaces.
11. Comparing temperatures is not difficult if you remember that zero *degrees Celsius* is the same as thirty-two *degrees Fahrenheit*, twenty-two *degrees Celsius* is equal to seventy-two *degrees Fahrenheit*, and that both *temperatures* are the same at minus forty *degrees.*
12. Tank *serial number* 2821 holds one thousand *litres* of oil, which is equivalent to 264 *United States gallons* or 220 *imperial gallons.*

Refer to the guidelines earlier in this chapter, or turn directly to the Glossary

Exercise 11.6

Correct any of the following sentences that are not complete or have not been punctuated properly.

1. We have examined your bubble-jet printer, repairs will cost $148.00 plus tax, shall we go ahead?

2. The hard disk crashed at 4:15 p.m. Before I had time to copy today's work onto a safety disk.

3. Do you have a Hama A16 Electronic Camera Flash in your department, you can recognize it by the words *Sorte mit Automatik* on the base.

4. With effect from October 31 we will assign Myra Weiss and Dan Helwig to your Division for six months, this confirms our telephone conversation of October 17.

5. The safety label warned: "Toxic solution, handle with great care."

6. Progressive corrosion inside the pipes has reduced liquid flow by 21% since 1989. A condition which, if not corrected, could cause system shutdown in less than 12 months.

7. Work on the Feldstet contract was completed on February 16 three days ahead of the February 19 scheduled completion date, a cause for celebration.

8. When the digital exchange was installed at Multiple Industries, eight lines were left unused for anticipated staff expansion. Also for providing dedicated lines for a planned facsimile transmission network between branches.

9. The overhead steam pipe ruptured at 10:10 a.m., fortunately the office was empty as everyone had gone down to the cafeteria for coffee break. Although damage was effected to the computer equipment and oak furniture.

10. May I have company approval to attend the course "Wireless Communication," to be held at the University of Toronto May 16–20. And as attendance is limited may I have a reply by April 28.

11. Your request to attend the May course on "Wireless Communication" as requested in your memo dated April 10 was approved at the April 15 meeting of the Executive Committee, Francine Williams who attended the same course in January was particularly helpful in recommending that you attend.

12. Before submitting purchase requests for new or replacement equipment, ensure that
    - manufacturer details are complete,
    - price quotations are attached.
    - The appropriate specifications are quoted, and
    - An alternative supplier is listed.
    - The divisional manager approves the request.

Exercise 11.7

Select the correctly spelled words among the choices offered below.

1. The (computer/computor) is supplied with a built-in 99-year (calendar/calender).
2. A (coarse/course)-grained (aggregate/agreggate) is used as a base before pouring the concrete.
3. Profits in the (forth/fourth) quarter increased by (forty/fourty) percent.
4. The (affluent/effluent) produced by the paper mill is (enviromentally/environmentaly/environmentally) sound.
5. The preface to a book or report is sometimes called a (forward/foreward/foreword).
6. The (cite/site) is (inaccessable/inaccessible) except by helicopter.
7. Version 6.0 of *4Tell for Windows 95* (supercedes/supersedes) version 5.5.
8. The tests show that the materials have (similar/similiar) properties.
9. To a young business owner seeking a cash flow loan, the (colatteral/collatteral/collateral) demanded by the bank may seem (exhorbitant/exorbitant).
10. When Multiple Industries bought all the outstanding shares of Torrance Electronics, the latter company became a (wholely/wholly)-owned subsidiary.
11. Its better to edit your own writing on (hard copy/hardcopy) rather than (online/online/on line).
12. Silica gel is a drying agent, or (desiccant/dessiccant/dessicant), that is packed with electronic equipment before shipment.
13. Software designers who have (entepreneurial/entrepeneurial/entrepreneurial) drive do not (necessarily/neccessarily) have good management expertise.
14. After (lengthy/lengthly) deliberation, the executive committee admitted that Ken Wynne's innovative design was indeed (ingenious/ingenuous).
15. After we have (accumulated/accummulated) all the results from product tests, we (prescribe/proscribe) definitive purchase specifications.
16. An (auxiliary/auxilliary) heater cuts in when temperature drops below 3°C.
17. The (eigth/eighth) test demonstrated that the process is (feasable/feasible).
18. The incandescent lamps have been replaced with (flourescent/fluorescent/fluourescent) lamps.
19. The sales manager was (embarassed/embarrassed/embarrased) that customers were being (harassed/harrassed/harrased) by overly zealous sales staff.

Well? How good are you at spelling?

All these words are in the Glossary

20. Well? How many words did you (mispell/misspell)?

Exercise 11.8

Describe why gender-specific terms are inappropriate in the letters and reports you write.

Exercise 11.9

Improve the following passages so they contain no gender-specific language and (where appropriate) are better conveyors of information:

1. Memo to all lab technicians:

   The Department of Defence has informed us that a D.O.D. inspector will visit our calibration lab on May 17. He will be evaluating our equipment and calibration hierarchy to determine whether our lab meets MIL-STD-202 specifications. Please extend him every courtesy and your cooperation.

2. To: Andy Rittman:

   Please inform each field technician that from April 1 he will be covered by company-sponsored travel insurance arranged through ManSask Assurance Corporation. They won't have to pay for it, but belonging to the scheme won't be automatic, they have to apply for it.

Correcting gender-specific terms can be more subtle than is immediately evident

   I suggest you write a personal memo to each technician and enclose a copy of the enclosed application form and explanatory brochure. In each case be sure to remind him to apply by March 25, otherwise he'll have to wait until May 1 for his coverage to start.

3. From a company notice board:

Christmas Cheer is About to Start!

Are you planning to attend the Christmas Party? Tell your most significant other it will be at O'Halloran's on December 15 and to expect a royal feast!

Line up a babysitter today and tell her to expect to stay late...!

Call Dave Michaelson at extension 207 if you want to announce your intentions or to reserve a table, or if you want more information.

Exercise 11.10

Assume that you are employed by H L Winman and Associates and that you have been selected to be sent overseas in six weeks time, as part of a team of five engineers and technologists who will be working on a project in _____ *(your instructor will tell you which country)*. From your local library, research information on customs that will influence how you write, speak, and deport yourself in the host country. Write three short paragraphs describing the factors you will have to consider when you

Some research is needed here

1. write letters to people in the host country,
2. speak to your hosts at a conference, and
3. attend meetings or business lunches.

Exercise 11.11

Rewrite the following letters and memorandums to improve their effectiveness.

1. Customer Service Manager

   Emerald Air Express

   An overwritten letter

   To whomever it may concern:

   I am writing with reference to a shipment one of your drivers delivered to me on August 17. This particular shipment was an envelope containing eight papers that I needed urgently. For your records, your waybill number 7284 06292 36 is in reference to this shipment. The documents originated in Poughkeepsie, New York, and the package was picked up by one of your drivers and delivered to your Poughkeepsie office on August 13. The shipping envelope was marked "Next Morning Delivery."

   Your driver—here, the one who delivered the package to me— insisted that the shipment had been sent C.O.D., and I had to pay $48.76 for it before he would hand over the envelope. Only after the driver had gone did I discover that the shipment had been travelling for 4 days!!!

   What I want to know is this: Why was I charged such an outrageous sum for documents that were time-sensitive and that, by the time it reached me, was of no value? I look forward to receiving your cheque at your earliest convenience.

   I remain (annoyed)

   Peter LeMay

2.   Dear Mr Shasta

I am in receipt of your letter of Nov 17, you'll be glad to hear the problem you outlined is under consideration. A defective chip has been discovered in the output stage. Replacements are hard to get, I phoned around but no one has one locally. So I've ordered one directly from the manufacturer—Mansell Microprocessors—and asked them to ship it air express. I'll telephone you immediately it comes in. But it won't be for three weeks, too many are back-ordered, they can't ship before Dec 12.

Sincerely

T L Pedersen

3.   Dear Mr Reimer

It is with the sincerest regret that Vancourt Computers Inc has to inform you that there will be an unfortunate delay of about three weeks in filling your order. (Ref. your P.O. 2863 dated April 29) Due to measures entirely beyond our control the ship carrying a shipment of 1100 portable model 7000 computers from Nabuchi Electronics in Taiwan developed engine trouble in mid-Pacific and had to limp back to its home port. Your 3 computers are among the 1100. The shipping company has recently informed us that the ship will sail on June 6 and arrive in Vancouver on June 19. We will air express your 3 units as soon as they have cleared customs and absorb the additional expense ourselves. We hope this will help you understand our position.

Yours very truly

Vern Kerpov

4.   To: All H L Winman staff, Calgary

From: Tanys Young

Date: August 10

Ref: New facility

H L Winman and Associates is pleased to announce that construction will start shortly in room B101, which is to become a Child Care Centre that will accommodate 20 children between 1 and 5. It will open on November 1 and will be open *only* to children of H L Winman employees.

If you have preschoolers and would like to take advantage of this unusual new in-house service, you are invited to procure form CCC01 from Rick Davis in Personnel. It is our expectation that the CC Centre will be subject to oversubscription, hence we suggest your application be placed early because we will evaluate them in the order in which they are received.

Additionally, we will be looking for three experienced child care workers to man the facility. If you or your spouse knows of someone who might be suitable, ask her to telephone Rick for information and an application form.

5. Dear Ms Sorchan

Re: Inspection of your residential lot at 2127 Victoria Street. My survey shows your neighbor's fence is 62 mm inside your property (your neighbor to the south, that is, at #2123). Fence to the north is okay: its yours and its 13 mm inside your property. Does your neighbor at #2123 know about this? You have basis for watertight legal action if that's the rout you want to take. Our survey report and invoice #236 is enclosed.

Regards

Wilton Candrow

A discontinuous, under-written letter

6. To: Annette Lesk

   From: Mark Hoylan

   Date: 09/07/97

   Subject: Progress at MMW

Chatty and friendly; but are there too many words?

As you are well aware, Ken Poitras and I have been at Morriss Machine Works for two and a half weeks now, where we are shoring up the flooring for the NCR machine to be installed this week, and have had to build a 3.8 metre extention along the width of the north wall.

All this work is now complete and we should've been heading back to the office by now except theres a problem: the NCR machine arrived today and instead of being installed its sitting outside under a tarp. Why? Because no one seems to of calculated that a 1.26 metre wide machine (which is it's narrowest dimention) can't be greased through a .92 metre wide door!

So...Mr Grindelbauer who is the Machine Works manager (actually, he's the owner) has asked Ken and yours truly to stay behind and tear out part of the wall to make the door wider (which we're doing now) so the machine can be put in tomorrow, and then for us to rebuild the wall and reinstal the door which I reckon will take two extra days. I tried to explain to him there would be an extra charge for doing all this as its not in the contract, and he said for you to call him. Will you do that? Thanks.

Mark

P.S.   I reckon the additional time will be 32 hours and there will be two additional nights accomodation and per diem, plus some extra lumber and wallboard at a cost that will probably come in at about $135.

PPS   We'll be back in the office on the tenth. OK?

Exercise 11.12

Rewrite this one-paragraph notice and use information design to make it clearer, more personal, and more likely to encourage readers to do as it requests.

PROCEDURE RE EXPENSE CLAIMS

You can use your imagination when revising this one!

Expense claims must be handed to the Accounts Section before 10:30 a.m. on Wednesday for payment on Friday. Personnel failing to hand in their forms at the proper time, may do so at any time until 4:30 p.m. on Thursday but must wait until Monday for payment. Under no account will a late claim be paid in the same week that it was filed. Claims handed in after Thursday will be processed with the following week's claims and will be paid on the next Friday.

Online Technical Writing: Online Textbook
www.io.com/~hcexres/tcm1603/acchtml/acctoc.html

The online textbook of Austin Community College's Online Technical Writing course is available at this site. The book includes chapters on all aspects of technical writing and many examples. The appendices give additional information about audience analysis; libraries, documentation, and cross-referencing; common grammar, usage, and spelling problems; power-revision techniques; and related topics.

Self-Study Course for Professional Engineers and Scientists
darkstar.engr.wisc.edu/alley/professionals.html

This self-study course by Michael Alley of the Technical Communication Program of the University of Wisconsin is designed to help engineers and scientists make their scientific documents more efficient to write and to read. The site includes numerous online practice exercises.

Grammar, Punctuation, and Spelling: A Handbook for Technical Writers
sti.larc.nasa.gov/html/Chapt3/Chapt3-TOC.html

This handbook from the National Aeronautics and Space Administration describes a functional concept of punctuation.

Technical Writing: Books and References Sources
www.interlog.com/~ksoltys/twritres.html

This extensive bibliography of resources to do with technical communications is divided into 31 sections. Included are audience analysis, biomedical writing and editing, graphics, issues in technology, scientific and mathematical writing and editing, and many other topics.

So You Want to Be a Science Writer?
dspace.dial.pipex.com/town/square/ac073/writer.htm

Anthony Tucker, a former science editor of *The Guardian*, and the Association of British Science Writers produced this article on becoming a science writer. Although the focus of the article is British, readers in other countries will be interested in the general issues discussed.

Glossary of Technical Usage

A standard glossary of usage contains rules for combining words into compound terms, for forming abbreviations, for capitalizing, and for spelling unusual or difficult words. The glossary in *Technically-Write!* also offers suggestions for handling many of the technical terms peculiar to industry. Hence, it is oriented toward the technical rather than the literary writer.

The entries in the glossary are arranged alphabetically. Among them are words that are likely to be misused or misspelled, such as

- words that are similar and frequently confused with one another; e.g. *imply* and *infer*; *diplex* and *duplex*; *principal* and *principle*,
- common minor errors of grammar, such as *comprised of* (should be *comprises*), *most unique* (*unique* should not be compared), *liaise* (an unnatural verb formed from *liaison*),
- words that are particularly prone to misspelling, e.g. *desiccant, oriented, immitance,* and
- words for which there may be more than one "correct" spelling; e.g. *programer* or *programmer*.

Where two spellings of a word are in general use (e.g. *symposiums* and *symposia*), the glossary lists both and states which is preferred.

Definitions have been included when they will help you select the correct word for a given purpose, or to differentiate between similar words having different meanings. These definitions are intentionally brief and are intended only as a guide; for more comprehensive definitions, consult one of the authoritative dictionaries we recommend in Chapter 11 (pages 388 and 389).

All entries in the glossary are in lower case letters. Capital letters are used where capitals are recommended for a specific word, phrase, or abbreviation. Similarly, periods have been eliminated except where they form part of a specific entry. For example, the abbreviation for "inch" is *in.*, and the period that follows it is inserted intentionally to distinguish it from the word "in".

Finally, think of the glossary as a guide rather than a collection of hard and fast rules. Our language is continually changing, so that what was fashionable yesterday may seem pedantic today and a cliché tomorrow. We expect that in some cases your views will differ from ours. Where they do, we hope that the comments and suggestions we offer will help you to choose the right expression, word, abbreviation, or symbol, and that you will be able to do so both consistently and logically.

The Glossary

General abbreviations used throughout the glossary:

abbr	abbreviate(d); abbreviation	pl	plural
adj	adjective	pref	prefer; preferred; preference
Br	Britain; British	rec	recommend(ed)
Can.	Canada; Canadian	SI	International System of Units
def	definition	US	United States
lc	lower case		

A

a; an use *an* before words that begin with a silent *h* or a vowel; use *a* when the *h* is sounded or if the vowel is sounded as *w* or *y*; *an hour* but *a hotel, an onion* but *a European*

aberration

above- as a prefix, *above-* combines erratically: *aboveboard, above-cited, aboveground, above-mentioned*

abrasion

abscess

abscissa

absence

absolute abbr: abs

absorb(ent); adsorb(ent) *absorb* means to swallow up completely (as a sponge absorbs moisture); *adsorb* means to hold on the surface, as if by adhesion

abut; abutted; abutting; abutment

ac abbr for alternating current

accelerate; accelerator; accelerometer

accept; except *accept* means to receive (normally willingly), as in *he accepted the company's offer of employment*; *except* generally means exclude: *the night crew completed all the repairs except rewiring of the control panel*

access; accessed; accessible

accessory; accessories abbr: accy

accidental(ly)

accommodate; accommodation

account abbr: acct

accumulate; accumulator

achieve means to conclude successfully, usually after considerable effort; avoid using *achieve* when the intended meaning is simply to reach or to get

acknowledg(e)ment *acknowledgment* pref in US, rec in Can.; *acknowledgement* pref in Br

acquiesce def: agree to

acquire; acquisition

across not *accross*

actually omit this word: it is seldom necessary in technical writing

actuator

adapt; adept; adopt *adapt* means to adjust to; *adept* means clever, proficient; *adopt* means to acquire and use

adapter; adaptor *adapter* pref

adaptation pref spelling; *adaption* also common in US

addendum pl: *addenda*

adhere to never use *adhere by*

ad hoc def: set up for one occasion

adjective (compound) two or more words that combine to form an adjective are either joined by a hyphen or compounded into a single word; see page 387; abbr: adj

adsorb(ent) see absorb

advanced power manager abbr: apm (pref) or APM

advantageous

advice; advise use *advice* as a noun and *advise* as a verb: *the engineer's advice was sound; the technician advised the driver to take an alternative route*; spell: **adviser, advisable**

ae; e *ae* is pref in Br and common in Can., as in *aesthetic* and *anaemic*; *e* is pref in US and rec in Can., as in *esthetic* and *anemic*

aerate

aerial see antenna

aero- a prefix meaning of the air; it combines to form one word; *aerodynamics, aeronautical*; in some instances it has been replaced by *air*: *airplane, aircraft*

aesthetic; esthetic see ae

affect; effect *affect* is used only as a verb, never as a noun; it means to produce an effect upon or to influence (*the potential difference affects the transit time*); *effect* can be used either as a verb or as a noun; as a verb it means to cause or to accomplish (*to effect a change*); as a noun it means the consequences or result of an occurrence (as in *the detrimental effect upon the environment*), or it refers to property, such as *personal effects*

aforementioned; aforesaid avoid using these ambiguous expressions

after- as a prefix, usually combines to form one word: *afteracceleration, afterburner, afterglow, afterheat, afterimage*; but *after-hours*

agenda although plural, *agenda* is generally treated as singular; *the agenda is complete*

aggravate the correct definition of *aggravate* is to increase or intensify (worsen) a situation; try not to use it when the meaning is *annoy*

aggregate

aging; ageing *aging* pref in US, rec in Can.; *ageing* pref in Br

agree to; agree with to be correct, you should *agree to* a suggestion or proposal, but *agree with* another person

air- as a prefix, normally combines to form one word: *airborne, airfield, airflow*; exceptions: *air-condition(ed) (er) (ing), air-cool(ed) (ing)*

air horsepower abbr: **ahp**

airline; air line an *airline* provides aviation services; an *air line* is a line or pipe that carries air

algorithm

alkali; alkaline pl: *alkalis* (pref) or *alkalies*

allot(ted)

all ready; already *all ready* means that all (everyone or everything) is ready: *already* means by this time: *the samples are all ready to be tested; the samples have already been tested*

all right def: everything is satisfactory; never use *alright*

all together; altogether *all together* means all collectively, as a group; *altogether* means completely, entirely: *the samples have been gathered all together, ready for testing; the samples are altogether useless*

allude; elude *allude* means refer to; *elude* means avoid

almost never contract *almost* to *most*; it is correct to write *most of the software has been tested*, but wrong to write *the software is most ready*

alphanumeric def: in alphabetical, then numerical sequence

alternate; alternative *alternate(ly)* means by turn and turn about: *the inspector alternated among the four construction sites; alternative(ly)* should offer a choice between only two things: *the alternative was to return the samples*; however, it is becoming common for *alternative* to refer to more than two, as in *there are three alternatives*

alternating current abbr: **ac**

alternator

altitude abbr: **alt**

a.m. def: before noon (*ante meridiem*)

amateur

ambient abbr: **amb**

ambiguous; ambiguity

American standard code for information exchange abbr: **ASCII**

American Wire Gauge abbr: **AWG**

among; between use *among* when referring to three or more items; use *between* when referring to only two; avoid using *amongst*

amount; number use *amount* to refer to a general quantity: *the amount of time taken as sick leave has decreased*; use *number* to refer to items that can be counted: *the number of applicants to be interviewed was reduced to six*

ampere(s) abbr: **A** (pref) or **amp**; other abbr: **kA, mA, µA, nA, pA, A/m** (amperes per minute)

ampere-hour(s) abbr: **Ah** (pref) or **amp-hr** (more common)

amplitude modulation abbr: **AM**

an see **a**

anaemic; anemic see **ae**

anaesthetic; anesthetic see **ae**

analog(ue) *analog* pref in US, rec in Can.; *analogue* pref in Br

analogous

analogy

analyse; analyze *analyse* pref in Br, rec in Can.; *analyze* pref in US; also: **analyser**

AND-gate

and/or avoid using this term; in most cases it can be replaced by either *and* or *or*

ångström abbr: **A**

anion def: negative ion

anneal; annealed; annealing

annihilate; annihilated; annihilation

anonymous ftp

antarctic see **arctic**

ante- a prefix that means before; combines to form one word: *antecedent, anteroom*

ante meridiem def: before noon; abbr: **a.m.**; can also be written as *antemeridian*

antenna the proper plural in the technical sense is *antennas; antennae* should be limited to zoology; *antenna* has generally replaced the obsolescent *aerial*

anti- a prefix meaning opposite or contradictory to; generally combines to form one word: *antiaircraft, antiastigmatism, anticapacitance, anticoincidence, antisymmetric*; if combining word starts with i or is a proper noun, insert a hyphen: *anti-icing, anti-American*

antimeridian def: the opposite of meridian (of longitude); e.g. the antimeridian of 96°30′W is 83°30′E

anybody; any body *anybody* means any person; *any body* means any object: *anybody can attend; discard the batch if you find any body containing foreign matter*

anyone; any one *anyone* means any person; *any one* means any single item: *you may take anyone with you; you may take any one of the samples*

anyway; any way *anyway* means in any case or in any event; *any way* means in any manner: *the results may not be as good as you expect, but we want to keep them anyway; the work may be done in any way you wish*

apm advanced power manager

apparatus; apparatuses

apparent; apparently

appear(s); seem(s) use *appears* to describe a condition that can be seen: *the equipment appears to be new*; use *seems* to describe a condition that cannot be seen: *the computer seems to be fast*

appendix def: the part of a report that contains supporting data; pl: *appendices* (pref) or *appendixes*

applets def: JAVA computer programs

approximate(ly) abbr: **approx**; but *about* is a better word

arbitrary

arc; arced; arcing

Archie

architect; architecture

arctic capitalize when referring to a specific area: *beyond the Arctic Circle*; otherwise use lc letters: *in the arctic*; never omit the first *c*

area the SI unit for area is the *hectare* (abbr: **ha**)

areal def: having area

around def: on all sides, surrounding, encircling

arrester, arrestor *arrester* pref

artwork

as avoid using when the intended meaning is *since* or *because*; to write *he could not open his desk as he left his keys at home* is incorrect (replace *as* with *because*)

ASCII American standard code for information exchange

as per avoid using this hackneyed expression, except in specifications

asphalt *asfalt* also used in US, but less pref

assembly; assemblies abbr: **assy**

assure means to state with confidence that something has been or will be made certain; it is sometimes confused with *ensure* and *insure*, which it does not replace; see **ensure**

as well as avoid using when the meaning is *and*

asymmetric; asymmetrical

asynchronous

atmosphere abbr: **atm**

atomic weight abbr: **at. wt**

attenuator

atto def: 10^{-18}; abbr: **a**

audible; audibility

audio frequency abbr: **af** (pref) or **a-f**

audiovisual

audit; auditor

aural def: that which is heard; avoid confusing with *oral*, which means that which is spoken

author; writer avoid referring to yourself as *the author* or *the writer*; use *I*, *me*, or *my*

authoritative

authorize; authorise *authorize* pref

auto- a prefix meaning self; combines to form one word: *autoalarm, autoconduction, autoionization, autoloading, autotransformer*

automatic frequency control abbr: **afc** (pref) or **AFC**

automatic volume control abbr: **avc** (pref) or **AVC**

auxiliary abbr: **aux**

average see **mean**

avocation def: an interest or hobby; avoid confusing with *vocation*

ax; axe *axe* pref in Br, rec in Can.; *ax* pref in US; pl: *axes*

axis the plural also is *axes*

azimuth abbr: **az**

B

back- as a prefix normally combines into one word: *backboard, backdate(d), backlog, backup*

balance; reminder use *balance* to describe a state of equilibrium (as in *discontinuous permafrost is frozen soil delicately balanced between the frozen and unfrozen state*), or as an accounting term; use *remainder* when the meaning is the rest of: *the remainder of the shipment will be delivered next week*

ball bearing

bandwidth

bare; bear *bare* means barren or exposed; *bear* means to withstand or to carry (or a wild animal)

barometer abbr: **bar.**

barrel; barrel(l)ed; barrel(l)ing *ll* pref in Br, rec in Can., single *l* pref in US; the abbr of *barrel(s)* is **bbl**

barretter

barring def: preventing, excepting

bases this is the plural of both *base* and *basis*

basically

basic input/output system abbr: **bios** (pref) or **BIOS**

baud; baud rate

ba(u)lked *baulked* pref in Br, rec in Can., *balked* pref in US

because; for use *because* when the clause it introduces identifies the cause of a result: *he could not open his desk because he left his keys at home*; use *for* when the clause introduces something less tangible: *he failed to complete the project on schedule, for reasons he preferred not to divulge*

becquerel def: a unit of activity of radionuclides (SI): abbr: **Bq**; other abbr: **PBq, TBq, GBq, kBq**; in SI, the *becquerel* replaces the *curie*

behavio(u)r *behavior* pref in US and rec in Can.; *behaviour* pref in Br

benefit; benefit(t)ed; benefit(t)ing single *t* pref

beside; besides *beside* means alongside, at the side of; *besides* means as well as

between see **among**

bi- a prefix meaning two or twice; combines to form one word: *biangular, bidirectional, bifilar, bilateral, bimetallic, bizonal*

biannual(ly); biennial(ly) *biannual(ly)* means twice a year; *biennial(ly)* means every two years

bias; biased; biases; biasing

billion def: 10^9 (Can. and US); 10^{12} (Britain)

billion electron volts although the pref abbr is **GEV**, *beV* and *bev* are more commonly used

Bill of Materials abbr: **BOM**

bimonthly def: every two months

binary

binaural

bioelectronics

bionics def: application of biological techniques to electronic design

bios basic input/ouput system

birdseye (view)

Bitnet

bits per second abbr: **bps**

biweekly def: every two weeks

blow- as a prefix combines to form one word: *blowhole, blowoff, blowout*

blueprint

blur; blurred; blurring; blurry

board feet abbr: **fbm** (derived from *feet board measure*)

boiling point abbr: **bp**

borderline

bps bits per second

brakedrum; brake lining; brakeshoe

brake horsepower; brake horsepower-hour abbr: **bhp, bhp-hr**

brand-new

break- when used as a prefix to form a compound noun or adj, *break* combines into one word: *breakaway, breakdown, breakup*; in the verb form it retains its single-word identity: *it was time to break up the meeting*

bridging

Brinell hardness number abbr: **Bhn**

British thermal unit abbr: **Btu**

buoy; buoyant

burned; burnt *burned* pref in Can. and US; *burnt* pref in Br

bur(r) *burr* pref

buses; bused; busing; bus bar

business; businesslike; businessperson avoid using *businessman* or *businesswoman* unless referring to a specific male or female person

by- as a prefix, *by-* normally combines to form one word: *bylaw, byline, bypass, byproduct*

byte abbr: **kbyte** and **Mbyte** (pref), or **kb** and **Mb**

C

cache memory

calendar; calender; colander a *calendar* is the arrangement of the days in a year; *calender* is the finish on paper or cloth; a *colander* is a sieve

calibre; caliber *calibre* pref in Br and rec in Can.; *caliber* pref in US

calk see **caulk**

cal(l)iper *caliper* pref

calorie abbr: **cal**

calorimeter; colorimeter a *calorimeter* measures quantity of heat; a *colorimeter* measures color

cancel(l)ed; cancel(l)ing *ll* pref in Br, rec in Can.; single *l* pref in US; **cancel** always has single *l*; **cancellation** always has *ll*

candela def: unit of luminous intensity (replaces *candle*); abbr. **cd**; recommended abbr for candela per square foot and square metre are **cd/ft^2** and **cd/m^2**

candlepower; candlehour(s) abbr: **cp, c-hr**

candoluminescence

cannot one word pref; avoid using *can't* in technical writing

canvas; canvass *canvas* is a coarse cloth used for tents; *canvass* means to solicit

capacitor

capacity for never use *capacity to* or

capacity of

capillary

capital letters abbr: **caps.**

car- as a prefix normally combines to form one word: *carload, carlot, carpool, carwash*

carburet(t)or *carburetor* pref; a third, seldom used spelling is *carburetter*; also: **carburetion**

carcino- as a prefix combines to form one word

case- as a prefix normally combines to form one word: *casebook, caseharden, casework(er)*; exceptions: *case history, case study*

cassette

caster; castor use *caster* when the meaning is to swivel freely, and *castor* when referring to castor oil, etc

catalog(ue) *catalogue(d)* and *cataloguing* pref in Can, and Br; *catalog(ed)* and *cataloging* pref in US

catalyst; catalytic

cathode-ray tube abbr: **crt** (pref) or **CRT** (commonly used)

cation def: positive ion

ca(u)lk *caulk(ed), caulking* pref in Br, rec in Can.; *calk(ed), calking* pref in US

CD-ROM

-ceed; -cede; -sede only one word ends in *-sede*: *supersede*; only three words end in *-ceed*: *exceed, proceed, succeed*; all others end in *-cede*: e.g. *precede, concede*

centre; center *centre, centred, centring, central* pref in Br, rec in Can.; *center, centered, centering, central* pref in US

Celsius abbr: **C**; see **temperature**

centreline abbr: **₵** (pref) or **CL**

central processing unit abbr: **cpu** (pref) or **CPU**

centre-to-centre abbr: **c-c**

centi- def: 10^{-2}, as a prefix combines to form one word: *centiampere, centigram*; abbr: **c**; other abbr:

centigram	cg
centilitre	cL
centimetre	cm
centimeter-gram-second	cgs
centimetres per second	cm/s
square centimetre	cm^2

centigrade abbr: **C**; in SI, *centigrade* has been replaced by *Celsius*; see **temperature**

centri- a prefix meaning centre; combines to form one word: *centrifugal, centripetal*

cga color/graphics adapter

chairperson avoid using *chairman* or *chairwoman*

chamfer

changeable; changeover

channel(l)ed; channel(l)ing *ll* pref in Br, rec in Can.; single *l* pref in US

chargeable

chassis both singular and plural are spelled the same

check- as a prefix combines to form one word: *checklist, checkpoint, checkup* (noun or adjective)

checksum def: a term used in computer technology

cheque in US, spelled *check*

chisel(l)ed; chisel(l)ing *ll* pref in Br, rec in Can., single *l* pref in US

chrominance

chunk; chunking

cipher in Br also spelled *cypher*

circuit abbr: **cct**; also: **circuitous; circuit breaker**

cite def: to quote; see **site**

climate avoid confusing *climate* with *weather*; *climate* is the average type of weather, determined over a number of years, experienced at a particular place; *weather* is the state of the atmospheric conditions at a specific place at a specific time

clockwise (turn) abbr: **CW**

cmc computer-mediated communication

cmos complementary metal-oxide conductor

co- as a prefix, *co-* generally means jointly or together; it usually combines to form one word: *coexist, coequal, cooperate, coordinate, coplanar* (*co-worker* is an exception); it is also used as the abbr for *complement of* (an arc or angle): *codeclination, colatitude*

coalesce; coalescent

coarse; course *coarse* means rough in texture or of poor quality; *course* implies movement or passage of time; *a coarse granular material; the technical writing course*

coaxial abbr: **coax.**

coefficient abbr: coef

coerce; coercion

collaborate; collaborator avoid writing *collaborate together* (delete *together*)

collapsible

collateral

cologarithm abbr: colog

colon when a colon is inserted in the middle of a sentence to introduce an example or short statement, the first word following the colon is not capitalized; a colon rather than a semicolon should be used at the end of a sentence to introduce subparagraphs that follow; a hyphen should not be inserted after the colon

colo(u)r *color* pref in US, rec in Can.; *colour* pref in Br

color/graphics adapter abbr: cga (pref) or CGA

colorimeter see **calorimeter**

column abbr: col.

combustible

comma a comma normally need not be used immediately before *and, but,* and *or,* but may be inserted if to do so will increase understanding or avoid ambiguity

commence in technical writing, replace *commence* with the more direct *begin* or *start*

commit; commitment; committed; committing

committee

communicate it is vague to write *I communicated the results to the client*; use a clearer verb: *I emailed/faxed/ wrote/telephoned*

compare; comparable; comparison; comparative use *compared to* when suggesting a general likeness; use *compared with* when making a definite comparison

compatible; compatibility

complement; compliment *complement* means the balance required to make up a full quantity or a complete set; to *compliment* means to praise; *in a right angle, the complement of 60° is 30°; Mr Perchanski complimented Janet Rudman for writing a good report*

complementary metal-oxide conductor abbr: cmos (pref) or CMOS

composed of; comprising; consists of all three terms mean "made up of" (specific items); if any one of these terms is followed by a list of items, it implies that the list is complete; if the list is not complete, the term should be replaced by *includes* or *including*

compound terms two or more words that combine to form a compound term are joined by a hyphen or are written as one word, depending on accepted usage and whether they form a verb, noun, or adjective; the trend is toward one-word compounds; (see page 387)

comprise; comprised; comprising to write *comprised of* is incorrect, because the verb comprise includes the preposition *of*

computer-mediated communication abbr: cmc (pref) or CMC

concur; concurred; concurrent; concurring

condenser

conductor

conform use *conform to* when the meaning is to abide by; use *conform with* when the meaning is to agree with

connection; connexion *connection* pref in Can. and US; *connexion* is alternative spelling in Br

conscience

conscious

consensus means a general agreement of opinion; hence to write *consensus of opinion* is incorrect; e.g. write: *the consensus was that a further series of tests would be necessary*

consistent with never *consistent of*

consists of; consisting of see **composed of**

contact *contact* should not be used as a verb when *write, visit, speak, fax,* or *telephone* better describes the action taken

continual; continuous *continual(ly)* means happens frequently but not all the time: *the generator is continually being overloaded* (is frequently overloaded); *continuous(ly)* means goes on and on without stopping: *the noise level is continuously at or above 100dB* (it never drops below 100 dB)

continue(d) abbr: cont

continuous wave abbr: cw

contra- as a prefix normally combines into a single word

contrast when used as a verb, *contrast* is followed by *with*; when used as a noun, it may be followed by either *to* or *with* (*with* pref)

control; controlled; controlling; controller

conversant with never *conversant of*

converter; convertible

conveyor

cooperate

coordinate; coordinator

copyright not *copywrite*

corollary

correlate

correspond *to correspond to* suggests a resemblance; *to correspond with* means to communicate in writing

corroborate

corrode; corridible; corrosive

cosecant abbr: csc (pref) or cosec

cosine abbr: cos

cotangent abbr: cot

coulomb def: a quantity of electricity, electric charge (SI); abbr: C; other abbr: kC, mC, μC, nC, pC, C/m^2

counter- a prefix meaning opposite or reciprocal; combines to form one word: *counteract, counterbalance, counterflow, counterweight*

counterclockwise (turn) abbr: CCW

counterelectromotive force abbr: cemf; also known as *back emf*

counts per minute abbr: cpm

course see **coarse**

cpu central processing unit

criteria; criterion the singular is *criterion*, the plural is *criteria*: e.g. *one criterion; seven criteria*

criticism; criticize; critique

cross- as a prefix combines erratically: *cross-check, crosshatch, crosstalk, cross-purpose, cross section*

cross-refer(ence) abbr: x-ref

crt cathode-ray tube

cryogenic

cryptic

crystal abbr: xtal

crystalline; crystallize

cubic abbr: **cu** or ^{3}; other abbr:

  cubic centimetre(s) cm^3 (pref); **cc**

  cubic decimetre(s) dm^3

  cubic foot (feet) ft^3 (pref); **cu ft**

  cubic feet per minute   **cfm** (pref); ft^3/min

  cubic feet per second   **cfs** (pref); ft^3/sec

  cubic inch(es)   **in.**^{3} (pref); **cu in.**

  cubic metre(s) m^3

  cubic millimetre(s) mm^3

  cubic yard(s) yd^3 (pref); **cu yd**

curb; kerb *curb* pref; *kerb* is Br

curie abbr: **Ci**; other abbr: **mCi, µCi**; in SI the *curie* is replaced by the *becquerel*

curriculum pl: *curriculums* (pref) or *curricula*

cursor

cw continuous wave

cycles per minute abbr: **cpm**

cycles per second abbr: **cps**; although occasionally used, this term has been replaced by **hertz**

cylinder; cylindrical abbr: **cyl**

D

daraf def: the unit of elastance

data def: gathered facts; although *data* is plural (derived from the singular *datum*, which is rarely used), it is more acceptable to use it as a singular noun: *when all the data has been received, the analysis will begin*

dateline

date(s) avoid vague statements such as "last month" and "next year" because they soon become indefinite; write a specific date, using day (in numerals), month (spelled out), and year (in numerals): *January 27, 1998* or *27 January 1998* (the latter form has no punctuation); to abbreviate, reduce month to first three letters and year to last two digits: *Jan 27, 98* or *27 Jan 98*

day- as a prefix, generally combines to form one word: *daybook, daylight, daytime, daywork*

days days of the week are capitalized: *Monday, Tuesday*

dc direct current

de- a prefix that generally combines to form one word: *deaccentuate, deactivate, decentralize, decode, deemphasize, deenergize, deice, derate, destagger*; an exception is *de-ionize*

dead- as a prefix combines erratically: *deadbeat, dead centre, dead end, deadline, deadweight, deadwood*

debug; debugged; debugging

decelerate def: to slow down; never use *deaccelerate*

decibel abbr: **dB**; the abbr for decibel referred to 1 mW is **dBm**

decimals for values less than unity (one), place a zero before the decimal point: *0.17, 0.0017*

decimate def: to reduce by one-tenth; can also mean to destroy much of

decimetre abbr: **dm**

declination abbr: **dec**

deductible

defective; deficient *defective* means unserviceable or damaged (generally lacking in quality); *deficient* means lacking in quantity (it is derived from *deficit*), and in the military sense incomplete: *a short circuit resulted in a defective transmitter; the installation was completed on schedule except for a deficient rotary coupler that will not be delivered until June 10*

defence; defense *defence* pref in Can. and Br; *defense* pref in US; **defensive** always has an *s*

defer; deferred; deferring; deferrable; deference

definite; definitive *definite* means exact, precise; *definitive* means conclusive, fully evolved; e.g. *a definite price* is a firm price; *a definitive statement* concerns a topic that has been thoroughly considered and evaluated

degree(s) abbr: **deg** (pref in narrative) or ° (following numerals); see **temperature**

demarcation

demi- a little-used prefix meaning half (generally replaced by *semi-*); combines to form one word: *demivolt*

demonstrate; demonstrator; demonstrable not *demonstratable*

depend; dependence; dependent; dependable

deprecate; depreciate *deprecate* means to disapprove of; *depreciate* means to reduce the value of: *the use of "as per" in technical writing is deprecated; the vehicles depreciated by 50% the first year and 20% the second year*

depth

desiccant, desiccate(d)

desirable

desktop

desktop publishing abbr: **dtp** (pref) or **DTP**

desktop video conference abbr: **dtv** (pref) or **DTV**

deter; deterred; deterrence; deterrent; deterring

deteriorate

develop not *develope*

device; devise the noun is *device*, the verb is *devise*: *a unique device; he devised a new software program*

dext(e)rous *dexterous* pref

diagram(m)ed; diagram(m)ing *mm* pref in Can. and Br; single *m* pref in US; **diagram** always has a single *m*; **diagrammatically** always has *mm*

dial(l)ed; dial(l)ing *ll* pref in Br, rec in Can.; single *l* pref in US

dialog(ue) *dialogue* pref

diameter abbr. **dia**

diaphragm

diazo

didn't never use this contraction in technical writing; use **did not**

dielectric

diesel; diesel-electric

dietitian

differ use *differ from* to demonstrate a difference; use *differ with* to describe a difference of opinion

different *different from* is preferred; *different to* is sometimes used; *different than* should never be used

diffraction

diffusion; diffusible

dilemma means to be faced with a choice between two unhappy alternatives; should not be used as a synonym for *difficulty*

diplex; duplex *diplex operation* means the simultaneous transmissions of two signals using a single feature, e.g. an antenna; *duplex operation* means that both ends can transmit and receive simultaneously

direct current   abbr: **dc**

directly   def: immediately; do not use when the meaning is as soon as

disassemble; dissemble   *disassemble* means to take apart; *dissemble* means to conceal facts or put on a false appearance

disassociate   see **dissociate**

disc; disk   *disc* pref in Br, rec in Can. (except in computer technology, where *disk* is more common); *disk* pref in US

discernible

discolo(u)r   *discolor* pref in US, rec in Can.; *discolour* pref in Br

discreet; discrete   *discreet* means prudent or discerning: *his answer was discreet*; *discrete* means individually distinctive and separate: *discrete channels*; **discretion** is formed from *discreet*, not from *discrete*

disinterested; uninterested   *disinterested* means unbiased, impartial; *uninterested* means not interested

disk cache

disk operating system   abbr: **DOS**

dispatch; despatch   *dispatch* pref

disseminate

dissimilar

dissipate

dissociate; disassociate   *dissociate* pref

distil(l)   *distil* pref in Can. and Br; *distill* pref in US; **distillate, distillation,** and **distilled** always have *ll*

distribute; distributor

don't; doesn't   such contractions should not appear in technical writing

donut; doughnut   for electronics/nucleonics, use *donut*

doppler   capitalize only when referring to the Doppler principle

DOS   disk operating system

double-   as a prefix combines erratically: *double-barrelled, doublecheck, double-duty, double entry, doublefaced*

down-   as a prefix combines into one word: *downgrade, downrange, downtime, downwind*

dozen   abbr: **doz**

drafting; draftsperson   avoid writing *draftsman* or *draftswoman*

drawing(s)   abbr: **dwg**

drier; dryer   the adjective is always *drier*; the pref noun is *dryer* in Can. and US, and *drier* in Br: *this material is drier; place the others back in the dryer*

drop; droppable; dropped; dropping

dtp   desktop publishing

dtv   desktop video conferencing

due to   an overused expression; *because of* pref

duo-   a prefix meaning two; combines to form one word: *duocone, duodiode, duophase*

duplex   see **diplex**

duplicator

E

each   abbr: **ea**

east   capitalize only if *east* is part of the name of a specific location: *East Africa*; otherwise use lc letters: *the east coast of Canada*; abbr: **E**; the abbr for *east-west* (control, movement) is **E-W**; *eastbound* and *eastward* are written as one word

eccentric; eccentricity

echo; echoes

economic; economical   use *economical* to describe economy (of funds, effort, time); use *economic* when writing about economics: *an economical operation* (it did not cost much); *an economic disaster* (it will have a major effect on the economy)

effect   see **affect**

efficacy; efficiency   *efficacy* means effectiveness, ability to do a job; *efficiency* is a measurement of capability, the ratio of work done to energy expended: *we hired a consultant to assess the efficacy of our training methods; the power house is to have a high-efficiency boiler*

e.g.   def: *for example*; avoid confusing with **i.e.**; no comma is ncessary after *e.g.*; may also be abbr **eg**

ega   enhanced graphics adapter

eighth

electric(al)   if in doubt, use *electric*; generally, *electric* means produces or

carries electricity, whereas *electrical* means related to the generation or carrying of electricity; abbr: **elec**

electro-   a prefix generally meaning pertaining to electricity; it normally combines to form one word: *electroacoustic, electroanalysis, electrodeposition, electromechanical, electroplate*; if the combining word starts with *o*, insert a hyphen: *electro-optics, electro-osmosis*

electromagnetic units   abbr: **emu**

electromotive force   abbr: **emf**

electronic(s)   use *electronic* as an adjective, *electronics* as a noun: *electronic counter-measures; your career in electronics*

electronic mail   abbr: **email**

electron volt(s)   abbr: **eV** (pref) or **ev**

electrostatic units   abbr: **esu**

elicit; illicit   *elicit* means to obtain or identify; *illicit* means illegal

eligible; eligibility

ellipse

email

embarrass; embarrassed; embarrassing; embarrassment

embedded

emigrate; immigrate   *emigrate* means to go away from; *immigrate* means to come into

emit; emitter; emittance; emission; emissivity

emulate; emulation

enamel(l)ed; enamel(l)ing   *ll* pref in Br, rec in Can.; single *l* pref in US

encase; incase   *encase* pref

encipher

enclose; inclose   *enclose* pref; *inclose* is used mainly as a legal term

enforce   not *inforce*

engineer; engineered; engineering

enhanced graphics adapter   abbr: **ega** (pref) or **EGA**

enquire; inquire   *enquire* pref

enrol; enroll   both are correct, but *enroll* pref; universal usage prefers *ll* for **enrolled** and **enrolling**, but a single *l* for **enrolment**

en route   def: on the road, on the way; never use *on route*

ensure; insure; assure use *ensure* when the meaning is to make certain of: *the new oscilloscope will ensure accurate calibration*; use *insure* when the meaning is to protect against financial loss: *we insured all our drivers*; use *assure* when the meaning is to state with confidence that something has been or will be made certain: *he assured the meeting that productin would increase by 8%*

entrepreneur; entrepreneurial

entrust; intrust *entrust* pref

envelop; envelope *envelop* is a verb that means to surround or cover completely; *envelope* is a noun that means a wrapper or a covering

environment; environmental; environmentally

EPROM def: abbr for erasable programmable read-only memory; can also be abbr as **eprom**

equal; equal(l)ed; equal(l)ing *equal, equality, equalize* always have single *l*; *equaled, equaling* pref in US, rec in Can; *equalled, equalling* pref in Br; *equalize* can be spelled *equalise* in Br

equi- a prefix that means equality; combines to form one word: *equiphase, equipotential, equisignal*

equilibrium; equilibriums

equip; equipped; equipping; equipment

equivalent abbr: equiv

erase; erasable

errata although *errata* is plural (from the singular *erratum*, which is seldom used), it can be used as a singular or plural noun: both the *errata are ready* and *the errata is complete* are acceptable

erratic

erroneous

escalator

especially; specially *specially* pref when it refers to an adjective, as in: *a specially trained crew*; *especially* should introduce a phrase, as in: *they were well trained, especially the computer technicians*

et al. def: and others; now rarely used

et cetera def: and so forth, and so on; abbr: etc; use with care in technical writing: *etc* can create an impression of vagueness or unsureness: *the*

transmitters, etc, were tested sounds much less definite than *the transmitter, modulator, and power supply were tested* (or, if to restate all the equipment is too repetitious, *the transmitting equipment was tested*)

everybody; every body *everybody* means every person, or all the persons; *every body* means every single body: *everybody was present*; *every body was examined for gun-powder scars*

everyone; every one *everyone* means every person, or all the persons; *every one* means every single item: *everyone is insured*; *every one had to be tested in a saline solution*

exa def: 10^{18}; abbr: E

exaggerate

exceed

excel; excelled; excellent; excelling

except def: to exclude; see **accept**

exhaust

exhibit; exhibitor

exorbitant

expedite; expediter (pref) or **expeditor**

explicit; implicit *explicit* means clearly stated, exact; *implicit* means implied (the meaning has to be inferred from the words): *the supervisor gave explixit instructions* (they were clear); *that the manager was angry was implicit in the words he used*

extemporaneous

extracurricular

extraordinary

extremely high frequency abbr: ehf

F

face- as a prefix normally combines to form one word: *facedown, facelift, faceplate*; exceptions: *face-saver, face-saving*

Fahrenheit abbr: F; see **temperature**

fail-safe

fallout one word in the noun form

familiarize; familiarization

farad def: a unit of electric capacitance; abbr: F; other abbr: µF, nF, pF

farfetched; far-out; far-reaching;

farseeing; farsighted

farther; further *farther* means greater distance: *he travelled farther than the other salespeople*; *further* means a continuation of (as an adjective) or to advance (as a verb): *the promotion was a further step in her career plan*, and *to further his education, he took a part-time course in industrial drafting*

fasten; fastener

faultfinder; faultfinding

favo(u)r *favor* pref in US and rec in Can.; *favour* pref in Br

feasible; feasibility

February

feet; foot abbr: ft; other abbr:

feet board measure (board feet)	fbm
feet per minute	fpm
feet per second	fps
foot-candle(s)	fc (pref); ft-c
foot-pound(s)	fp (pref); ft-lb
foot-pound-second (system)	fps system

femto def: 10^{-15}; abbr: f; other abbr:

femtoampere(s)	fA
femtovolt(s)	fV

ferri- a prefix meaning contains iron in the ferric state; combines to form one word: *ferricyanide, ferrimagnetic*

ferro- a prefix meaning contains iron in the ferrous state; combines to form one word: *ferroelectric, ferromagnetic, ferrometer*

ferrule; ferule a *ferrule* is a metal cap or lid; a *ferule* is a ruler

fewer; less use *fewer* to refer to items that can be counted; *fewer technicians than we predicted have been assigned to the project*; use *less* to refer to general quantities: *there was less water available than predicted*

fibre; fiber; fibrous *fibre, fibreglass* pref in Br, rec in Can.; *fiber, fiberglass, fiberoptic* pref in US; *Fiberglas* is a US trade name

field- as a prefix normally does not combine into a single-word or hyphenated form: *field glasses, field test, field trip*; but: *fieldwork(er)*

figure numbers in text, spell out the word *Figure* in full, or abbr it to

Fig.: *the circuit diagram in Figure 26* and *for details, see Fig. 7*; use the abbreviated form beneath the illustration; always use numerals for the figure number

file transfer protocol abbr: **ftp** (pref) or **FTP**

final; finally; finalize

fire- as a prefix combines erratically: *firearm, fire alarm, firebreak, fire drill, fire escape, fire extinguisher, firefighter, firepower, fireproof, fire sale, fire wall*

firmware

first to write *the first two...* (or three, etc) is better than to write *the two first...*; never use *firstly*; as a prefix, *first-* combines erratically: *first-class, firsthand, first-rate*

fix in technical usage, *fix* means to firm up or establish as a permanent fact; avoid using it when the meaning is to repair

flameout; flameproof

flammable def: easily ignited; see **inflammable**

flexible

flight- usually combines to form two words: *flight control, flight deck, flight plan*

flip-flop

flotation this is the correct spelling for describing an item that floats: *flotation gear*

flowchart

fluid abbr: **fl**; the abbr for fluid ounces is **fl oz**

fluorescence; fluorescent

fluorine; fluoridation

focus; focused; focusing; focuses pl: *focuses* (pref in Can. and US) or *foci* (pref in Br)

foot- as a prefix normally combines into one word: *footbridge, footcandle, foothold, footnote, footwork*; also see **feet**

for see **because**

forceful; forcible use *forceful* to describe a person's character; use *forcible* to describe physical force

fore- def: that which goes before; as a prefix normally combines into one word: *foreclose, forefront, foreground, foreknowledge, foremost, foresee,*

forestall, forethought, forewarn

forecast this spelling applies to both present and past tenses

foreman avoid using in a general sense, except when referring to a person specifically, as in *John Hayward, the foreman*; never use *forewoman* (a better word is *supervisor*)

foresee

forestall

foreword; forward a *foreword* is a preface or preamble to a book; *forward* means onward: *the scope is defined in the foreword to the book; he requested that we bring the meeting date forward*

for example abbr: **e.g.** (pref) or **eg**

former; first use *former* to refer to the first of only two things; use *first* if there are more than two

formula pl: *formulas* (pref in Can. and US) or *formulae* (pref in Br)

forty def: 40; it is not spelled *fourty*

fourth def: 4th; it is not spelled *forth*

fractions when writing fractions that are less than unity, spell them out in descriptive narrative: *by the end of the heat run, nine-tenths of the installation had been completed*; for technical details use decimals rather than fractions, as in *a flat case 0.75 m wide by 0.060 m deep*, except when a quantity is normally stated as a fraction (such as *3/8 in. plywood*)

free- as a prefix normally combines into one word: *freehand, freehold, freestanding, freeway, freewheel*

free from use *free from* rather than *free of*: *he is free from prejudice*

free on board abbr: **fob** (pref), **f.o.b.** (commonly used), or **FOB**

frequency abbr: **freq**

frequency modulation abbr: **FM**

ftp file transfer protocol

fulfil(l) *fulfil, fulfilment* pref in Br, rec in Can.; *fulfill, fulfillment* pref in US; fulfilled and fulfilling always have *ll*

funnel(l)ed; funnel(l)ing single *l* pref

further see **farther**

fuse as a verb, means join together or weld; as a noun, means a circuit protection device or (in Br) a detonation initiation device

fuselage

fuze def: a detonation initiation device

G

gage US alternative spelling of *gauge*

gallon gallons differ between US ($3.785\ dm^3$) and Britain ($4.546\ dm^3$); abbr: **gal**; other abbr:

gallons per day	**gpd**
gallons per hour	**gph**
gallons per minute	**gpm**
gallons per second	**gps**

gang; ganged; ganging

gas; gases; gassed; gassing; gasious; gassy

gauge *gauging* does not retain the *e*

gearbox; gearshift

geiger (counter)

gelatin(e) *gelatin* pref

geo- a prefix meaning of the earth; combines to form one word; *geocentric, geodesic, geomagnetic, geophysics*

giga def: 10^9, abbr: **G**; other abbr:

gigabecquerel(s)	**GBq**
gigahertz	**GHz**
gigajoule(s)	**GJ**
gigaohm(s)	**GΩ; Gohm**
gigapascal(s)	**GPa**
gigavolt(s)	**GV**

gimbal

glue; glueing; gluey

glycerin(e) *glycerine* pref in Br, rec in Can.; *glycerin* pref in US

gnd ground

Gopher def: an Internet search tool

government capitalize when referring to a specific government either directly or by implication; use lc if the meaning is government generally: *the Canadian Government; the Government specifications; no government would sanction such restrictions*

gram abbr: **g**; abbr for **gram-calorie** is **g-cal**

grammar; grammatical(ly)

grateful not *greatful*

gray def: absorbed dose of ionizing radiation (SI); abbr: **Gy**; other abbr: **mGy, µGy**; in SI, the *gray* replaces the *rad*

Greenwich mean time abbr: **GMT**

grey; gray def: a color; *grey* pref in Br, rec in Can.; *gray* pref in US

grill(e) when the meaning is a loudspeaker covering or a grating, *grille* pref

ground (electrical) abbr: **gnd**; **ground crew; ground floor**

guage wrongly spelled; the correct spelling is *gauge*

guarantee never *guaranty*; also **guarantor**

guesstimate

guideline

gyroscope abbr: **gyro**

H

half; halved; halves; halving as a prefix, *half* combines erratically; some common compounds are: *half-hourly, half-life, half-monthly, halftone, halfwave*; for others, consult your dictionary

hand- as a prefix normally combines to form one word: *handbill, handbook, handful, handfuls, handhold, handpicked, handset, handshake*

hangar; hanger a *hangar* is a large building for housing aircraft; a *hanger* is a supporting bracket

hard- as a prefix normally combines into one word: *hardbound, hardhanded, hardhat, hardware*; exceptions: *hard-earned, hard-hitting*

haversine abbr: **hav**

H-beam

head- as a prefix normally combines into one word: *headfirst, headquarters, headset, headstart, headway*

heat- as a prefix, *heat* combines erratically; some typical compounds are: *heatresistant, heat-run, heatsink, heat-treat*; for others, consult your dictionary

heavy-duty

hectare def: a large unit of area, used in surveying and agriculture; in SI, *hectare* replaces *acre*; abbr: **ha**

height (not *heighth*) abbr: **ht**; also: **heighten; heightfinder; heightfinding**

helix pl: *helices* (pref) or *helixes*

hemi- a prefix meaning half; combines

to form one word: *hemisphere, hemitropic*

henry def: a unit of inductance; abbr: **H**; other abbr: **mH, μH, nH, pH**

here- whenever possible avoid using *here-* words that sound like legal terms, such as *hereby, herein, hereinafter, hereof*; they make a writer sound pompous; as a prefix, *here-* combines to form one word

hertz def: a unit of frequency measurement (similar to *cycles per second*, which it replaces); abbr: **Hz**; other abbr: **THz, GHz, MHz, KHz**

heterodyne

heterogeneous; homogeneous *heterogeneous* means of the opposite kind; *homogeneous* means of the same kind

hexadecimal

high- as a prefix either combines into one word or the two words are joined by a hyphen: *highhanded, highlight*; as compound adj: *high-frequency, high-power, high-priced, high-speed*

high frequency abbr: **hf**

high-pressure (as an adjective) abbr: **h-p**

high voltage abbr: **hv** (pref) or **HV**

hinge; hinged; hinging

homogeneous see **heterogenous**

horizontal abbr: **hor**

horsepower abbr: **hp**; the abbr for horse-power-hour is **hp-hr**

hotkey

hour(s) abbr: **hr** or **h** (SI)

HTML hypertext markup language

hundred abbr: **C**

hundredweight def: 112 lb; abbr: **cwt**

hybrid

hydro- a prefix meaning of water; combines to form one word: *hydroacoustic, hydroelectric, hydromagnetic, hydrometer*

hyper- a prefix meaning over; combines to form one word; *hyperacidity, hypercritical*

hyperbola the plural is *hyperbolas* (pref) or *hyperbolae*

hyperbole def: an exaggerated statement

hyperbolic cosine, sine, tangent abbr:

cosh, sinh, tanh

hypertext

hypertext markup language abbr: **HTML**

hyphen in compound terms you may omit hyphens unlesss they need to be inserted to avoid ambiguity or to conform to accepted usage; e.g. *preemptive* is preferred without a hyphen, but *photo-offset* and *re-cover* (when the meaning is *to cover again*) both require one; refer to individual entries in this glossary

hypothesis pl: *hypotheses*

I

I-beam

ibid. def: Latin abbr for *ibidem*, meaning in the same place; used in footnoting, but becoming obsolete

icon

ID card

i.e. def: *that is*; avoid confusing with *e.g.*; no comma is necessary after *i.e.*; may also be abbr **ie**

ignition abbr: **ign**

ill- as a prefix combines into a hyphenated expression: *ill-advised, ill-defined, ill-timed*

illegible

im- see **in-**

imbalance this term should be restricted for use in accounting and medical terminology; use *unbalance* in other technical fields

immalleable

immaterial

immeasurable

immigrate see **emigrate**

immittance

immovable

impasse

impel; impelled; impelling; impeller

imperceptible

impermeable

impinge; impinging

imply; infer speakers and writers can *imply* something; listeners and readers *infer* from what they hear or read: *in* *his closing remarks, Mr Smith implied that further studies were in order; the technician inferred from the report that his work was better than expected*

impracticable; impractical
impracticable means not feasible; *impractical* means not practical; a less-preferred alternative for impractical is *unpractical*

in; into *in* is a passive word; *into* implies action; *ride in the car; step into the car*

in-; im-; un- all three prefixes mean not; all combine to form one word: *ineligible, impossible, unintelligible*; if you are not sure whether you should use *in-, im-,* or *un-,* use *not*

inaccessible

inaccuracy

inadmissible

inadvertent

inadvisable; unadvisable *inadvisable* pref

inasmuch as a better word is *since*

inaudible

incalculable not *incalculatable*

incandescence; incandescent

incase *encase* pref

inch(es) abbr: **in.**; other abbr:
inches per second	**ips** (pref); **in./s**
inch-pound(s)	**in.-lb**

inclose see **enclose**

includes; including abbr: **incl**; when followed by a list of items, *includes* implies that the list is not complete; if the list is complete, use *comprises* or *consists of*

incomparable

incompatible

incur; incurred; incurring

index pl: *indexes* pref, except in mathematics (where *indices* is common)

indicated horsepower abbr: **ihp**; the abbr for indicated horsepower-hour is **ihp-hr**

indifferent to never use *indifferent of*

indiscreet; indiscrete *indiscreet* means imprudent; *indiscrete* means not divided into separate parts

indispensable

indorse *endorse* pref

industrywide

ineligible

inequitable

inessential; unessential both are correct; *unessential* pref

inexhaustible

inexplicable

infallible

infer; inferred; inferring; inference also see **imply**

inflammable def: easily ignited (derived from *inflame*); *flammable* is a better word: it prevents readers from mistakenly thinking the *in* of *inflammable* means *not*

inflexible

infrared

infrastructure

ingenious; ingenuous *ingenious* means clever, innovative; *ingenuous* means innocent, naive; *ingenuity* is a noun derived from ingenious

in-house

inoculate

inoperable not *inoperatable*

input/output abbr: **I/O**

inquire; enquire *enquire* pref in Br, rec in Can.; *inquire* pref in US; also: *enquiry*

insanitary; unsanitary both are correct; *insanitary* pref

inseparable

inside diameter abbr: **ID**

in situ def: in the normal position

instal(l) *install, installed, installer, installing, installation* pref; but *instalment* pref in Br, rec in Can.; *installment* pref in US

instantaneous

instrument

insure the pref def is to protect against financial loss; can also mean make certain of (particularly in US); see **ensure**

integer

integral; integrate; integrator

intelligence quotient abbr: **IQ**

intelligible

inter- a prefix meaning among or between; normally combines to form one word: *interact, intercarrier, interdigital, interface, intermodulation, interoffice*

intermediate-pressure (as an adjective) abbr: **i-p**

intermittent

internal abbr: **int**

Internet

interrupt

into see **in**

intra- a prefix meaning within; normally combines to form one word: *intranet, intranuclear*; if combining word starts with *a*, insert a hyphen: *intra-atomic*

Internet relay chart abbr: **irc** (pref) or **IRC**

intrust *entrust* pref

I/O input/output

IQ

irc Internet relay chart

irrational

irregardless never use this expression; use *regardless*

irrelevant frequently misspelled as *irrevelant*

irreversible

iso- a prefix meaning the same, of equal size; normally combines to form one word: *isoelectronic, isometric, isotropic*; if combining word starts with *o*, insert a hyphen: *iso-octane*

its; it's *its* means belonging to; *it's* is an abbr for it is: *the transmitter and its modulator; if the fault is not in the remote equipment, then it's most likely in master control*; in technical wirting *it's* should seldom be used: replace with *it is*

J

jobholder; jobseeker; job lot

joule def: a unit of energy, work, or quantity of heat (SI); abbr: **J**; other abbr: **TJ, GJ, MJ, kJ, mJ, J/m³, J/K** (joule(s) per kelvin), **J/kg, J/mol** (joule(s) per mole)

journey; journeys

judg(e)ment *judgment* pref in US and Can.; *judgement* pref in Br

judicial; judicious *judicial* means related to the law; *judicious* means sensible, discerning

K

kelvin def: the SI unit for thermodynamic temperature; abbr: **K**

kerb Br equivalent of *curb*

key- as a prefix normally combines to form one word: *keyboard, keypunch, keying, keystroke;* but *key word*

kilo def: 10^3, abbr: **k**; other abbr:

kiloampere(s)	**kA**
kilobecquerel(s)	**kBq**
kilobyte(s)	**kbyte** (pref) or **kb**
kilocalorie(s)	**kcal**
kilocoulomb(s)	**kC**
kilogram(s), (see **kilogram**)	**kg**
kilohertz	**kHz**
kilohm(s)	**kΩ; kohm**
kilojoule(s)	**kJ**
kilolitre(s)	**kL**
kilometre(s)	**km**
kilometres per hour	**km/h**
kilomole(s)	**kmol**
kilonewton(s)	**kN**
kilopascal(s)	**kPa**
kilosecond(s)	**ks** (pref); **ksec**
kilosiemens	**kS**
kilovolt(s)	**kV**
kilovolt-ampere(s)	**kVA**
kilovolt-ampere(s) reactive	**kVAr**
kilowatt(s)	**kW**
kilowatthour(s)	**kWh** (pref); **kw-hr**

kilogram def: the SI unit for mass; abbr: **kg**; other typical abbr: **Mg, g, mg, µg**; also:

kilogram-calorie(s)	**kg-cal**
kilogram(s) per metre	**kg/m**
kilogram(s) per square metre	**kg/m²**
kilogram(s) per cubic metre	**kg/m²**
kilogram metre(s) per second	**kg · m/s**

knockout as noun or adjective, one word

knot abbr: **kn**

knowledge; knowledgeable

L

label(l)ed; label(l)ing *ll* pref in Br, rec in Can.; single *l* pref in US

laboratory abbr: **lab**

labo(u)r *labor* pref in US, rec in Can.; *labour* pref in Br; also: laborsaving

lacquer

lambert abbr: **L**; use the abbr L with care: it is also the SI abbr for *litre*

lampholder

large scale integration abbr: **lsi** (pref) or **LSI**

last; latest; latter *last* means final; *latest* means most recent; *latter* refers to the second of only two things (if more than two, use *last*); it is better to write *the last two* (or *three*, etc) than *the two last*

lath; lathe a *lath* is a strip of wood; a *lathe* is a machine

latitude abbr: **lat** or **Φ**

lay- as a prefix generally combines to form one word (as noun or adj): *layoff, layout, layover*

lcd liquid crystal display

learned; learnt *learned* pref

led light emitting diode

left-hand(ed) abbr: **LH**

length the SI unit of length is the *metre*, expressed in multiples and submultiples of *kilometres* (**km**), *metres* (**m**), and *millimetres* (**mm**)

lengthy not *lengthly*

less see **fewer**

letter- as a prefix combines erratically: *letterhead, letter-perfect, letter writer*

letter of intent; letter of transmittal pl: *letters of intent, letters of transmittal*

level(l)ed; level(l)er; level(l)ing *ll* pref in Br, rec in Can.; single *l* pref in US

liable to means under obligation to; avoid using as a synonym for *apt to* or *likely to*

liaison liaison is a noun; it is sometimes used uncomfortably as a verb: *liaise*

libel(l)ed; libel(l)ing *ll* pref in Br, rec in Can.; single *l* pref in US

licence; license in Can. and Br, the noun is *licence* and the verb is *license;* in US, *licence* is pref for both noun and verb

light- as a prefix *light-* generally combines to form one word: *lightface* (type), *lightweight;* but *light-year;* the past tense is *lighted,* but *lit* is common in Can. and Br

light emitting diode abbr: **led** (pref) or **LED**

lightening; lightning *lightening* means to make lighter; *lightning* is an atmospheric discharge of electricity

linear abbr: **lin**; the abbr for lineal foot is **lin ft**

lines of communication not *line of communications*

liquefy; liquefaction

liquid abbr: **liq**

liquid crystal display abbr: **lcd** (pref) or **LCD**

Listserv

litre; liter the SI spelling is **litre** (pref in Can. and Br), but in US liter is more common; abbr: **L**; other abbr: **kL, mL, µL**; the abbr for *litre(s) per day/hour/minute/second* are **L/d, L/h, L/m, L/s**

lock- as a prefix combines into a single word: *locknut, lockout, locksmith, lockstep, lockup, lockwasher*

locus pl: *loci*

logarithm abbr: (common) **log**; (natural) **ln**

logbook

logistic(s) use *logistic* as an adjective, *logistics* as a noun: *logistic control; the logistics of the move*

long- as a prefix normally combines into a single word or is hyphenated: *long-distance, longhand, longplaying, long-term, long-winded;* but *long shot*

longitude abbr: **long.** or **λ**

lose; loose *lose* is a verb that refers to a loss; *loose* is an adjective or a noun that means free or not secured: *three loose nuts caused us to lose a wheel*

louvre; louver *louvre* pref in Can. and Br; *louver* pref in US

low frequency abbr: **lf**

low-pressure (as an adjective) abbr: **l-p**

lsi large scale integration

lubricate; lubrication abbr: **lub**

lumen def: a unit of luminous flux (SI); abbr: **lm**; other abbr:

lumen-hour(s)	**lm-h** (pref); **lm-hr**
lumens per square foot	**lm/ft²**
lumens per square metre	**lm/m²**
lumens per watt	**lm/W**

lumens-second(s)	lm · s
microlumen(s)	µlm
millilumen(s)	mlm

luminance; luminescence; luminosity; luminous

lux def: a unit of illuminance (SI); abbr: **lx**; other abbr: **klx**

M

Mach

macro- a prefix meaning very large; combines to form one word: *macroscopic; macroview*

magneto pl: *magnetos*; as a prefix normally combines to form one word: *magnetoelectronics, magnetohydrodynamics, magnetostriction*; if combining word starts with *o* or *io*, insert a hyphen: *magneto-optics, magneto-ionization*

magneton; magnetron a *magneton* is a unit of magnetic moment; a *magnetron* is a vacuum tube controlled by an external magnetic field

maintain; maintained; maintenance

majority use *majority* mainly to refer to a number, as in a *majority of 27*; avoid using it as a synonym for many or most; e.g. do not write *the majority of technicians* when the intended meaning is *most*

make- as a prefix normally combines to form one word: *makeshift, makeup*

malfunction

malleable

manage; managed; manageable; managing

manoeuvre; maneuver *manoeuvre, manoeuvred, manoeuvring* pref in Br, rec in Ca.; *maneuver, maneuvered, maneuvering* pref in US

manufacturer abbr: **mfr**

marketplace

marshal(l)ed; marshal(l)er; marshal(l)ing *ll* pref in Br, rec in Can.; single *l* pref in US

mass see **kilogram**

material; materiel *material* is the substance or goods out of which an item is made; when used in the plural, it describes items of a like kind, such as *writing materials*; *materiel* are all the

equipment and supplies necessary to support a project or undertaking (a term commonly used in military operational support)

matrix pl: *matrices* (pref) or *matrixes*

maximum pl: *maximums* (pref) or *maxima*; abbr: **max**; like *minimize, maximize* can be used as a verb

maybe; may be *maybe* means "perhaps": *maybe there is a second supplier*; the verb form *may be* means "perhaps it will be" or "possibly there is": e.g. *there may be a second supplier*

mda monochrome display adapter

mean; median the *mean* is the average of a number of quantities; the *median* is the midpoint of a sequence of numbers; e.g. in the sequence of five numbers 1, 2, 3, 7, 8, the mean is 4.2 and the median is 3

mean effective pressure abbr: **mep**

mean sea level a bbr: **msl** (pref) or **MSL**

mediocre

medium when *medium* is used to mean substances, liquids, materials, or communication or advertising, the pref plural is *media*; in all other senses the pref plural is *mediums*

mega def: 10^6, abbr: **M**; other abbr:

megabyte(s)	**Mbyte** (pref) or **Mb**
megacoulomb(s)	**MC**
Megaelectronvolt(s)	**MeV**
megahertz	**MHz**
megajoule(s)	**MJ**
meganewton(s)	**MN**
megaohm(s)	**MΩ; Mohm**
megapascal(s)	**MPa**
megavolt(s)	**MV**
megawatt(s)	**MW**

memorandum pl: *memorandums* (pref in US, rec in Can.) or *memoranda* (pref in Br); abbr: **memo** (singular), **memos** (pl)

merit; merited; meriting

metal(l)ed; metal(l)ing *ll* pref in Br, rec in Can.; single *l* pref in US; **metallic** and **metallurgy** always have *ll*

meteorology; metrology *meteorology* pertains to the weather; *metrology* pertains to weights, measures, and calibration

meter def: a measuring instrument (noun) or to measure out (verb)

metre; meter def: metric unit of length;

the SI spelling is *metre* (pref in Can. and Br), but in US *meter* is more common; abbr: **m**; other typical abbr:

square metre(s)	**m²**
cubic metre(s)	**m³**
metres per second	**m/s**
newton-metre(s)	**N · m**
newtons per square meter	**N/m²**
kilogram(s) per cubic metre	**kg/m³**

micro def: 10^{-6}; abbr: **µ** (pref) or **u**; other abbr:

microampere(s)	**µA**
microcoulomb(s)	**µC**
microfarad(s)	**µF**
microgram(s)	**µg**
microgray(s)	**µGy**
microhenry(s)	**µH**
microhm(s)	**µΩ; µohm**
microlumen(s)	**µlm**
micromho(s)	**µmho**
micrometre(s)	**µm**
micromole(s)	**µmol**
micronewton(s)	**µN**
micropascal(s)	**µPa**
microsecond(s)	**µs** (pref); **µsec**
microsiemens	**µS**
microtesla(s)	**µT**
microvolt(s)	**µV**
microwatt(s)	**µW**

micro- as a prefix meaning very small, normally combines to form one word: *microammeter, micrometre, microorganism, microprocessor, microswitch, microview, microwave*; the term *micromicro-* (10^{-12}) has been replaced by **pico**

microphone abbr: **MIC** (pref) or **mike**

Microsoft disk operating system abbr: **MS-DOS**

mid- a prefix that means in the middle of; generally combines into one word: *midday, midpoint, midweek*; if used with a proper noun, insert a hyphen: *mid-Atlantic*

mile the word mile is generally understood to mean a statute mile of 5280 ft (1609 m), so the statement *I drove 326 miles* implies statute miles; when referring to the *nautical mile* (6080 ft; 1853 m), always identify it as such: *the flight distance was 4210 nautical miles* (or *4210 nmi*); abbr:

statute mile(s)	**mi**
nautical mile(s)	**nmi** (pref) or **n.m.**

miles per gallon   mpg
miles per hour   mph

mileage; milage *mileage* pref

milli def: 10^{-3}; abbr: **m**; other abbr:
  milliampere(s)   **mA**
  millicoulomb(s)   **mC**
  millicurie(s)   **mCi**
  millifarad(s)   **mF**
  milligram(s)   **mg**
  milligray(s)   **mGy**
  millihenry(s)   **mH**
  millijoule(s)   **mJ**
  millikelvin(s)   **mK**
  millilitre(s)   **mL**
  millilumen(s)   **mlm**
  millimetre(s)   **mm**
  millimho(s)   **mmho**
  millimole(s)   **mmol**
  millinewton(s)   **mN**
  milliohm(s)   **mΩ; mohm**
  millipascal(s)   **mPa**
  milliroentgen(s)   **mR**
  millisecond(s)   **ms** (pref); **msec**
  millisiemens   **mS**
  millitesla(s)   **mT**
  millivolt(s)   **mV**
  milliwatt(s)   **mW**
  milliweber(s)   **mWb**

milli- as a prefix, combines to form one word: *milliammeter, milligram, millimicron*

millibar def: a unit of pressure (= 100 Pa); abbr: **mbar**

mini- as a prefix combines to form one word: *minicomputer, minireport*

miniature; miniaturization

minimum pl: *minimums* (pref) or *minima*; abbr: **min**; also **minimize**

minority use mainly to refer to a number, as in *a minority by 2*; avoid using it as a synonym for several or a few; to write *a minority of the technicians* is incorrect when the intended meaning is *a few technicians*

minuscule not *miniscule*; def: very small

minute abbr:
  time   **min**
  angular measure   °

mis- a prefix meaning wrong(ly) or bad(ly); combines to form one word: *misalign, misfired, mismatched, misshapen*

miscellaneous

miscible

misspelled; misspelt *misspelled* pref (note double *s*)

mitre; miter *mitre* pref in Br, rec in Can.; *miter* pref in US

mnemonic

model(l)ed; model(l)er; model(l)ing *ll* pref in Br, rec in Can.: single *l* pref in US

mole def: the SI unit for amount of substance; abbr: **mol**; other abbr: **kmol**, mmol, µmol, mol/m^3

momentary; momentarily both mean *for a moment*, not *in a moment*

money- as a prefix normally combines to form one word: *moneymaking, moneysaving*

mono- a prefix meaning one or single; combines to form one word: *monopulse, monorail, monoscope*

monochrome display adapter abbr: **mda** (pref) or **MDA**

months the months of the year are always capitalized: *January, February,* etc; if abbr, use only the first three letters: *Jan, Feb,* etc; the abbr for *month* is **mo**

mortice; mortise *mortise* pref

mosaic

most never use as a short form for *almost*; to say *most everyone is here* is incorrect

motherboard

mo(u)ld *mould* pref in Br, rec in Can.; *mold* pref in US

movable; moveable *movable* pref

Mr; Ms address men as *Mr* and women as *Ms*; use *Miss* or *Mrs* only if you know the person prefers to be so addressed; the period (punctuation) may be inserted after *Mr* and *Ms*

MS-DOS Microsoft disk operating system

multi- a prefix meaning many; combines to form one word: *multiaddress, multicavity, multielectrode, multistate*

municipal; municipality

N

NAND-gate

nano def: 10^{-9}; abbr: **n**; other abbr:

nanoampere(s)   **nA**
nanocoulomb(s)   **nC**
nanofarad(s)   **nF**
nanohenry(s)   **nH**
nanometre(s)   **nm**
nanosecond(s)   **ns** (pref); **nsec**
nanotesla(s)   **nT**
nanovolt(s)   **nV**
nanowatt(s)   **nW**

naphtha(lene)

national information infrastructure abbr: **NII**

nationwide

nautical mile def: 1853 m (6080 ft); abbr: **nmi** (pref) or **n.m.**; see **mile**

navigate; navigator; navigable

NB means note well, and is the abbr for *nota bene*; it is more common to use the word *NOTE*

NC abbr for *normally closed* (contacts)

nebula pl: *nebulae* (pref in Br, rec in Can.) or *nebulas* (pref in US)

negative abbr: **neg**

negligible

nevertheless

newsgroup

newton def: a unit of force (SI); abbr: **N**; other abbr: **MN, kN, mN, µN, N · m** (newton metre), **N/m** (newtons per metre)

next write the *next two* (or *next three,* etc) rather than *the two next* (etc)

nickel

night never use *nite*; write *nighttime* as one word

NII national information infrastructure

nineteen; ninety; ninth all three are frequently misspelled

NO abbr for *normally open* (contacts)

No. abbr for **number**

noise-cancel(l)ing *ll* pref; see **cancel**

nomenclature

nomogram; nomograph *nomogram* pref

non- as a prefix meaning not or negative, normally combines to form one word: *nonconductor, nondirectional, nonnegotiable, nonlinear, nonstop*; if combining word is a proper noun, insert a hyphen: *non-Canadian*; avoid forming a new word with *non-* when a similar word that serves the same purpose already exists (i.e. you should not form *nonaudible* because *inaudible* already exists)

none when the meaning is "not one," treat as singular; when the meaning is "not any," treat as plural: *none* (not one) *was satisfactory; none* (not any) *of the receivers were repaired*

no one two words

NOR-gate

norm def: the average or normal (distribution, situation, or condition)

normalize

normally closed; normally open (contacts) abbr: **NC, NO**

normal to def: at right angles to

north abbr: **N**; other abbr:
northeast	**NE**
northwest	**NW**
north-south	
(control, movement)	**N-S**

northbound and *northward* are written as one word; for rule on capitalization, see **east**

notable

not applicable abbr: **N/A**

note well abbr: **NB** (derived from *nota bene*), but *NOTE* is more common

NOT-gate

notice; noticeable; notification

not to exceed an overworked phrase that should be used only in specifications; in all other cases use *not more than*

nth (harmonic, etc)

nuclear frequently misspelled

nucleus the plural is *nuclei* (pref) or *nucleuse*

null

number although **no.** would appear to be the logical abbr for number (and is pref), **No.** is much more common (the symbol # is no longer used as an abbr for number); the abbr *no.* or *No.* must always be followed by a quantity in numerals: it is incorrect to write *we have received a No. of shipments*; for the difference in usage between *amount* and *number*, see **amount**

numbers (in narrative) as a general rule, spell out up to and including nine, and use numerals for 10 and above; for specific rules, see pages 398 and 399

O

oblique; obliquity

obsolete; obsolescent

obstacle

obtain; secure use *obtain* when the meaning is simply to get; use *secure* when the meaning is to make safe or to take possession of (possibly after some difficulty); *we obtained four additional samples; we secured space in the prime display area*

occasional; occasionally

occur; occurred; occurrence; occurring

o'clock avoid using; see **time**

OCR optical character recognition

odo(u)r *odor* pref in US, rec in Can.; *odour* pref in Br

off- as a prefix either combines into one word, or a hyphen is inserted: *offset, offshoot, off-centre(d), off-scale, off-the-shelf*

offence; offense *offence* pref in Can. and Br; *offense* pref in US; **offensive** always has an *s*

offline

off of an awkward construction; omit the word *of*

ohm def: a unit of electric resistance; abbr: Ω or **ohm**; other abbr: $G\Omega$, Gohm, $M\Omega$, Mohm, $k\Omega$, kohm, $m\Omega$, mohm, $\mu\Omega$, uohm; abbr for ohm-centimetre(s) is **ohm-cm**; *ohmmeter* has *mm*

oilfield; oil-filled

omit; omitted; omission

omni- a prefix meaning all or in all ways; combines to form one word: *omnibearing, omnidirectional, omnirange*

on; onto *on* means positioned generally; *onto* implies action or movement: *the report is on Mr Cord's desk; the speaker stepped onto the platform*

one- as a prefix mostly combines with a hyphen: *one-piece, one-sided, one-to-one, one-way*; but *oneself* and *onetime*

online

onward(s) *onward* pref

opaque; opacity

op. cit. def: Latin abbr for *opere citato*, meaning the work cited; used in footnoting, now obsolescent

operate; operator; operable not *operatable*

optical character recognition abbr: **OCR**

optimum pl: *optima* (pref), and sometimes *optimums*; also: **optimal**

oral def: spoken, avoid confusing with *aural*

orbit; orbital; orbited; orbiting

organize; organizer; organization

OR-gate

orient; orientation the noun form is *orientation*; the pref verb form is *orient, oriented, orienting*

orifice

origin; original; originally

oscillate

oscilloscope slang abbr: **scope**

ounce(s) abbr: **oz**; other abbr:
ounce-foot	**oz-ft**
ounce-inch	**oz-in.**

out- as a prefix normally combines to form one word: *outbreak, outcome, outdistance*; when *out-* is followed by *of*, insert hyphens if used as a compound adjective (as in *an out-of-date list*), but treat as separate words when used in place of a noun (as in *the printing schedule is out of phase*)

outside diameter abbr: **OD**

outward(s) *outward* pref

over- as a prefix meaning above or beyond, normally combines to form one word: *overbunching, overcurrent, overdriven, overexcited, overrun*; avoid using as a synonym for more than, particularly when referring to quantities: *more than 17 were serviceable* is better than *over 17 were serviceable*

overage means either too many or too old

overall an overworked word; as an adjective it often gives unnecessary additional emphasis (as in *overall impression*) and should be deleted; avoid using as a synonym for *altogether, average, general,* or *total*

oxidation; oxidization *oxidation* pref in US, rec in Can.; *oxidization* pref in Br; also: **oxidize**

oxyacetylene

P

pacemaker; pacesetter

page; pages abbr: **p; pp**

paid not *payed,* when the meaning is to spend

pair(s) abbr: **pr**

pamphlet

panel(l)ed; panel(l)ing *ll* pref in Br, rec in Can.; single *l* pref in US

paper- as a prefix mostly combines to form one word: *paperback, paperbound, paperwork;* but *paper-covered, paper-thin*

parabola; parabolas; parabolic; paraboloid

paragraph(s) abbr: **para**

parallax

parallel; paralleled; paralleling; parallelism; parallelogram both *parallel to* and *parallel with* are correct

paralysis pl: *paralyses;* also: **paralyze**

parameter; perimeter *parameter* means a guideline; *perimeter* means a border or edge

paraplegic

paraprofessional

parenthesis the pl is *parentheses;* in Can., the term **bracket(s)** is more common

parity

particles

partly; partially use *partly* when the meaning is "a part" or "in part"; use *partially* when the meaning is "to a certain extent," or when preference or bias is implied

parts per million abbr: **ppm**

part-time

pascal def: a unit of pressure (SI); abbr: **Pa;** other abbr: **Gpa, Mpa, kPa, mPa, µPa, pPa, Pa • s** (pascal second)

pass- as a prefix normally combines to form one word: *passbook, passkey, passport, password*

passed; past as a general rule, use *passed* as a verb and *past* as an adjective or noun: *the test equipment has passed quality control inspection; past experience has demonstrated a tendency to fail at low temperature; in the past...*

pay- as a prefix normally combines to form one word: *paycheque, payload, payroll;* but *pay day*

pcb printed circuit board

PCMCIA Personal Computer Memory Card International Association

pel

pencil(l)ed; pencil(l)ing *ll* pref in Br, rec in Can.; single *l* pref in US

pendulum pl: *pendulums*

penultimate def: the next to last

people; persons *people* pref: *all the people were present;* use *persons* to refer only to small numbers of people: *one person was interviewed; seven people failed the test*

per in technical writing it is acceptable to use *per* to mean either *by, a,* or *an,* as in *per diem* (by the day) and *miles per hour;* avoid using *as per* in all writing

per cent; percent *percent* is pref in US and common in Can., where it is replacing *per cent;* abbr: **%;** use **%** only after numerals: 42%; use *percent* after a spelled-out number: *about forty percent;* avoid using the expression *a percentage of* as a synonym for *a part of* or *a small part;* also: **percentage** and **percentile**

perceptible

peripheral

permeable; permeameter; permeance

permissible

permit; permitted; permitting; permittivity; permit-holder

perpendicular abbr: **perp**

persevere; perseverance

persistent; persistence; persistency

personal; personnel *personal* means concerning one person; *personnel* means the members of a group, or the staff: *a personal affair; the personnel in the powerhouse*

Personal Computer Memory Card International Association abbr:

peta def: 10^{15}; abbr: **P;** other abbr: **PBq** (petabecquerel)

pharmacy; pharmacist; pharmaceutical

phase in the nonelectric sense, *phase* means a stage of transmission or development; it should not be used as a synonym for aspect; it is used correctly in *the second phase called for a detailed cost breakdown*

phase-in; phaseout but use two words for the verb forms: *to phase in, to phase out*

phenolic

phenomenon pl: *phenomena*

photo- as a prefix, normally combines to form one word: *photoelectric, photogrammetry, photoionization, photomultiplier;* if combining word starts with *o,* insert a hyphen: *photo-offset*

pico def: 10^{-12}, abbr: **p;** other abbr:

picoampere(s)	**pA**
picocoulomb(s)	**pC**
picofarad(s)	**pF**
picohenry(s)	**pH**
picosecond(s)	**ps** (pref); **psec**
picowatt(s)	**pW**

piecemeal; piecework

piezoelectric; piezo-oscillator

pilot; piloted; piloting

pipeline

pixel

plagiarism def: to copy without acknowledging the original source

plateau pl: *plateaus* (pref in US, rec in Can.) or *plateaux* (pref in Br)

plug; plugged; plugging

plumbbob; plumb line

p.m. def: after noon (post meridiem)

pneumatic

polarize; polarizing; polarization

poly- a prefix meaning many; combines to form one word: *polydirectional, polyethylene, polyphase*

polyvinyl chloride abbr: **pvc**

pop-up window

positive abbr: **pos**

post- a prefix meaning after or behind; combines to form one word:

postacceleration, postgraduate, postpaid; but *post office*

post meridiem def: after noon; abbr: **p.m.**; can also be written as *post-meridian* (less pref)

potentiometer abbr: **pot.**

pound(s) (weight) abbr: **lb**; other abbr:

pound-foot	**lb-ft**
pound-inch	**lb-in.**
pounds per square foot	**psf** (pref); **lb/ft**^{2}
pounds per square inch	**psi** (pref); **lb/in**^{2}
pounds per square inch, absolute	**psia**

power factor abbr: **pf** or spell out

powerhouse; power line; powerpack

practicable; practical these words have similar meanings but different applications that sometimes are hard to differentiate; *practicable* means feasible to do: *it was difficult to find a practicable solution* (one that could reasonably be implemented); *practical* means handy, suitable, able to be carried out in practice: *a practical solution would be to combine the two departments*

practice; practise the noun always is *practice*; the verb is *practise* (pref in Br, rec in Can.), but can also be *practice* (pref in US)

pre- a prefix meaning before or prior; normally combines to form one word: *preamplifier, predetermined, preemphasis, preignite, preset*; if combining word is a proper noun, insert a hyphen; *pre-Roman*

precede; proceed *precede* means to go before; *proceed* generally means carry on or continue: *a brief business meeting preceded the dinner* (the meeting occurred first); *after dinner, we proceeded with the annual presentation of awards*; see **proceed**

precedence; precedent *precedence* means priority (of position, time, etc): *the pressure test has precedence* (it must be done first); a *precedent* is an example that is or will be followed by others: *we may set a precedent if we grant his request* (others will expect similar treatment)

précis

predominate; predominant; predominantly

prefer; preferred; preference; preferable avoid overstating *preferable*, as in *more preferable* and *highly preferable*

prescribe; proscribe *prescribe* means to state as a rule or requirement; *proscribe* means to deny permission or forbid

presently use *presently* only to mean soon or shortly; never use it to mean *now* (use *at present* instead)

pressure-sensitive

prestigious

pretence; pretense *pretence* pref in Can. and Br; *pretense* pref in US

preventive; preventative *preventive* pref

previous def: earlier, that which went before; avoid writing *previous to* (use *before*); see **prior**

principal; principle as a noun, *principal* means (1) the first one in importance, the leader; or (2) a sum of money on which interest is paid: *one of the firm's principals is Martin Dawes; the invested principal of $10 000 earned $950 in interest last year*; as an adjective, *principal* means most important or chief: *the principal reason for selecting the Arrow microprocessor was its low capital cost; principle* means a strong guiding rule, a code of conduct, a fundamental or primary source (of information, etc): *his principles prevented him from taking advantage of the error*

printed circuit board abbr: **pcb**

printout (noun and adj form)

prior; previous use only as adjectives meaning earlier: *he had a prior appointment*, or *a previous commitment prevented Mr Perchanski from attending the meeting*; write *before* rather than *prior to* or *previous to*

proceed; proceeding; procedure use *proceed to* when the meaning is to start something new; use *proceed with* when the meaning is to continue something that was started previously

producible

program(me) single *m* pref in US, rec in Can.; *mm* pref in Br; **program(m)ed, program(m)er, program(m)ing** always have *mm* in Can. and Br, and usually in

US (where single *m* also is used but is less pref)

prohibit use *prohibit from*; never *prohibit to*

prominent; prominence

proofread

propel; propelled; propelling; propellant (noun); **propellent** (adjective); **propeller**

prophecy; prophesy use *prophecy* only as a noun, *prophesy* only as a verb

proposition in its proper sense, *proposition* means a suggestion put forward for argument; it should not be used as a synonym for *plan, project*, or *proposal*

pro rata def: assign proportionally; sometimes used in the verb form as **prorate**: *I want you to prorate the cost over two years*

prospectus; prospectuses

protein

protocol

proved; proven use *proven* only as an adjective or in the legal sense; otherwise use *proved: he has been proven guilty; he proved his case*

provincewide

psycho- as a prefix normally combines to form one word: *psychoanalysis, psychopathic, psychosis*; if combining word starts with *o*, insert a hyphen: *psycho-organic*

purge; purging

pursuant to avoid using this wordy expression

Q

quality control abbr: **QC**

quantity; quantitative the abbr of quantity is **qty**

quart abbr: **qt**

quasi- a prefix meaning seemingly or almost; insert a hyphen between the prefix and the combining word: *quasi-active, quasi-bistable, quasi-linear*

question mark insert a question mark after a direct question: *how many booklets will you require?*; omit the question mark when the question posed is really a demand: *may I have your* *decision by noon on Monday*

questionnaire

quick- as a prefix normally combines with a hyphen: *quick-acting, quick-freeze, quick-tempered*; exceptions: *quicklime, quicksilver*

quiescent; quiescence

R

rack-mounted

racon def: a radar beacon

radian def: a unit of angular measurement; abbr: **rad**

radiator

radio- as a prefix, combines to form one word: *radioactive, radiobiology, radioisotope, radioluminescence*; if combining word starts with *o*, omit one of the *o's*: *radiology, radiopaque*; in other instances *radio* may be either combined or treated as a separate word, depending on accepted usage; typical examples are *radio compass, radio countermeasures, radio direction-finder, radio frequency* (as a noun), *radio-frequency* (as an adj), *radio range, radiosonde, radiotelephone*

radio frequency abbr: **rf**

radio frequency interference abbr: **rfi** (pref) or **RFI**

radius pl: *radii* (pref) or *radiuses*

radix pl: *radices* (pref) or *radixes*

rain- as a prefix normally combines to form one word: *raincoat, rainproof, rainwear*; exception: *rain check*

rally; rallied; rallying

RAM def: random access memory

ramdrive

range; ranging; rangefinder; range marker

rare; rarely; rarity; rarefy; rarefaction

ratemeter

ratio; ratios

rational; rationale *rational* means reasonable, clear-sighted: *John had a rational explanation for the error*; *rationale* means an underlying reason: *Tricia explained the company's rationale for diversifying the product line*

re def: a Latin word meaning in the case of; avoid using *re* in technical writing, particularly as an abbr for *regarding, concerning, with reference to*

re- a prefix meaning to do again, to repeat; normally combines to form one word: *reactivate, rediscover, reemphasize, reentrant, reignition, rerun, reset*; if the compound term forms an existing word that has a different meaning, insert a hyphen to identify it as a compound term, as in *re-cover* (to cover again)

reaction use *reaction* to describe chemical or mechanical processes, not as a synonym for *opinion* or *impression*

reactive kilovolt-ampere; reactive volt-ampere see **kilo** or **volt**

readability

read only memory abbr: **ROM**

readout (noun and adj form)

realize; realization

real-time chat; real-time transmission

recede

receive; receiver; receiving; receivable

rechargeable

recipe; receipt often confused; *recipe* means cooking instructions; *receipt* means a written record that something has been received

recommend; recommendation

reconcile; reconcilable

reconnaissance

recover; re-cover *recover* means to get back, to regain; *re-cover* means to cover again

recur; recurred; recurring; recurrence these are the correct spellings; never write *reoccur* (etc)

recycle; recyclable

reducible

reenforce; reinforce *reenforce* means to enforce again; *reinforce* means to strengthen: *Rick Davis reenforced his original instructions by circulating a second memorandum; The Artmo Building required 34 750 tonnes of reinforced concrete*

refer; referred; referring; referral; referee; reference

referendum Br and US dictionaries list *referenda* as the pref pl, but *referendums* is much more commonly used

reiterate def: to say again

relaid; relayed *relaid* means laid again, like a carpet; *relayed* means to send on, as a message would be relayed from one person to another

remit; remitted; remitting; remitter; remittance

remodel; remodelled; remodelling see **model**

removable

rent-a-car

reoccur(rence) never use; see **recur**

repairable; reparable both words mean in need of repair and capable of being repaired; *reparable* also implies that the cost to repair the item has been taken into account and it is economically worthwhile to effect repairs

repellant; repellent use *repellant* as a noun, *repellent* as an adjective; **repeller**

replaceable

reproducible

rescind

reservoir

reset; resetting; resettability

resin; rosin these words have become almost synonymous, with a preference for *resin*; use *resin* to describe a gluey substance used in adhesives, and *rosin* to describe a solder flux-core

respective(ly) this overworked word is not needed in sentences that differentiate between two or more items; e.g. it should be deleted from a sentence such as: *pins 4, 5, and 7 are marked R, S, and V respectively*

restart

resume the correct spelling is *résumé* (with two accents), but the single accent (*résume*) or no accent has become standard

retrieve; retrieval

retro- a prefix meaning to take place before, or backward; normally combines to form one word: *retroactive, retrofit, retrogression*; if combining word starts with *o*, insert a hyphen: *retro-operative*

reverse; reverser; reversal; reversible

revolutions per minute; revolutions per second abbr: **rpm; rps**

rfi radio frequency interference

rheostat

rhombus pl: *rhombuses* (pref) or *rhombi*

rhythm; rhythmic; rhythmically

richochet; richocheted; richocheting

right-hand(ed) abbr: **RH**

rigo(u)r *rigor* pref in US, rec in Can.; *rigour* pref in Br; *rigorous* is standard spelling in Can., US, and Br

RISC

rivet; riveted; riveter; riveting

road- as a prefix normally combines to form one word: *roadblock, roadmap, roadside*

roentgen abbr: **R**

role; roll a *role* is a person's function or the part that he or she plays (in an organization, project, or play); a *roll*, as a technical noun, is a cylinder; as a verb, it means to rotate: *the technician's role was to make the samples roll toward the magnet*

ROM def: read only memory

root mean square abbr: **rms**

rosin see **resin**

rotate; rotator; rotatable; rotary

round def. circular; in Br, *round* is commonly used in place of *around*, but this is not rec in Can.

ruggedize

rustproof; rust-resistant

S

salvageable

same avoid using *same* as a pronoun; to write *we have repaired your receiver and tested same* is awkward; instead, write *we have repaired and tested your receiver*

satellite

saturate; saturation; saturable

save; savable

sawtooth; saw-toothed

scalar; scaler *scalar* is a quantity that has magnitude only; *scaler* is a measuring device

scarce; scarcity

sceptic(al); skeptic(al) *sceptic(al)* pref in Br, rec in Can.; *skeptic(al)* pref in US

schedule

schematic although really an adjective (as in *schematic diagram*), in technical terminology *schematic* can be used as a noun (meaning a *schematic drawing*)

science; scientific(ally); scientist

screwdriver; screw-driven

seamweld

serial input/output abbr: **sio** (pref) or **SIO**

seasonal; seasonable *seasonal* means affected by or dependent on the season; *seasonable* means appropriate or suited to the time of year; *a seasonal activity; seasonable weather*

seasons the seasons are not capitalized: *spring, summer, autumn* or *fall, winter*

secant abbr: **sec**

secede; secession

second as a prefix, *second-* combines erratically: *second-class, second-guess, secondhand, second-rate, second sight*; the abbr for *second* (time) is **sec**, and for *second* (angular measure) it is **"**

secure see **obtain**

-sede *supersede* is the only word to end with *-sede*; others end with *-cede* or *-ceed*

seem(s) see **appear(s)**

self- insert a hyphen when used as a prefix to form a compound term: *self-absorption, self-bias, self-excited, self-locking, self-setting*; but there are exceptions: *selfless, selfsame*

semi- a prefix meaning half; normally combines to form one word: *semiactive, semiannually* (every six months), *semiconductor, semimonthly* (half-monthly), *semiremote, semiweekly* (half-weekly); if combining word starts with *i*, insert a hyphen: *semi-idle, semi-immersed*

separate; separable; separator; separation all are frequently misspelled

sequence; sequential

serial number abbr: **ser no.** or **S/N**

series-parallel

serrated

serviceable

serviceperson avoid using *serviceman* or *servicewoman*

servo- as a prefix, combines to form one word: *servoamplifier, servocontrol, servosystem*; as a noun, *servo* is an abbr for *servomotor* or *servomechanism*

sewage; sewerage *sewage* is waste matter; *sewerage* is the drainage system that carries away the waste matter

SGML standard generalized markup language

shall *shall* is rarely used in technical writing (*will* is pref), except in specifications when its use implies that the specified action is mandatory

short- as a prefix, may combine with a hyphen, as in *short-circuit, short-form* (report), *short-lived, short-term*; in some cases it may combine into one word, as in *shorthand* (writing), *shorthanded, shortcoming, shortsighted*

shrivel(l)ed; shrivel(l)ing *ll* pref in Br, rec in Can.; single *l* pref in US

sic a Latin word which means a quotation has been copied exactly, even though there was an error in the original; e.g. *the report stated: "Our participation will be an issential (sic) requirement."*

siemens def: a unit of electric conductance (SI); abbr: **S**; other abbr: **kS, mS, μS**

sight def: the ability to see; see **site**

signal(l)ed; signal(l)er; signal(l)ing *ll* pref in Br, rec in Can.; single *l* pref in US

signal-to-noise (ratio)

silhouette

silverplate; silver-plate use *silverplate* as a noun or adjective; *silver-plate* as a verb

similar not *similiar*

sine abbr: **sin**

singe; singeing the *e* must be retained to avoid confusion with *singing*

singlehanded

siphon not *syphon*

sirup; syrup *syrup* pref

site; sight; cite three words that often are misspelled; a *site* is a location: *the construction site; sight* implies the ability to see: *mud up to the axles became a familiar sight; cite* means quote: *I cite the May 17 progress report as a typical example of good writing*

siz(e)able *sizable* pref in US, rec in

Can.; *sizeable* pref in Br

skeptic(al) see **sceptic(al)**

skil(l)ful *skilful* pref in Br, rec in Can.; *skillful* pref in US; note that this is contradictory to most *l* and *ll* situations listed in this glossary

slip- usually combines into a single word: *slippage, slipshod, slipstream*; but *slip ring(s)*

smelled; smelt *smelled* pref in Can. and US; *smelt* pref in Br

smo(u)lder *smolder* pref in US, rec in Can.; *smoulder* pref in Br

soft key; software

solder

solely

someone; some one *someone* is correct when the meaning is any one person; *some one* is seldom used

some time; sometimes *some time* means an indefinite time: *some time ago; sometimes* means occasionally: *he sometimes works until after midnight*

sound- combines irregularly: *sound-absorbent, sound-absorbing, sound-powered, soundproof, sound track, sound wave*

south abbr: **S**; other abbr:

southeast	**SE**
southwest	**SW**

southbound and *southward* are written as one word; for rule on capitalization, see **east**

space- as a prefix normally combines to form one word: *spacecraft, spaceflight*

spare(s) can be used as a noun meaning spare part(s)

specially see **especially**

specific gravity abbr: **sp gr**

specific heat abbr: **sp ht**

spectro- as a prefix, combines to form one word: *spectrometer, spectroscope*; if combining word starts with *o*, omit one *o: spectrology*

spectrum pl: *spectra* (pref) or *spectrums*

spelled; spelt *spelled* pref in Can. and US; *spelt* pref in Br

spilled; spilt *spilled* pref in Can. and US; *spilt* pref in Br

spiral(l)ed; spiral(l)ing *ll* pref in Br, rec in Can.; single *l* pref in US

split infinitive to split an infinitive is to insert an adverb between the word

to and a verb: *to really insist* is a split infinitive; although grammarians used to claim that you should never split an infinitive, they now suggest you may do so if rewriting would result in awkward construction, ambiguity, or extensive rewriting

spoiled; spoilt *spoiled* pref in Can. and US; *spoilt* pref in Br

spotweld

square abbr: **sq** or **2**; other abbr:

square foot/feet	ft^2 (pref); sq ft
square inch(es)	in.^{2} (pref); sq in.
square metre(s)	m^2
square centimetre(s)	cm^2
square millimetre(s)	mm^2
curies per square metre	Ci/m^2
milliwatts per square meter	mW/m^2

standard generalized markup language abbr: **SGML**

standby; standoff; standstill all combine into one word when used as noun or adjective

standing-wave ratio abbr: **swr**

state-of-the-art

stationary; stationery *stationary* means not moving: *the vehicle was stationary when the accident occurred; stationery* refers to writing materials: *the main item in the October stationery requisition was an order for one thousand writing pads*

statute mile def: 1609 m (5280 ft); see **mile**

statutory

stencil(l)ed; stencil(l)ing *ll* pref in Br, rec in Can.; single *l* pref in US

stereo- as a prefix, combines to form one word: *stereometric, stereoscopic; stereo* can be used alone as a noun meaning multi-channel system

stimulus pl: *stimuli*

stocklist, stockpile

stop- as a prefix usually combines to form one word: *stopgap, stopnut, stopover* (when used as noun or adjective); but *stop payment, stop watch*

stoppage

strato- a prefix that combines to form one word: *stratocumulus, stratosphere*

stratum pl: *strata*

structural

stylus dictionaries list *styli* as the pref pl, but *styluses* is much more commonly used

sub- a prefix generally meaning below, beneath, under; combines to form one word: *subassembly, subcarrier, subcommittee, subnormal, subpoint*

subparagraph abbr: **subpara**; abbr for *subsubparagraph* is **subsubpara**

subpixel

subtle; subtlety; subtly

succinct

sufficient in technical writing, *enough* is a better word than *sufficient*

sulphur; sulfur *sulphur* pref in Can. and Br; *sulfur* pref in US; the prefix **sulph-** (or **sulf-**) combines to form one word: *sulphanilamide* (or *sulfanilamide*)

summarize

super- a prefix meaning greater or over; combines to form one word: *superabundant, superconductivity, superregeneration*

superhigh frequency abbr: **shf**

superimpose; superpose *superimpose* means to place or impose one thing generally on top of another; *superpose* means to lay or place exactly on top of, so as to be coincident with

supersede see **-sede**

supra- a prefix meaning above; normally combines to form one word: *supramolecular*; if combining word starts with *a*, insert a hyphen: *supra-auditory*

surfeit def: to have more than enough

surveillance

surveyor

susceptible

switch- *switchboard, switchbox, switchgear*

swivel(l)ed; swivel(l)ing *ll* pref in Br, rec in Can.; single *l* pref in US

syllabus pl: *syllabuses* (pref) or *syllabi*

symmetry; symmetrical

symposium pl: *symposia* (pref) or *symposiums*

synchro as a prefix combines to form one word: *synchromesh, synchronize,* *synchronous, synchroscope*; *synchro* can also be used alone as a noun meaning synchronous motor

synonymous use *synonymous with*, not *synonymous to*

synopsis pl: *synopses*

synthesis pl: *syntheses*

synthetic

syphon *siphon* pref

syringe

syrup; syrupy

systemwide

T

tail- as a prefix normally combines to form one word: *tailboard, tailless, tailwind*; but *tail end, tail fin*

take- *takeoff; takeover; takeup*; as nouns and adjectives these terms all combine into a single word

tangent abbr: **tan**

tangible

tape deck

taxable; tax-exempt; taxpayer

tci/ip transmission control protocol/Internet protocol

teamwork

technician

tele- a prefix meaning at a distance; combines to form one word: *teleammeter, telemetry, telephony, teletype (writer)*

telecom; telecon *telecom* is the abbr for *telecommunication(s)*; *telecon* is the abbr for *telephone conversation*

television abbr: **TV**

Telnet

temperature abbr: **temp**; combinations are *temperature-compensating* and *temperature-controlled*; when recording temperatures, the abbr for *degree* (deg or °) may be omitted: *an operating temperature of 85C; the water boils at 100C or 212F*; the pref (SI) unit for temperature is the degree Celsius (°C)

tempered

template; templet both spellings are correct; *template* pref in Can. and Br

temporary; temporarily

tenfold

tensile strength abbr: **ts**

tentative; tentatively

tenuous

tera def: 10^{12}; abbr: **T**; other abbr:

terabecquerel(s)	**TBq**
terahertz	**THz**
terajoule(s)	**TJ**
terawatt(s)	**TW**

terminus pl: *termini* (pref) or *terminuses*

tesla def: a unit of magnetic flux density, magnetic inductance (SI); abbr: **T**; other abbr: **mT, μT, nT**

that is abbr: **i.e.** (pref) or **ie**

their; there; they're the first two of these words are frequently misspelled, more through carelessness than as an outright error; *their* is a possessive meaning belonging to them: *the staff took their holidays earlier than normal*; *there* means in that place: *there were 18 desks in the room*, or *put it there*; *they're* is a contraction of *they are* and should not appear in technical or business writing

there- as a prefix combines to form one word: *thereafter, thereby, therein, thereupon*

therefor(e) *therefore* pref

thermo- a prefix generally meaning heat; combines to form one word: *thermoammeter, thermocouple, thermoelectric, thermoplastic*

thermodynamic temperature the SI unit is the kelvin (abbr: **K**), expressed in degrees Celsius (°C)

thesis pl: *theses*

thousand abbr: **k**

thousand foot-pound(s)	**kip-ft**
thousand pound(s)	**kip**

three- when used as a prefix, a hyphen normally is inserted between the combining words: *three-dimensional, three-phase, three-ply, three-wire*; exceptions are *threefold* and *threesome*

threshold

through never use *thru*

tieing, tying *tying* pref

timber; timbre *timber* is wood; *timbre* means tonal quality

time always write time in numerals, if possible using the 24-hour clock: *08:17*

or *8:17 a.m., 15:30 or 3:30 p.m.*; 24-hour times may be written as *20:45* (pref), *20:45 hr*, or *20:45 hours*; never use the term "o'clock" in technical writing: write *15:00 or 3 p.m.* rather than *j o'clock*

time- typical combinations are *time base, time-card, time clock, time constant, time-consuming, time lag, timesaving, timetable, time-wasting*

tinplate; tin-plate use *tinplate* as a noun, *tin-plate* as a verb or adjective

to; too; two frequently mispelled, most often through carelessness; *to* is a preposition that means in the direction of, against, before, or until; *too* means as well; *two* is the quantity 2

today; tonight; tomorrow never use *tonite*

tolerance abbr: **tol**

ton; tonne the US ton is 2000 lb and is known as a *short ton*; the Br ton is 2240 lb and is known as a *long ton*; the metric ton is 1000 kg (2204.6 lb) and is known as a *tonne* (abbr: **t**); other terms: *tonmile* and *tonnage*

toolbox; toolmaker; toolroom

top- top-heavy, top-loaded, top-up

torque; torqued; torquing

total(l)ed; total(l)ing *ll* pref in Br, rec in Can.; single *l* pref in US

touch-tone (dialing)

toward(s) *toward* pref

traceable

trade- as a prefix combines most often into a single word; *trademark, tradeoff*; but *trade-in, trade name*, and *trade show*

trans- a prefix meaning over, across, or through; it normally combines to form one word: *transadmittance, transcontinental, transship*; if combining word is a proper noun, insert a hyphen; *trans-Canada* (an exception is *transatlantic*, which through common usage has dropped the capital A and combines into one word); *transonic* has only one *s*

transceiver def: a transmitter-receiver

transfer; transferred; transferring; transferable; transference

transmission control protocol/Internet protocol abbr: **tci/ip** (pref) or **TCI/IP**

transmit; transmitted, transmitting;

transmittal, transmitter; transmission

transverse; traverse *transverse* means to lie across; *traverse* means to track horizontally

travel(l)ed; travel(l)er; travel(l)ing *ll* pref in Br, rec in Can.; single *l* pref in US

tri- a prefix meaning three or every third; combines to form one word: *triangulation, tricolor, trilateral, tristimulus, triweekly*

triple- all compounds are hyphenated: *triple-acting, triple-spaced*

trouble-free; troubleshoot(ing)

truncated

tune; tunable; tuneup (noun or adjective)

tunnel(l)ed; tunnel(l)ing *ll* pref in Br, rec in Can.; single *l* pref in US; **tunnel** has only one *l*

turbo- a prefix meaning turbine-powered; combines to form one word: *turboelectric, turboprop*

turbulence; turbulent

turn- *turnaround* and *turnover* form one word when used as adjectives or nouns; *turnstile* and *turntable* always form one word; *turns-ratio* is hyphenated

two- when used as a prefix to form a compound term, a hyphen normally is inserted: *two-address, two-phase, two-ply, two-position, two-wire*; an exception is *twofold*

type- as a prefix normally combines into one word: *typeface, typeset(ting)*

U

ucd user-centred drive

ultimatum pl: ultimatums

ultra- a prefix meaning exceedingly; normally combines to form one word: *ultrasonic, ultraviolet*; if combining word starts with *a*, insert a hyphen: *ultra-audible, ultra-audion*

ultrahigh frequency abbr: **uhf**

un- a prefix generally meaning not or negative; normally combines to form one word: *uncontrolled, undamped, unethical, unnecessary*; if combining word is a proper noun, or if term combines to form an existing word that

has a different meaning, insert a hyphen: *un-Canadian, un-ionized* (meaning not ionized); if uncertain whether to use *un-, in-,* or *im-*, try using *not*

unadvisable; inadvisable both are correct; *inadvisable* pref

unbalance; imbalance for technical writing, *unbalance* pref; see **imbalance**

unbiased

under- a prefix meaning below or lower; combines to form one word: *underbunching, undercurrent, underexposed, underrated, undershoot, undersigned*

underage means a shortage or deficit, or too young

unequal(l)ed *unequaled* pref; see **equal**

unessential; inessential *unessential* pref

unforeseen; unforeseeable

uni- a prefix meaning single or one only; combines to form one word: *uniaxial, unidirectional, unifilar, univalent*

uninterested def: not interested; avoid confusing with *disinterested*

unionized; un-ionized *unionized* refers to a group of people who belong to a union; *un-ionized* means not ionized

unique def: the one and only, without equal, incomparable; use with great care and never in any sense where a comparison is implied; you cannot write *this is the most unique design*; rewrite as *this design is unique*, or (if a comparison must be made) *this is the most unusual design*

universal resource locator abbr: **url** (pref) or **URL**

unmistakable

unnavigable

unparalleled

unpractical *impractical* pref

unsanitary

unserviceable abbr: **u/s**

unstable but *instability* is better than *unstability*

untraceable

up- as a prefix combines to form one word: *update, upend, upgrade, uprange, upswing*

upper case def: capital letters; abbr: **uc**

uppermost

up-to-date

url universal resource locator

use; usable; usage; using; useful

Usenet

user-centred drive abbr: **ucd** (pref) or **UCD**

utilize; utilise *utilize* pref; avoid using *utilizes* when *uses* or *employs* would be a better word

V

vacuum

valance; valence *valance* means a cover over a drapery track; *valence* is an electronic or nucleonic term, as in *valence electron*

valve-grind(ing)

vari- as a prefix meaning varied, combines into one word: *varicolored, variform*

variance write *at variance with*, never *at variance from*

varimeter; varmeter def: a meter for measuring reactive power; *varimeter* pref

vdisk virtual disk

vehicle, vehicular

vender; vendor *vendor* pref

versed sine abbr: **vers**

versus def: against; abbr: **vs**

vertex def: top; pl: *vertices* (pref) or *vertexes*; avoid confusing with *vortex*

very high frequency abbr: **vhf**

vga video graphics locator

vice; vise def: a clamping device; *vice* pref in Can. and Br; *vise* pref in US

vice versa def: in reverse order

video- as a prefix normally combines to form one word: *videocast, videocassette, videotape*; the abbr for videocassette recorder is **VCR** or **vcr**

video frequency abbr: **vf**

video graphics array abbr: **vga** (pref) or **VGA**

viewfinder; viewpoint

virtual disk abbr: **vdisk** (pref) or **VDISK**

vise see **vice**

visor; vizor *visor* pref

viz def: namely; this term is seldom used in technical writing

vocation; avocation *vocation* is a trade or calling; *avocation* means an interest or hobby

voice-over; voiceprint

volatile memory

volt def: electric potential or potential difference; abbr: **V**; other abbr: **MV, kV, mV, μV, nV**; also

volt-ampere(s)	**VA**
volt-ampere(s), reactive	**VAr**
volts, alternating current	**Vac**
volts, direct current	**Vdc**
volts, direct current, working	**Vdcw**
volts per metre	**V/m**

volt- combines into one word: *voltammeter, voltohmyst*

volume abbr: **vol**

vortex def: spiral; pl: *vortices* (pref) or *vortexes*; avoid confusing with *vertex*

VU-meter

W

WAIS wide area information server

waive; waiver; waver *waive* and *waiver* mean to forgo one's claim or give up one's right; *waver* means to hesitate, to be irresolute

walkie-talkie

war- as a prefix, combines to form one word: *warfare, wartime*

warranty

waste; wastage

water- combines irregularly: *water-cool(ed), water cooler, waterflow, water level, waterline, waterproof, water-soluble, watertight*

watt def: a unit of power, or radiant flux (SI); abbr: **W**; other abbr: **TW, GW, MW, kW, mW, μW, nW, pW, W/m²**; the abbr for *watt-hour(s)* is **Wh** (pref) or **W-hr**; as a prefix, *watt-* forms *watthourmeter* and *wattmeter*

wave- normally combines to form one word: *waveband, waveform, wavefront, waveguide, wavemeter, waveshape*; exceptions are *wave angle* and *wave-swept*

wavelength abbr:

waver see **waiver**

wear and tear *no* hyphens

weather use only as a noun; never write *weather conditions*; avoid confusing with *climate* and *whether*

weatherproof

weber def: a unit of magnetic flux (SI); abbr: **Wb**; other abbr: **mWb**

Wednesday often misspelled

week(s) abbr: **wk**

weekend

weight abbr: **wt**

well- as a prefix normally combines with a hyphen: *well-adjusted, well-defined, well-timed*

west abbr: **W**; *westbound* and *westward* are written as one word; for rule on capitalization, see **east**

where- as a prefix combines to form one word: *whereas, wherein*; when combining word starts with *e*, omit one *e*: *wherever*

whether; weather *whether* means if; *weather* has to do with rain, snow, sunshine, etc

while; whilst *while* pref

whoever

wholly not *wholely*

wide; width abbr: **wd**

wide area information server abbr: **WAIS**

wideband; widespread

wirecutter(s); wire-cutting; wirewound

withheld; withhold

word processor; word processing abbr: **WP**; as an adj, insert a hyphen: *the word-processing software*

words per minute abbr: **wpm**

work- as a prefix usually combines to form one word: *workbench, workflow, workload, workshop*; but *work force* and *work station*

working volts, dc abbr: **Vdcw**

worldwide

World Wide Web abbr: **www** or **The Web**

wrap; wrapped; wrapping; wraparound

writeoff; writeup both combine into one word when used as noun or adjective

writer see **author**

writing only one *t*

www World Wide Web

wysiwyg def: What You See Is What You Get

X

x- *x-axis, X-band, x-particle, x-radiation, x-ray*

Xerox

X-Y recorder

Y

y- *Y-antenna, y-axis, Y-connected, Y-network, Y-signal*

yard(s) abbr: **yd**

yardstick

year(s) abbr: **yr**; typical combinations are *year-end* and *year-round*

your; you're *your* means belonging to or originating from you: *I have examined your prototype analyser*; *you're* is a contraction of *you are* and should not appear in technical or business writing

Z

z-axis

zero pl: *zeros* (pref) or *zeroes*; typical combinations are *zero-access, zero-adjust, zero-beat, zero-hour, zero level, zero-set, zero reader*

zoology; zoological

Index

MARKING CONTROL CHART

This control chart will show you which aspects of your writing need attention and, as time progresses, whether you have successfully corrected your most predominant faults. As each assignment is returned to you, count up the errors indicated as marginal notations by your instructor and enter them on the chart. For instance, if on assignment 1 your instructor enters "F" and "U" once, and "S" three times, in the margin, you are being told you have used the wrong format (F), your work is untidy (U), and you have three spelling errors (S). In column 1 of the chart enter "1" in the squares opposite F and U, and "3" in the square opposite "S".

ASSIGNMENT NUMBER

1	2	3	4	5	6	7	8	9	10	11	12	13	14	15

A – Awkward construction

B – Brevity overdone; too few details

C – Continuity weak; paragraph lacks coherence/unity

D – Development inadequate; support your argument

E – Error! Check your facts, data, information

F – Format incorrect

G – Grammar fault

H – Heavy going; dull; uninteresting

I – Illogical or irrelevant (correct or omit)

J – Jumpy — too many short sentences, reads like primary reader

K – King-size paragraph or sentence (shorten it)

L – Low Information Content words or phrase (delete)

M – Missing words or information

N – No! Never do this; never use slang, contractions, unexplained abbreviations, etc.

O – Organization poor

P – Punctuation error, or punctuation missing

Q – Query: what does this mean? not understood; can't read your writing

R – Repetition

S – Spelling error

T – Tone wrong

U – Untidy, messy, or careless work (improve "presentation")

V – Vague; ambiguous; not clear enough

W – Wishy-washy; weak argument; unconvincing

X – X-out (delete, omit) this unnecessary statement

Y – Yak! Yak! Yak! — too wordy; too many generalities

Z – Lacks continuity; needs better transitions

– Numbers wrongly presented

// – Use parallel construction